THE DYNAMICS OF CITIES

THE DYNAMICS OF CITIES

Ecological determinism, dualism and chaos

Dimitrios Dendrinos

London and New York

First published 1992
by Routledge
11 New Fetter Lane, London EC4P 4EE

Simultaneously published in the USA and Canada
by Routledge
a division of Routledge, Chapman and Hall, Inc.
29 West 35th Street, New York, NY 10001

Typeset in September by
Leaper & Gard Ltd., Bristol
Printed and bound in Great Britain by
Biddles Ltd, Guildford and King's Lynn

A catalogue reference for this title is available from the British Library
ISBN 0-415-07721-4

Library of Congress Cataloging in Publication Data
has been applied for.

To
Haralambos S. Dendrinos

So maybe what we call imaginary time is really more basic, and what we call real is just an idea that we invent to help us describe what we think the universe is like.

Stephen W. Hawking, *A Brief History of Time*

CONTENTS

List of Figures xi
List of Tables xiii
Preface xv
Acknowledgements xxi
Abbreviations xxiii

INTRODUCTION
Some central issues 1
Structure of the book 7
A note on social science theories, data, and their markets 8

1 GLOBAL INTERDEPENDENCIES
Introduction 16
Macrodynamics of the world's largest urban areas: order in chaos 18
Interdependencies and ecological determinism 24
Conclusions 54

2 NATIONS, CITIES, INDUSTRIES, AND THEIR CONNECTANCE
Introduction 56
Instability from international exchanges 58
Two national economies; their internal growth and foreign trade 62
Unstable urban macrodynamics 82
Currencies and dualisms 95
Anatomy and evolution of some very large urban economies 105
Macrodynamics, stability, cycles, and relative dynamics 130
The mathematical formulation of relative macrodynamics 148
Conclusions 157

3 THE DYNAMIC CODE OF A GLOBAL URBAN HIERARCHY
Introduction 160
Hyper-concentrations, hypo-concentrations, and agglomeration gradients 163
Aggregate development code, relative parity, and the empirical evidence 175
The urban sector of the world's four largest nations 208
A phase portrait of global hierarchical dynamics? 211
Conclusions 217

4 EPILOGUE
Some future scenarios 222
Observers of and participants in social dynamics 234

APPENDICES
1 Toward an ecological determinism
Introduction 243
Individual preferences and public policy markets 247
Dualisms 260

2 A brief review of major theories of development
Pure economic theories of development 271
Dependence theory 285
Keynesian macroeconomics, Marxism, laissez-faire, *and ecological determinism* 287
Public and private sectors: their ecological associations 292
Two theories from anthropology 298

3 Data and sources
Time series used 303
Data on cities and metropolitan areas from other Censuses 310
Some Chinese metropolitan regions and their internal structure 312
World population counts 319
Sources of data for the United States of America 323
Sources of data for economic indicators of the United States of Mexico 324

Notes 325
Bibliography 360
Index 375

FIGURES

1.1	The dynamic paths of the world's largest urban agglomerations, 1958–80	20
1.2	Various versions of the Malthusian trap dynamic model	51
2.1	The actual path of the New York MSA's relative dynamics, 1930–80	116
2.2	The relative macrodynamics for selected US MSAs, 1890–1980	120
2.3	The computer-simulated dynamic path of growth for the New York MSA relative to the US, 1890–2050	143
2.4	The US economy's macrodynamics within the framework of the world's market economies	155
3.1	The negative and positive effects of agglomeration versus distance	173
3.2	Urban macrodynamics and world population growth	178
3.3	The urban macrodynamics aggregate development code	182
3.4(a)	The phase portrait and field of motion in urban macrodynamics: case (a)	187
3.4(b)	The phase portrait and field of motion in urban macrodynamics: case (b)	187
3.5	(a) The relative urban population, x, and (b) relative per capita income, $\ln y$, distributions	216
3.6	The world's urban hierarchy	219
3.7	Schematic probability distribution for $\ln y$ at different urban sizes	219
4.1	A possible set of slow motions in the aggregate developmental code	227
4.2(a)	The phase portrait and field of motion in urban macrodynamics under an upwardly sloping developmental threshold	228

4.2(b) The phase portrait and field of motion in urban macrodynamics under a downwardly sloping developmental threshold 230
A.2.1 A four-phase model of public–private sector interaction 295
A.3.1 Location of the twenty largest metropolitan regions in the People's Republic of China, 1983 316

TABLES

2.1 US GNP by industry at first digit SIC, 1960 and 1980 65
2.2 US manufactures, current value added, third digit SIC: ranking of selected industries 66
2.3 US foreign exports and imports by major category, 1967, 1970 and 1981 67
2.4 US international trade, fourth digit SIC, 1981, ranked according to valuation 68
2.5 US foreign trade, third digit SIC, 1960, ranked according to valuation 69
2.6 US foreign trade in 1959 in major categories of merchandise, by location 70
2.7 US foreign trade in 1980 in major categories of merchandise, by location 71
2.8 US trade with selected countries in 1958, 1970 and 1981 72
2.9 Mexico's gross domestic product by industry, first digit SIC share, 1910, 1960 and 1980 74
2.10 Mexico's employment share by industry at the first digit SIC, 1960 and 1979 75
2.11 Value of output in manufacturing for the four major categories, third digit SIC, 1982–83 76
2.12 Employment in manufacturing for major categories, 1982–83 76
2.13 Mexico's foreign trade, 1982 77
2.14 Mexico's foreign trade with selected nations, 1982 77
2.15 US exports and imports: share by major customs region, 1965 and 1981 114
2.16 The New York MSA: employment by industry type, first digit SIC, 1960 and 1980 117

2.17 The New York MSA: employment for selected industrial categories, third digit SIC, 1960 and 1980 118
2.18 The Los Angeles MSA: employment by industry type, first digit SIC, 1960 and 1980 119
2.19 The Los Angeles MSA: employment for selected industrial categories, third digit SIC, 1960 and 1980 119
2.20 The Federal District of Mexico: employment share by industrial category, first digit SIC, 1960, 1970 and 1980 126
N.1 The fuzzy population count of the world's largest urban areas, 1985 339
N.2 Taipei's monthly and yearly population counts, 1974–85 352
A.2.1 Exports and imports as a percentage of gross domestic product for selected developed countries, 1950–83 281
A.3.1 Relative population and per capita product (income) of nineteen urban agglomerations, 1950–80 303
A.3.2 City size of fifteen large urban areas of the globe, 1980–82 310
A.3.3 Urban metropolitan size for thirteen large centers of the globe, 1980–82 311
A.3.4 The twenty largest metropolitan regions in China, ranked by urban population size, 1983 314
A.3.5 Per capita value of output produced in the twenty largest Chinese urban regions, 1983 317
A.3.6 Growth rates in absolute population for seven large Chinese metropolitan regions, 1981–83 318
A.3.7 The industrial base of Wuxi City at the third digit SIC code, 1980 320
A.3.8 The world population, 1950–87 321
A.3.9 The world per capita product, 1974–85 322

PREFACE

The event described here has lasted for at least a quarter century and is still ongoing. It might initially seem to the reader to be rather esoteric or even obscure, too insignificant to write a book about. This event, however, has a story of potentially significant proportions to tell, and with far-reaching repercussions for numerous social science fields. It touches fundamental issues in social evolution. But thoroughly addressing these fundamental issues is not the central concern here. Of interest instead are the inner dynamics of this event which reveal the existence of comprehensive linkages among the urban and rural settings of the globe, giving rise to global dynamic instability and a host of spatial disparities.

Among the broad implications of this event and its dynamics are the numerous components which make up the whole, and one could dwell on each of these. The main goal of this book, however, is to supply the reader with a macrodynamic theory of *interdependent evolution* of the world's largest urban agglomerations. This goal and the manner in which it is pursued makes rather unusual reading. Written in an extensively modified Malthusian growth tradition,[1] the book aims at producing an *interdisciplinarian* social science and public policy analysis framework, and addresses some fundamental issues in *population growth and decline.* The term 'interdisciplinarian' is used here partly to indicate that the material presented does not rely on any specific current social science or public policy field. Thus, the subject matter of the book does not fall under any disciplinarian field of inquiry and cannot be considered as an extension of any particular social science theory or field.

By focusing on urban population and per capita income dynamics, the book sets the stage for analyzing possible futures open to the giant cities of the world. It draws extensively from available empirical evidence in testing certain aggregate theories, to the degree possible at this stage. There are three salient features to the book's main goal: to supply the basic elements for a modified Malthusian demographic and economic theory of aggregate macrolevel population and per capita income growth and decline, linking it to the recent theory of chaos; to outline a speculative theory of individual and collective action and policy making, not necessarily based on economic principles of efficiency; and to state the central components of a general theory of social science theories.

Two major findings are documented in the book: that instability is the central ingredient of urban evolution at the global scale; and that the future of the world's largest urban agglomerations, in particular those of the Third World, is bleak under a wide range of scenarios.

From a *methodological* standpoint, an analytical integrating and synthesizing mechanism is employed which extensively relies on non-linear interactive dynamics, a novel branch of mathematics. Numerous rich insights on topics of evolution emerge from bifurcation, chaos, and catastrophe theories. Certain key developments are exposed as currently understood from these advances, making it worthwhile revisiting some long-established and extensively analyzed branches of social science. Selectively mentioned are specific social science theories, including Keynesian macroeconomics, neoclassical economic growth theory, theory of speculation, social choice theory (from political science), Marxist cultural and structural anthropology, Malthusian population geography and demography, dependence theory in political economy, and mathematical ecology. It is shown that selective elements from various theories in these fields and disciplines can be combined, extended, and possibly unified under a higher level of theory referred to as 'ecological.'[2]

A central *substantive* component of this book is that of global spatial (urban) interdependencies. Under an analytical (mathematical) ecological perspective, and by utilizing some empirical evidence from publicly available information on certain worldwide urban and national population and per capita income dynamic patterns, issues of global spatial (urban, regional) interdependencies can be addressed. The growth or decline of the world's largest urban agglomerations is the central empirical focus of the analysis. Although only urban population and per capita income (product) dynamics are studied – these being referred to as urban macrodynamics – the basic ideas emerging from this analytical perspective are much broader in scope.

Undertaking such an ambitious integrating task has obvious theoretical and possibly practical benefits. In an era where disciplinary boundaries are weakened, an interdisciplinary approach is certainly helpful. Particularly when such an approach does not involve the complexity of individual disciplines but instead supplies a general, simple model of the socio-spatial world which is none the less rich in dynamical qualities. Benefits are, however, necessarily countered by equivalent costs. The brief coverage of numerous theories and their branches and the occasionally rough impressions conveyed by such coverage constitute its main costs. The open-minded social science reader – that is, the reader who does not worship or swear allegiance to any orthodoxies – will have to ultimately decide whether the derived benefits from such an attempt toward integration and synthesis can and will, in the long and arduous road toward what might be construed as 'understanding' in the social sciences, outweigh the costs of the brevity and uniqueness of the attempt.

The exposition of the subject matter moves along four broadly defined dimensions, more or less simultaneously:

1 Certain socio-spatial *events* are looked at and analyzed in some detail. Among the geographical phenomena examined is 'spatial dualism,' a phenomenon which contains, but is not limited to, the sharp differences in population density among regions. Also addressed are its allied events of urban and regional hyper- and hypo-concentration, i.e., the very high and extremely low relative concentration of population and other socio-economic activities at core and peripheral regions respectively, a phenomenon often referred to in urban and regional analysis as 'primacy.'[3] Economic and demographic cycles, i.e., the regular or non-regular periodic movement in the size of various socio-economic stocks or indices, the persistent sharp differences in per capita wealth (the phenomenon of relative poverty in space–time) are events also addressed here.
2 In addressing these events, a general *analytical* perspective of *socio-spatial dynamics* is developed. Its mainstays are certain recently obtained insights from bifurcation, chaos, and catastrophe theories as well as mathematical ecology, which are transferred onto the study of growth or decline in human populations. Issues which are explored in some detail include a pervasive dynamic instability leading to the explosive growth or decline and eventual extinction of stocks in space–time and dualism; turbulence, involving periodic and non-periodic (seemingly chaotic but none the less order following) motion; determinism versus stochasticity; individual and

collective speculative dynamics and their interactions, to mention only a few of the qualitative dynamics featured.

3 Instead of employing any specific disciplinarian theory from economics, demography, sociology, anthropology, geography, psychology or any other social science discipline an interdisciplinarian *theoretical* perspective is developed. This perspective is christened *macro spatio-temporal ecological determinism.* It is argued that within this framework the combined effects of all these disciplinarian forces are included. The fallacy of exclusively relying on any discipline and its method to understand social action is avoided. Analytical and theoretical tools are constructed which account for the multifaceted structure of social dynamic behavior.
4 Finally, the issue of public actions and policies is addressed in reference to the specific large-scale social events studied here. The subject is approached from a speculative public (collective) investment perspective. In particular, the role and function of social data, models, theories, and the perceived outcomes they contain in terms of social action are viewed from a speculative angle. An options-based *theory of social science theories and data* is outlined. The role of observer and participant (player, actor) type models is examined. Models are looked at as mathematical expressions of social science theories. It is argued that these roles and the theories underlying model building principles are linked directly through the formation of expectations to the evolution of socio-spatial systems and the events they entail.

Social events, analytical qualitative properties of social events, theoretical perspectives, and the markets where these perspectives are exchanged constitute four strongly connected components-dimensions. Making progress in advancing a socio-spatial topic along any one of them requires that all must be taken under consideration simultaneously. Put differently, the event and its path in time, the story told about the ongoing event, the story teller, the story teller's audience, and the language used to tell it are all intricately related.

Obviously, this book cannot claim that it does provide the complete statement on this integration. Such a task must combine the resources of many scholars over a long time-frame. It might trigger the mechanism for such an integration to commence, by setting the stage for the central components of the discussion to emerge and unfold within a disciplined frame of reference. Preliminary evidence supplied and argumentation presented here may provide enough justification to restructure the manner in which socio-spatial events can be approached for analysis and intervention by the public sector.

Although the main components of the book draw from non-linear dynamics it is not written primarily for the applied mathematician or the natural scientist, although both could draw some points of potential interest if their subject of analysis happened to be mathematical socio-spatial science. Its intended main audience is the social scientist. In drawing references and verbal definitions from the field of non-linear dynamics a number of key notions are supplied, particularly in the notes to Chapters 1 and 2. Since this book does not, however, supply a full and self-contained exposition of all these advanced and technical notions, it is required that the reader have some prior exposure to the subject of non-linear dynamics. The reader is also expected to have had some exposure to the variety of social science fields and theories on which the exposition relies. A broadly based book such as this cannot be self-contained; indeed it is arguable whether any book can be claimed to be self-contained – particularly one in the field of mathematical social science.

Parts of this book could be of interest to individuals working with various agencies for international development. Further, presented in the form of an extensive research program – and to the extent that its main hypotheses make sense, can be analytically followed up, and are attractive enough to the reader – the material is intended as a resource and stimulus for future academic work among social scientists from all fields of social science. To the extent that it contains empirical findings, it can be used secondarily in its present form for instructional purposes in certain social science fields with emphasis on urban (intra- or international) development. Specifically, it can be recommended for advanced level undergraduate or entry level graduate courses in urban and world geography, regional and urban policy sciences, urban planning, urban sociology, demography, anthropology, and urban and regional economics.

ACKNOWLEDGEMENTS

A great deal of the data collection and formation of ideas contained in this book took place during my sabbatical leave from the University of Kansas in the Fall of 1985. I wish to acknowledge the contribution of the following departments, research institutes, and universities for providing me with valuable information and for hosting me at various lengths during the 1985–86 academic year and during various trips since then: the Department of Geography at the University of Bombay; the Department of Architecture and the Department of Geography at the University of Singapore; the Center for Urban Studies and Urban Planning at the University of Hong Kong; the Building Construction Institute in Cairo; the Commonwealth Scientific and Industrial Research Organization, Division of Building Research in Melbourne; the Department of Architecture the University of Indonesia in Jakarta; the National Chiao Tung University's School of Transportation Engineering and Management in Hsinchu and Taipei; the Department of Urban Planning at the National Cheng Kung University in Tainan; the Department of Regional Development at Chung-Ang University in Seoul; the Department of Economics of Gakushuin University in Tokyo; the Institute for Socio-Economic Planning at the University of Tsukuba; the Department of Urban Engineering of the University of Tokyo; the Urban Construction Institute in Wuhan; the School of Architecture at Tsinghua University in Beijing; the Demography Institute of Fudan University in Shanghai; the Department of Urban Planning at Tong-ji University in Shanghai; and the Department of Geography at East China Normal University in Shanghai.

I am grateful to my colleagues Peter Nijkamp, Emilio Casetti, Borje Johansson, J. Barkley Rosser Jr, Daniel Vining, and the late Henry Mullally who have commented on parts of various drafts of this book and contributed many and diverse inputs and invaluable insights. I do not expect them to agree with all my opinions or findings, and I certainly do not hold them responsible for any errors in fact or judgement I may have committed in the writing of this manuscript.

I have run up a great intellectual debt to all the authors whose work I cite, even when doing so in a critical manner. They have supplied me with the stock essential for the development of these ideas. Some of these authors rank among the great minds of social sciences, and when compared to their contributions my efforts pale.

The seeds of this book were planted in 1983 when its nucleus was formed, while the computer work on the dynamics of the nineteen largest urban agglomerations was initiated in the spring of 1984. Through the course of the following two years the central ideas were more concretely shaped. A number of my students and research assistants worked on this project during the 1984–88 period. Jude Jimoha collected and coded the data sets; Saj Jivanjee drew the hand-drawn figures, while Jian Zhang assisted me with the computer-drawn graphs; Mehrdad Givechi carried out the statistical analyses included here. The main body of the book was initially written in the 1984–87 period, and hence it draws from data and sources available as of 1987. During the 1987–91 period numerous major and minor revisions of the text were undertaken while refining the central ideas. The book went to press in August 1991 during the aftermath of the significant events which had taken place in the USSR. These turbulent events will no doubt have implications with regard to issues covered in this book, in the light of which some sections will need to be revised appropriately when future trends become clearer.

Acknowledgement is made to Pion Limited for permission granted to reproduce Figures 2(b) and 3(b) from my article 'The decline of the U.S. economy: a perspective from mathematical ecology,' *Environment and Planning*, A, vol. 16, 651–62 (1984).

I wish to thank my editor at Routledge, Tristan Palmer, for his support in publishing this book and the editorial and production staff for their excellent work in producing this volume. Finally, I must add that no financial support has been sought from, and no part of this research effort has been sponsored by, any funding agency or any other source outside the University of Kansas, which is of course gratefully acknowledged. The importance of this will become clear during the reading of the book.

D.S.D.
August 1991

ABBREVIATIONS

ADC	aggregate developmental code
B	biological force
c.i.f.	cost, insurance, freight
E–D	economic–demographic force
f.o.b.	freight on board
LDC	less developed country
MCMR	Mexico City Metropolitan Region
MDC	more developed country
MSA	metropolitan statistical area
NGR	natural growth rate
PCGDP	per capita Gross Domestic Product
PCGNP	per capita Gross National Product
PRB	Population Reference Bureau
RIC	recently industrialized country
SAUS	Statistical Abstract of the US
SIC	Standard Industrial Classification
SPP	Secretaria de Programacion y Presupuesto

INTRODUCTION

SOME CENTRAL ISSUES

Undoubtedly, a major event of the twentieth century has been the acceleration in the increase of the globe's human population, when compared with that of previous centuries. Population grew from approximately 1.6 billion in 1900[1] to about 5 billion in 1987.[2] When plotting the world's population in time and going back to approximately 8,000 BC one sees an exponential growth curve to the present.[3] Clearly, the event is of considerable interest to all social sciences. Accompanying this event has been the considerable growth in the population of the world's largest urban agglomerations, to current sizes not even considered in theoretical studies only a century ago.[4]

Per capita income (or product) has experienced significant changes over this period as well. Far less precise to define and measure than population, its recording on a global scale has been very recent with somewhat reliable counts made available only in the 1970s. It has been estimated at approximately US$1,360 for 1973,[5] and to have reached about US$3,330 in 1987[6] in current US$. How it grew over the first half of the twentieth century, or during prior centuries or millennia, is largely unknown at present.

Income originates at a variety of sources, and comes in a multiplicity of ways. What people did with it in the past, what they do with it at present, and how they plan to expend it in the future have significant repercussions on the past, current, and expected future level of welfare and population stock size. At the same time, current and future population size affects the manner in which current and future income is generated and spent.

Multiple forces drove and drive the growth of these two variables globally. A host of local and global factors fuel changes in them locally, i.e., at various regions on the globe. Similarly, numerous factors affect the manner in which income is distributed to various population groups, inter-regionally and intra-regionally. What these factors are, have been, and are likely to be at the various regions of the world or *in toto* globally is largely unknown at present and a matter of continuous and intensive theoretical speculation and social debate. What is apparent is that the spatial distribution of these two variables is not uniform, instead demonstrating sharp spatio-temporal disparities or dualisms.

It is also safe to assert that there are multifaceted interactions linking various regions of the globe, at least in terms of population and per capita income. Through speculative in time and flowing in space individual and collective actions, behavior at one point in space–time affects behavior at another. Can the actions of a small vendor in Kalemie, Zaire today affect the fortunes of Paris next month? Maybe. But this turns out to be a question one wishes to ask, when looking at the global interdependencies linking the urban settings of the world, in the light of recent developments in non-linear dynamics and chaos theory. In addressing it, significant insights are obtained.

It is rather safe to assert that the factors which affect the dynamic behavior of these two variables are changing in time and differ over space. Further, it seems reasonable to hypothesize that the dynamics of these two variables over the long history of the human race have not been calm or stable. But their instability has not been unruly, for the human race would not have been able to survive. *Disciplined chaos* has prevailed over the history of mankind, permitting it to evolve and at the same time giving us some hints as to what is likely to occur in the future. This is what the study here will try to document, as far as this is possible. It will show that out of this limited chaotic dynamics a pattern emerges which allows one to outline certain major scenarios of evolution for the cities and regions of the globe.

The main focus of this book and the central socio-spatial event documented in this work is the *dynamic instability* in population and per capita product growth experienced individually by the largest urban agglomerations of the globe during the third quarter of the twentieth century. Instability has also prevailed in global (collective) dynamics. Dynamic instability is a complex notion requiring a definition under the various conditions in which it arises. At this stage, it is enough to define it as a state in which population and per capita income do not remain constant for relatively prolonged time-periods.[7]

As mentioned, urban population and per capita income are functions of a multiplicity of factors. Their unstable dynamics are the direct result of global interdependencies governing and linking the evolution of metropolitan areas. Individual city and collective urban dynamic patterns emerging from these interdependencies are large in scale and of a long-term nature. It is suggested that the interdependencies among the urban areas of the globe constitute a *macro spatio-temporal ecological determinism.* This general principle is useful in understanding growth and decline processes, not only of present-day major cities and regions but also of cities and regions of the past. Often, the argument is made by historians that studying the past enables one to better understand the present. Here, the reverse is also argued: namely, that studying present-day dynamics enables us to better understand the past.

Interdependencies are spatio-temporal; they are broadly defined as multifaceted associations among many diverse socio-economic stocks located in space.[8] These associations exist because they entail a set of perceived individual and collective multifaceted benefits and costs in space–time. Interdependencies cause diverse interactions (economic, social, political, cultural, environmental, etc.) which are manifested as intra- and inter-national as well as inter-temporal flows. Population migration, capital flows, information exchange, and commodities trade constitute the core of such spatio-temporal interactions. Other movements in space–time include the spatial spreading of diverse elements like technological innovation, social norms, ideas, theories, or disease. All these are examples of obvious spatio-temporal interactions. Movements in space and time of unemployment, poverty, environmental pollution, depletion of natural resources, political dominance (or subordination), power, cultural traits and demographic shifts, are a few other examples of less apparent inter-national and intra-national interaction.

The end effect(s) of multifaceted interpendencies and their embedded interactions can be conveniently and effectively captured by certain central variables, among them population and per capita income, in a *relative* framework.[9] These variables can be normalized by using either an aggregate (total), or a background prevailing average count. Employing relative rather than absolute dynamics may reveal qualitatively different conclusions to the analyst.

Spatio-temporal interactions (flows) keep various stocks found at any point in space and time at levels which are not haphazard or random. To an observer, spatial interactions result from a social (individual and collective) effort toward achieving a perceived (by the observer) *composite relative parity in net attraction* among locations. The principle of composite relative parity is a central component of the theory of macro spatio-temporal ecological determinism. Like a valve, this effort toward a relative parity in net attraction principle is perceived by an observer to regulate the size, direction and specific path followed by flows during the spatio-temporal dynamics of the various stocks located at different points in space–time.

From an observer's viewpoint these flows intend to act as equalizers of the perceived current bundle type net attraction among locations at any point in time.[10] Among other variables, the various urban conglomerates' current population stock and per capita income levels obey this composite relative parity principle. Their size in space–time is affected by spatial attractivity differentials, and simultaneously their changes affect these differentials in a non-linear dynamic manner. This non-linear interdependency is one of the central components of dynamic instability. Understanding it allows one to comprehend why, even though the system in aggregate is driven by this relative parity principle, sharp spatio-temporal dualisms prevail.

The strength, extent and selectivity of interspatial interactions are also components which directly affect the dynamic stability or instability of spatial population agglomerations as well as of other stocks. This finding, obtained from extensive theoretical work in mathematical ecology, has significant implications in the evolution of human settlements. Weak, extensive and largely random interactions produce dynamically unstable patterns whereas, strong, few and highly selective linkages underlie stable dynamics.[11] It is demonstrated here, through a look at international trade as a case study, that certain recorded economic interactions among globally large urban agglomerations and nations are highly selective and rather weak when viewed in absolute terms and from a worldwide perspective. Collectively, however, the diverse socio-economic spatio-temporal interactions may not be selective and instead exhibit relative strength – particularly when they are expressed in relative terms in reference to a worldwide environment.

Hypotheses offered on the total population size dynamics of the world's largest urban agglomerations draw from the fields of economics demography and ecology, as well as dynamical analysis. They employ the central features of non-linear dynamics including discontinuities,[12] periodic and non-periodic cycles,[13] phase transitions,[14] irreversibility, the slaving (i.e., domination) principle[15] whereby fast and small in spatial scale movements are determined by slow and large in scale dynamics, and other features.

All these notions emerge from the richness in the methods and phenomena recent mathematical theories invoke – particularly catastrophe theory,[16] bifurcation theory and the theory of turbulence and chaos.[17] Elements from the economic theory of *speculative behavior*, involving futures markets,[18] are also contained and form the backbone of a macro ecological determinism. Theoretical perspectives provided here are at variance with some long-standing assumptions, theories and models used in the field of urban, regional and national development, at a micro or macroscale. They differ from conventional analyses in that they mainly focus on multiplicity of state, dynamics, disequilibrium, and discontinuities rather than on uniqueness, statics, equilibrium, and continuities. Consequently, they differ as to the meaning and use as well as in the message they contain for models of policy making and public action.

Methods used here and associated findings involve 'observer' type large-scale and long-term deterministic dynamic models not directly appropriate for or open to social action or public policy making, simple in form but rich in outcome. To an extent, they provide material for reflection at the pre-policy or ideas-formation phase. The message they convey, however, goes far beyond this realization as the main intent here is to define the constraints set on urban policy making rather than to expand the realm of such policies. These models expose the severe limitations, ill definition, and often misguided action of public intervention.

Individual (private sector) and collective (public sector) determinism is contained in the models presented. The presence of multiple, integrated, aggregate, and deterministic, as opposed to a few, fragmented, disaggregate, and stochastic[19] private and public markets of exchange is stressed within a local and global environment. The multiple and severe market imperfections[20] are highlighted. Highly imperfect markets in – among other factors – population, capital, information, and commodities are seemingly overwhelmed by macrolevel dynamical determinism. A complex ecology of factors affects the evolution of and interactions among the globe's giant urban agglomerations under a long-term time horizon. There is no intention here to place judgement, or go beyond the analytical aspects of the urban macrodynamics these markets entail, largely because such judgements may be irrelevant. For in these macrodynamics the effects of various judgement markets may already have been discounted.

It is certainly an understatement to remark that the socio-spatial system under investigation is very large and complex in its inner structure. In spite of its scale and complexity, however, in the span of a quarter century the macrodynamics exhibit certain simple and persistent unstable patterns. That a complex behavior by a very large number of individuals and collectives on the globe can in aggregate produce simple outcomes, and that such complexity can be captured by seemingly simple mathematical constructs at a more or less random moment in the history of social evolution, is certainly of methodological and substantive interest.

Central to the specific findings is the documentation of the hypothesis that each large metropolitan area's evolution is determined by an empirically imputed *code*. This code is the outcome of all other large urban settings (and the variety of agents they contain) and individual and collective speculative fast[21] dynamic behavior. It is hypothesized that an intra- and inter-nationally prevailing relative parity in composite net attraction is the underlying principle giving rise to this code. To an observer, under a continuous effort toward adaptation to a worldwide environment to achieve this relative parity in net attraction, the socio-economic structure of these urban agglomerations rapidly evolves, in a relatively short time horizon (a ten- to twenty-year time-period). In the longer term, the level of this relative parity undergoes slow[22] dynamics. This book focuses primarily on fast dynamics. Some inevitable and inevitably speculative statements regarding possible slow motion in the code are also supplied in Chapter 4.

As a result of these interdependencies and relative parity, *relative development thresholds* which constrain the individual dynamics of the world's largest urban agglomerations are identified. Further, the existence of 'relative population size' and 'relative per capita product' dynamic distributions can be detected. The two distributions, in combination, point to a worldwide urban dynamic hierarchical scheme defined over these two macrovariables. Among other findings, and beyond the implications they hold for each individual giant urban agglomeration, these distributions allow also for an estimate to be made of the total current (as of 1980) number of urban areas on the globe, and the current average urban area's size.

Various data sets are used in this study, some extended over a century or so. The central findings, however, utilize available numerical evidence covering approximately a 25-year period of fast dynamics. Although only data from the largest cities of the third quarter of the twentieth century are employed, there are indications to support two broader hypotheses. First, these urban fast macrodynamics must have been in place at much earlier chronological periods having affected a variety of cities. Second, that irreversible slow dynamics of a much longer time horizon must have been in place, impacting the current large urban agglomerations. Certain implications of these inter-temporal comparisons and irreversible dynamics are addressed in Chapter 4.

The world's greatest cities are identified here merely as points on a phase portrait of relative population size and per capita product paths. Sharp focus is drawn on the documented analytical properties of their aggregate growth or decline as traced in a worldwide environment. Thus, this book does not intend to enrich one's knowledge of the detailed and varied past or current conditions (or predicaments) of these agglomerations. Nor is it a substitute for the host of publications, scientific or not, which collectively describe their inner structure. In center stage here is the purpose of learning more about their macrodynamics rather an elucidation where the reader is informed of the richness of their individual microstructure.

STRUCTURE OF THE BOOK

Briefly, the book is divided into four chapters and a set of appendices. In Chapter 1 and its associated appendices the central empirical finding is presented, setting the stage for the analysis on global interdependencies and ecological determinism to follow. The key notions in relative urban macrodynamics are defined, and the presence of chaotic dynamics within bounds is documented. Two fundamental forces are identified and extensively discussed. Appendix 1 elaborates on the notion of ecological determinism, and the nature of individual and collective speculative behavior is expanded. A brief survey into the current major theories of development is carried out in Appendix 2, with an eye on how these theories consider the behavior of this book's central macrovariables: relative population and per capita product.

Chapter 2 deals with aspects of instability in the dynamics of cities and nations. These instabilities are due to the extent, strength, and degree of linkages among spatially distributed stocks. Linkages between nations and their urban areas are studied in both absolute and relative terms by means of foreign trade patterns. Case studies dealing with the US and Mexico are presented, together with a glance into the economy of a number of major metropolitan areas of these two nations. In a relative lens, the concept of a worldwide urban ecology is expanded. The notion of a relative inter- and intra-national parity in net composite attractions is defined and initially presented. It is argued that it governs urban development as an organizing principle.[23] For an observer, this parity regulates the flow of spatio-temporally liquid socio-economic stocks.

Chapter 3 presents most of the empirical work. It addresses the unstable dynamics in the formation of spatio-sectoral hyper-concentrations and its allied concept of hypo-concentration. It demonstrates the individual relative macrodynamics of nineteen large urban agglomerations over a 25-year period and detects the presence of a relative developmental threshold. The relative parity principle is analytically presented from the point of view of an observer of the socio-spatial system, and it is argued that it can be used to describe the observed urban macrodynamics. This observer's principle is juxtaposed to the principles underlying participant type behavior, i.e., social (collective and individual) spatial action which is intractable to any observer.

An effort is made to derive a worldwide urban hierarchy, allowing for an estimate of the total number (about 130,000) of urban settings (with a population greater than 2,500 inhabitants) on the globe as of 1980. Deterministic and stochastic aspects of this urban hierarchy over the two relative macrovariables are examined. Chapter 4 follows and contains some rather highly speculative statements on alternative urban futures. It discusses certain methodological aspects of modeling socio-spatial dynamics, and makes reference to 'observer' and 'participant' type models.

A NOTE ON SOCIAL SCIENCE THEORIES, DATA, AND THEIR MARKETS

Next, a few remarks are made on the subject of social science data and theories. These remarks bear directly on the main subject of this book: how data are perceived and how theories are formulated regarding past, present, or expected future social events in the social sciences impact the manner in which the social system behaves. In social sciences, action, empirical evidence, theoretical speculation, and interpretations can never be fully identified and separated. This is the nature of social theory and action. Social analysts can use this condition profitably for getting at the core of what knowledge and understanding is in the social sciences.

Accuracy and reliability of the available numerical empirical evidence, as opposed to the verbal and more descriptive record mostly used in social sciences in general and development theory in particular, is an issue worth addressing at the outset. The discussion offers an opportunity to touch upon fundamental notions in: (i) the nature proper of social science data, and events; (ii) social science theory construction; and (iii) theories about past, present, and future events and current social action. Only a few observations are offered, since a more complete treatment of the topic would require one to step deep into a formal theory of social action, social science epistemology, the study of history, and arguments found in formal logic and theoretical linguistics.[24]

Socio-spatial dynamics are sequences of social events involving individuals or collectives and defined over space. Social events are multifaceted, in that many variables from a variety of social science disciplines can be used by different observers to record and report them.[25] They are fuzzy[26] and as a result they are open to multiple interpretations. Events occupy time, lasting different periods; some are temporally sharp (almost instantaneous at one extreme), whereas other events are more durable, lasting relatively long time-periods. Further, events occupy space; some again are spatially sharp (highly localized or of limited spatial influence), whereas other events enjoy a relatively large spatial expansion, possibly having a global influence.

A social event can be thought of as the state of a social system. The 25-year dynamics analyzed in this book can be construed as a socio-spatial event, or the state of the socio-spatial system under review over a 25-year period. The past and present state of a social system are never sharp in space–time. Both are always subject to a degree of ambiguity. Contradictory, conflicting and antithetical interpretations are obtained from the multiple recordings of social events, past and present. In the social sciences one must frequently wrestle with multiple viewpoints even on a particular individual's theory about a social system, a branching scheme that could go on *ad infinitum*. The vast diversity in Marxist analysis is a good case in point.

One could argue that these attributes are necessary for social evolution, and for theory formation and development. Not only are these multiplicities in recording and interpretation not impeding the performance of the system, but instead they may facilitate it by providing efficient means for it to function in space–time.

Fuzzy measurement is a key attribute of multiple interpretations and their associated complexities.[27] Numerical evidence, being unambiguous and sharp in nature (as opposed to the ambiguous and fuzzy verbal description), may not be as efficient a description of social events as the verbal description might be. Thus one might argue that numerical (analytical) treatment of socio-spatial dynamics is inefficient, despite the fact that numerical evidence may be more precise. Verbal description may carry more information regarding the event than the numerical evidence can. To accommodate the fuzziness of social events – for example, the dynamic behavior of metropolitan areas' population size and per capita product – social agents (for example, governments) provide multiple recordings of these variables.

Fuzzy recordings may also be the product of turbulent and chaotic motions of socio-economic variables in the phase portrait, which may not allow for precise estimation and measurement. The paths of socio-economic stocks in particular may be chaotic, impeding inadvertently or intentionally their accurate measurement at any point in space–time. Population and per capita income dynamics of any urban area, present or past, may have been blurry in the phase portrait and statistically difficult to plot precisely. In addition, chaotic behavior significantly impedes one's ability to estimate the model's parameters with precision. This is a new source of concern, raised by chaos theory, as a great deal of time series data – far beyond our ability to collect, now or in the future – are necessary.

Intricately related to this methodological question and difficulty is another source of ambiguity. The macrovariables under study here, as well as other socio-economic variables (and their associated theories), are periodically revised by the various agencies recording and distributing their counts. This periodic revision is an economic process containing economic and political profit. Data which have direct impact upon economic performance are subject to betting by economic agents. In fact, when data supplied by governments are regularly revised, the bets are on the size of the expected revisions. More broadly, social agents react to social science data recordings, on both their current counts and the expected revisions. Inexactness is an element which affords the opportunity for interested (directly or indirectly affected) agents to take bets. A 'quantum type' theory of social evolution might be constructed on the premise that social recordings always fall anywhere within bounds, in a manner so that no one can exactly pinpoint; but such a theory will not be elaborated here.

A fuzzy measurement is necessarily provided on otherwise sharp social science variables, depicting these well-defined but ill-recorded social science events. The need for such ambiguity is found in the nature of social science data markets. General statements about data and their markets may provide insights into the reasons why the historical record, recent and past, is incomplete and fuzzy, and why it will and must remain incomplete, ill-recorded, and fuzzy as well.

Data are, among other things, economic commodities.[28] They are one of the many input factors demanded by and supplied to the planning, management, and policy-making processes involved in social (individual and collective) action. They also figure in the demand for and supply of recording, theorizing about, explaining, and describing past, present, or expected future social events. At the time of their collection and distribution, social science data have a social, political, and economic price, i.e., an expected benefit and cost to their suppliers and buyers. As any other economic commodity produced, they require social, political and economic resources as inputs.

Such input factors used for their collection, processing and distribution, include capital, labor, time and technology, and a social, institutional and political infrastructure to handle them. Their supply is associated with a multifaceted expected profit to their collectors and sellers, normally governments. Their demand is linked to a multifaceted expected reward to their buyers and users. Of interest is the fact that, like all markets, data markets are imperfect. Rarely do they clear, as disequilibrium (excess demand or supply of data) may prevail. A variety of rationing schemes are used to allocate data (as well as theories) when excess demand conditions prevail. Data production and distribution markets are characterized by monopsonies or oligopsonies. Social science data are traded in both formal and informal data markets. Data have vintage and consequently their value depreciates (appreciating only rarely) over time, while new or existing data stocks accumulate.

Data are traded in highly complex dynamic seasonal options markets,[29] when time–place and holder of the primary data sets carries a premium. Some data sets are durable, i.e., they last relatively long, without being significantly changed (converted). The vast majority of social science data, however, are frequently revised non-durable commodities, as is the case with census data for example. All that and a bit more is well known by social scientists as well as by traders at various stock exchanges.[30]

What has not been widely discussed and carefully analyzed is the existence of markets for interpretations of social events, i.e., social science knowledge and understanding resulting from processing social science data sets.[31] Demand for these interpretations, and their supply and prices (i.e., of theories, knowledge, understanding, descriptions or explanations) are the central elements of these markets, being strongly linked to the data markets themselves. The year 1992 marks the quincentenary of the first voyage by Christopher Columbus to America. This social event, and its multiple current and past as well as possibly future revisions, interpretations and recordings, points to the core of the arguments in this book: the existence of a demand for and the supply of various interpretations of social events, and the concomitant net benefits accruing to the buyers and sellers of these interpretations.

Some of these interpretation markets (which may involve theories, fragments of knowledge, or partial understanding) are formal, although most are informal. Formal markets exist when a multiplicity of interpretations are made available by a large number of suppliers of such interpretations to a large number of adopters of interpretations in formal forms of exchange. The academic environment is one forum in which multiple interpretations are exchanged. Governmental policy-making agencies and the political arena are other means where presumably such exchange occurs. Absence of formal markets gives rise to informal trade of interpretations.

Present and past social event interpretation markets are not efficient – indeed, they are highly imperfect. In particular, they are imperfect because they are very thin, as a very large number of social events took and are taking place. Consequently they provide grounds for many theories to be drawn up and in turn for many theories on such theories to appear and coexist. Very rarely does one find a large number of suppliers of a particular social event interpretation, so that monopolies or oligopolies in interpretation markets may occur, together with many theories supplied in excess of demand.

Thinness in adoption of a particular viewpoint is accentuated by the fact that information diffusion (i.e., spatio-temporal theory-data arbitrage) on a particular social event is not always very fast; and it may entail a variety of transaction and transmission costs and market imperfections including a host of externalities. Individuals or agencies close to the source of information may enjoy significant comparative advantages against those that are remote in space–time from it. The latter example is particularly obvious in the case of certain governmental agencies holding on to, and delaying the publication of, various census reports for a host of reasons.

Timing of publication and the form of the published information is subjected to a 'rational' decision-making process by the suppliers of data, particularly governments. Here, the term 'rational' implies that some objectives are to be satisfied subject to constraints in the release and distribution of data sets. It directly impacts upon the supply of interpretations, social actions, and their multifaceted prices (i.e., expected benefits and costs). Presumably this is why the reporting of politically and economically sensitive data is subject to strict regulatory controls in many instances. None the less, insider trading involving data and theories is widespread in social sciences, public policy and social action.

Although at any point in space–time a number of interpretations may be traded in interpretation exchanges of a particular social event, interest among them is rarely evenly distributed. This is because at present the currently perceived and discounted multiple future benefits and costs are unevenly distributed in space–time both to their suppliers, distributors and adopters, as well as to the recipients of actions resulting from such interpretations. Among other reasons, social conflicts may arise as a result of different interpretations competing for dominance, as well as the varying expectations regarding the possible prices (in terms of benefits and costs) these interpretations may convey to different social (and scientific) groups. Certainly, revolutions (scientific and social) fueled by ideology (interpretations) are a testament to this observation.

A type of dynamic options market of interpretations is thus present, clearing (but often failing to do so) demand for and supply of interpretations regarding past and present social events in space–time. It rarely performs efficiently, since it is rather thin and thus subject to distortions. Documentations, recordings and interpretations of past and present social events are intricately related to the occurrence and course of these events. Social systems and their development are not immune to the functioning of data and interpretation of options markets. Through arbitrage behavior by social (individual and collective) agents, social systems tend to react to the various recordings and interpretations being exchanged, when action vacuums are created and opportunities arise to those enjoying a competitive advantage.

Theories and data (information) offered and adopted, particularly the heavily traded ones, directly affect the dynamic behavior of the social system. This intrinsic interaction between interpretations of social events and events proper is uninterrupted. It continuously updates not only the currently perceived prices (the multifaceted bundle of expected socio-economic benefits and costs) of these options (odds and payoffs) on recordings and interpretations to their suppliers and users through some vague accounting method, but it also affects the speculative behavior and thus the dynamics proper of the social system at hand by altering the currently perceived odds.

Thus, an intricate linkage is established between the social event, the data recording it, the interpretations of the event, and the ongoing and future causes and effects of the event itself. Speculators and hedgers[32] of social action outcomes and their interpretations, together with spatio-temporal information brokers and arbitrageurs strongly link the subject of social science (theory construction) to its object (social action).

As the past and present are fuzzy, the future is even more unclear.[33]

The future's multiple possible perceptions, at present, allow for speculative behavior by social agents, thus affecting the future course of events. Positive (self-fulfilling) or negative (self-defeating) feedback between theory and action has a number of implications, particularly to the extent that forecasting of anticipated (public or private) social actions for purposes of management, planning or control is sought. Clearly, it severely limits the claim of objectivity or rationality placed on the sought-after forecasts. It demonstrates the fallacy of what is argued in economics to be 'rational expectations.'

These 'forecasts' (often termed 'projections' or 'predictions') are nothing more than simple expectations (beliefs or reflections) by the individual or collective supplying the particular interpretation made available for specific purposes in mind. These forecasts allow the social agents with interest and appropriate resources to take bets on them. They do not fall under the category of scientific predictions, in the sense that a natural scientist would refer to predictions (Popper 1957).[14]

Nowhere else are these comments more in evidence than in the case of census information (in the form of either 'actual counts' or estimates supplied by national governments) in both developed and developing nations. Population counts in particular are of extreme importance, as they affect resource allocations within national boundaries by various government agencies. This example is a clear-cut case showing how, in an interactive manner, the data recording affects social behavior and in turn social behavior affects data recording through their perceived rewards to the various agents involved in their markets and distribution.

In view of the above comments, and with regard to the models of this book, one may ask whether these models also obey the above-mentioned market determinism. The following two positions are taken regarding this book's models, which justify the reason why such a lengthy note on social science theories and data was provided:

1 An almost 'end effect' urban macrodynamic outcome is presented by these models. These macrodynamics have, to the maximum extent, discounted for the complex ecology of interactions among the multiplicity of speculative agents in this vast system of observers and participants they contain. The urban macrodynamics presented here have, at the utmost possible extent, discounted for the presence in space–time of the infinite (or at least a very large number of) data on and interpretations about past, present and anticipated future social events and actions.

2 Although the information from these models may be novel, it is of the utmost macroscale, thus minimizing the chance that any social agent (no matter the scale of the collective) could meaningfully affect it. Thus, the processes presented here allow for the most detached and exhaustive qualitative scenarios to be considered, without provoking reaction that would significantly alter the paths to be traced. To the extent possible any feasible and material reaction (i.e., relevant and significant) is already discounted in the recorded dynamics.

These two positions, however, do not necessarily imply that any of these scenarios can be quantitatively sharp, or prove accurate *ex ante.* They merely present a few possibly thinly traded options on the world's urban areas' futures. The code which is derived by the urban macrodynamics postulated here may not be subject to manipulation. Only by discounting the policy-making process in the mathematical form of the code can one construct models which might be considered *asymptotically approaching internal consistency in observer type models.* This consistency argument may only be valid (and feasible) at the most macroscale, long-term view one can obtain on urban macrodynamics. At the microscale and in the short-term social events, their recordings and their resulting interpretations may not afford the opportunity for such consistency.

In any event, given the sources of the data used here (mostly from the United Nations) some readers may question their relative quality (or accuracy), and as a result the relative reliability of the findings and conclusions drawn on urban macrodynamics. Even if one does not accept the consistency argument made earlier, this is not an appropriate (or relevant) concern. One always must extract the maximum quantity of information out of the current stock of data. Future additions to and modifications made on the current stock will confirm, modify or replace the existing theoretical propositions. This is the process which underlies all scientific development and evolution within the highly complex ecology and imperfect markets of social science information and theory. One must constantly keep in mind the above brief notes when reading this book.

1

GLOBAL INTERDEPENDENCIES

INTRODUCTION

Frequently, in many regions of the globe and through centuries of human history, the growth and decline of population and wealth in many human settlements has accompanied significant historical incidents. Landmark social economic and political events, ranging from wars to the rise and fall of empires to scientific breakthroughs, have been associated with changes in these two stocks. Such major events have at times triggered, often followed, occasionally coincided, and in other instances simply intermittently interrupted, critical phases in the dynamics of these two variables. Historians have spent significant efforts in attempting to detect such associations. Their emphasis has been, however, on chronicling major events rather than linking them to the underlying population and wealth dynamics in space–time.

To socio-spatial analysis, on the other hand, it is whether such events precede, succeed, coexist or intermittently occur in various identifiable phases in the dynamics of these two key variables which is of importance. When such phases are clearly evident in industries, cities, nations or empires, and contain significant modifications in the size of these two stocks of socio-spatially dominant entities, historians name these periods 'eras.'

Whether one confronts topics in paleontology, paleoanthropology, archeology or contemporary history, major events and the underlying population stock size (of animals, plants or humans) on the one hand, and the food resource base level (the intake nutrient and energy levels for animals and plants) or the income resources and wealth levels of humans on the other, are closely linked. It is surprising that no systematic study has been undertaken to date connecting such dynamics to the major socio-spatial events in history, let alone prehistory. Clearly, major events affect these dynamics and in turn these dynamics affect the occurrence of such major events in an interactive complex and non-linear manner.

Changes in population and in the wealth of cities, regions or nations are important causes of events in human history. They are also the outcome of a variety of forces which are at work within a specific social system at a particular location and time-period. Consequently, one cannot expect the dynamics of these two stocks at any point in space–time to be either uninformative or random. Instead, these dynamics must, in abstract and often succinct form, reveal a comprehensive story of fundamental change in the multifaceted phases of development and evolution of human societies and their settlements. A look at the recent history of these two variables at a number of locations may point to key processes underlying events of the past and foretell events at some time in the future.

That population and wealth are two central variables influencing and being influenced by one another and that in combination they efficiently record various socio-economic growth processes are hypotheses currently affecting theoretical developments in spatial dynamics.[1] Urban, regional, and national development paths are now being studied in view of a dynamic interdependency between population and wealth.

Studying this basic interdependency is nothing new to social science. Demographers,[2] economists,[3] geographers,[4] sociologists,[5] anthropologists[6] and other social scientists and historians have examined this interdependency over the past two centuries. Renewed interest in these two variables' dynamics has been injected by recent developments in the theory of non-linear difference and differential equations in mathematics. These advances significantly enlarge and possibly modify one's perspective on urban growth, as well as allowing one to reflect more broadly on subjects of socio-spatial development and evolution. They are the sources of some new insights into the study of history.

Briefly, three main themes seem to emerge out of non-linear theory. First, complex socio-spatial processes can be described by simple dynamical models. Second, out of very simple deterministic dynamic model specifications, over a few central variables, extraordinarily complex, powerful and insightful outcomes emerge. These outcomes shed light on the divergent, chaotic, or seemingly stochastic dynamic patterns which describe the qualitative features of real systems dynamics. Third, micro and macro socio-spatial systems alike are marred by an overwhelming likelihood for instability, this being argued here as the main factor contributing to the presence of extreme socio-spatial dualisms and disparities. Instability is caused by extensive socio-economic interdependencies and resulting interactions among social stocks in space–time. Key events in development[7] or evolution[8] are the by-products of this instability. Thus, the framework adopted does not require one to resort to complicated and at times convoluted and largely ideological theories (of Marxist or non-Marxist variety) to describe and account for the presence of such disparities. They are simply found to be inherent in any complex dynamical system.

With this environment as the background, considerable attention has been attracted to the fact that in 1987 the world's population toppled the 5 billion mark. An increase in interest has been shown by urban analysts with regard to the significant growth in absolute population size of the world's largest urban agglomerations and the various predicaments that this has resulted in during the past quarter century or so. Combined with the inevitable question of whether such growth rates are sustainable, one is motivated to explore these relatively recent dynamic events by utilizing the novel analytical methods made available from bifurcation theory. This study is a step toward this direction.

MACRODYNAMICS OF THE WORLD'S LARGEST URBAN AREAS: ORDER IN CHAOS

A principal finding of this book, and a key stimulus for the *deterministic ecological approach* theoretical framework adopted here, is that when the relative per capita gross product of the world's major urban areas, recorded as the ratio of their individual per capita product to the pervailing current average of the world's market economies, is plotted against these urban areas' share of the current world population over a time-period the emerging dynamic paths are not found to be random. Indeed, they turn out to reveal important clues in the study of the evolution of human settlements.

Data from nineteen metropolitan areas are presented in Figure 1.1, with population size greater than five million inhabitants as of 1975.[9] The dynamics of the world's largest urban agglomerations as shown in the phase portrait (the two-dimensional diagram which records their paths), demonstrate very informative and qualitatively quite robust dynamical features. In Figure 1.1, the observed share of a metropolitan area's population to the current world total (in 10^4) is recorded along the horizontal axis (x). In the upper diagram the vertical axis depicts ten times the current per capita domestic (or national) product to the world market economies' prevailing average, y, derived when both are expressed in current US$. Its logarithmic expression, $\ln y$, is shown in the upper diagram. The phase portrait contains the signatures from nineteen metropolitan agglomerations, expressed as the simultaneous dynamics of these two variables during the period 1958–80.

Notable is the decrease/increase in relative population above/below *a relative per capita product developmental threshold* level. Urban areas with a per capita product ratio higher than approximately 74 percent of the world market economies' current average per capita product, shown by the lines $\bar{y} = 7.4$(or in ln $\bar{y} = 2$) in Figure 1.1, exhibit declines in their share of the world's population. Urban areas below this threshold without exception increase their share of global population. This threshold exhibits a resilience throughout the admittedly short study period. However, it enables one to draw certain inferences and mildly strong conclusions about the settings' *macrodynamics*, in spite of the possible limitations characterizing United Nations data. The existence of a threshold seems to be confirmed by other researchers as well, although in absolute growth studies.[10]

One cannot fail to notice that nations whose urban areas lie underneath the threshold belong without exception to what the UN classify as less-developed countries (LDCs); whereas nations whose urban areas lie above the threshold (with the possible exception of Argentina) are classified by the UN as more developed countries (MDCs). Whether infinitesimal or significant changes in an urban area's per capita relative product have occurred over the 25-year period depends heavily on its past and current location relative to this threshold. Thus, this threshold seems to represent a lift-off level. Below it, urban wealth variations over time are almost negligible. The growth in a particular LDC metropolitan area's population share of the world's total dominates the urban area's dynamics. Above this threshold lie MDC metropolitan agglomerations with very pronounced changes in both population share and relative product, varying during the study period between .9 and 5.5 times the world's average per capita product.

The line $\bar{y} = 7.4$ seems to be a part of a code which separates two areas with distinctly different dynamics in the relative population–product space. Each metropolitan area's macrodynamics are dictated by its starting value, in reference to this line, on this phase portrait which contains the code. At this scale of analysis and at this point in time dynamic paths depicted in the phase portrait indicate a mild oscillatory motion and a strong indication of irreversibility in their qualitative properties. It is indeed quite interesting to recognize the persistence and non-randomness of the pattern formed in one quarter century as shown in Figure 1.1.

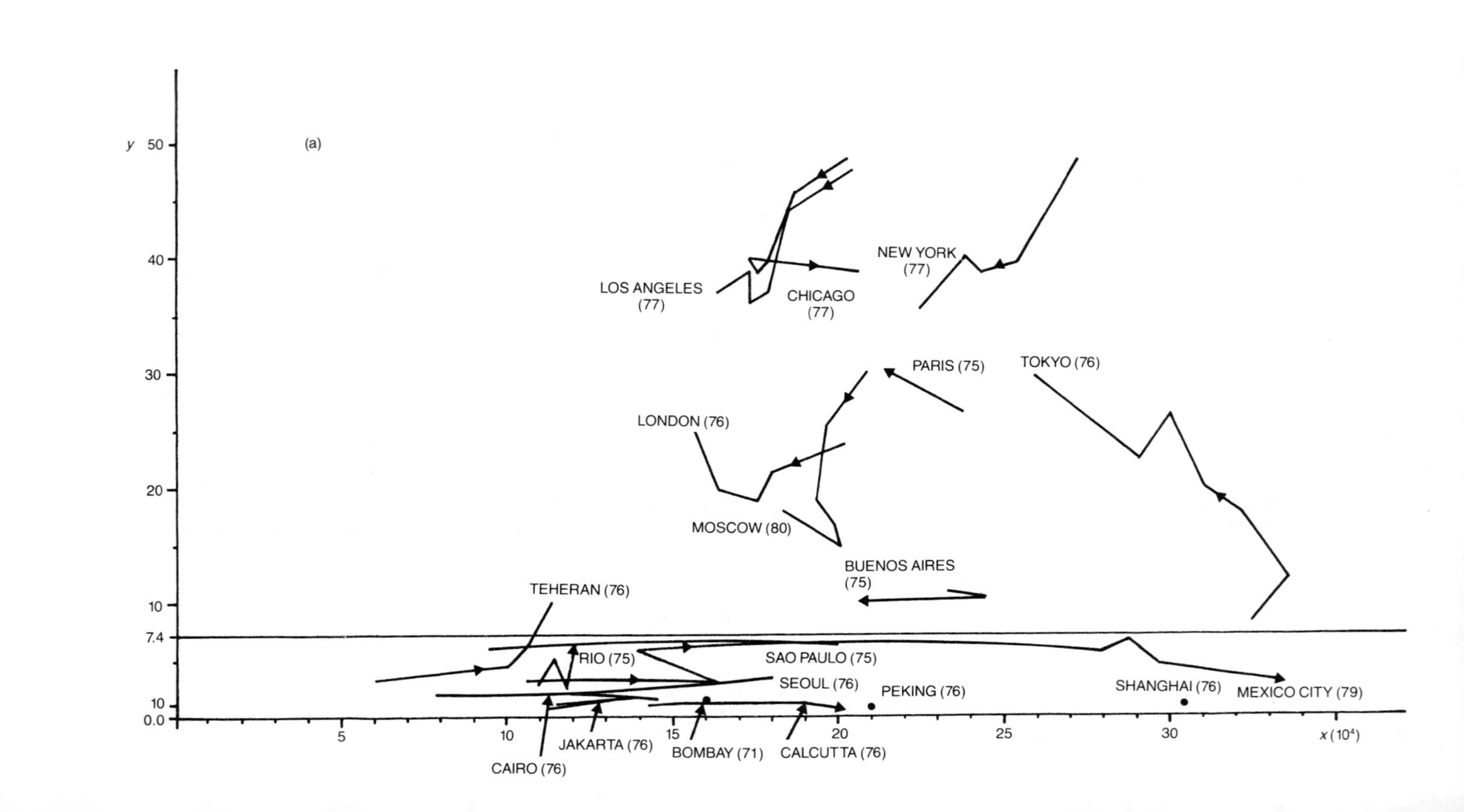

(a)
y
50
40
30
20
10
7.4
10
0.0
5
10
15
20
25
30
x (10⁴)
LOS ANGELES (77)
CHICAGO (77)
NEW YORK (77)
PARIS (75)
TOKYO (76)
LONDON (76)
MOSCOW (80)
BUENOS AIRES (75)
TEHERAN (76)
RIO (75)
SAO PAULO (75)
SEOUL (76)
PEKING (76)
SHANGHAI (76)
MEXICO CITY (79)
CAIRO (76)
JAKARTA (76)
BOMBAY (71)
CALCUTTA (76)

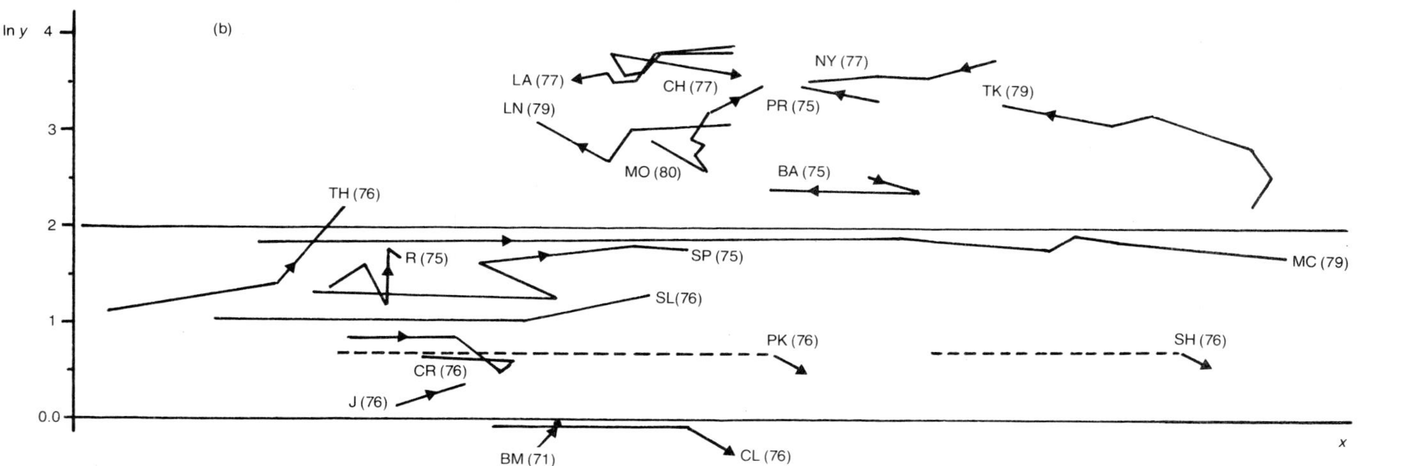

Figure 1.1 The dynamic paths of the world's largest urban agglomerations (more than 5 million inhabitants in 1975). On the vertical axis the actual ratio y of their per capita income to the world market economies' average is plotted (a), along with its logarithmic ln y transformation (b); on the horizontal axis their share of the world's population is recorded. The study period is 1958–80.
Key: BA (Buenos Aires), BM (Bombay), CH (Chicago), CL (Calcutta), CR (Cairo), J (Jakarta), LA (Los Angeles), LN (London), MC (Mexico City), MO (Moscow), NY (New York), PK (Peking), PR (Paris), R (Rio de Janeiro), SH (Shanghai), SL (Seoul), SP (São Paulo), TH (Teheran), TK (Tokyo)

In it many urban agglomerations are depicted as belonging to nations which differ significantly in many dimensions. These nations differ in their political structure, level of economic development, demographic ethnic, cultural and social structures, population and area size. The quantity and type of natural resource abundance varies greatly, their location spans the globe, and in general they differ in many other ways. They share only a few common elements in the historical period these recordings were taken.

During that quarter century, these cities experienced events which locally might be considered as being of large scale with long-term implications. The events, of great historical significance to these cities and the nations they are situated in, ranged from local conflicts involving war and revolution, to radical changes in their economic and political structure, to major socio-cultural transformations, to considerable technological innovation. In view of these myriad local events, and the effects of global events as well, the stability of the code during that period and the largely non-random motion of individual urban areas in the phase portrait of the code are remarkable. This resiliency indicates the presence of certain strong and global forces at work, governing the behavior of this macrosystem beyond the relative microevents just listed and having occurred at many locations at different points in time within the past quarter century.

As noted, looking at the phase portrait of Figure 1.1 one detects an *oscillatory movement* in almost all the urban settings's dynamic paths, these being confined however within their *particular field or domain of motion.* An obviously irregular path describes each urban area's dynamic within the phase portrait. None the less, each path's oscillations are *restrained* in their irregular motion. The apparently chaotic movement each trajectory exhibits is not totally random, but rather it is confined within a domain of motion obeying *floor* and *ceiling* (capped) dynamics. One does not see either unrestrained chaos or the semi-regularity of quasi-chaos or the strict regularity of periodic movement in these domains.[11] When a temporal (lagged or not) mean value is traced for each city's path, these averages seem to behave well within the phase portrait. The temporal average path for each city exhibits certain strong regularities in spite of the apparently chaotic motion that these trajectories collectively demonstrate. *Order in chaos* is obviously present in these dynamics, as is the case too for the dynamic paths of each among the nineteen cities. These trajectories are not rampantly unruly. Why? What is one to conclude from this? And what for?

Answered briefly, these motions seem to obey some scant regularity possibly because it might not be sustainable for these movements to be either totally chaotic (i.e., purely random), quasi-periodic, or periodic. The observed pattern allows for an ambiguous enough outcome to emerge, not too wildly unruly or fully predictable, so that the system can exist, i.e., survive and evolve, and also tell a number of stories about itself. Ordered chaos may be a prerequisite for socio-spatial evolution. This evolution might be allowed to occur along paths permitted by a code, which contains some flexibility in that some neighboring paths within a given domain assigned to a city may also be possible. The specific paths recorded might be slightly off the statistical averages of these bands in the phase portrait where the actual paths of these two fuzzily measured[12] variables are allowed to move.

It may be concluded that if real (as opposed to model) chaos is present in each city's trajectory then the actual chaotic movement of the two macrovariables' *macrodynamics*, i.e., the dynamics of these two aggregates, must also take place within bands for each city. Only the general neighborhood for these bands can be and is outlined, inside which these measurements and movements – based on the UN or any other data sources – are traced. In the phase portrait of Figure 1.1, where the domains of motion for cities belonging to the less and more developed nations converge to or are restrained by roughly identified regions, these latter may be construed respectively as possible 'strange attractors' or 'strange containers' of these macrodynamics.

Strange attractors (or containers) in non-linear dynamics are areas in the phase portrait where dynamic trajectories wander around (are attracted to or are confined by) in their unstable non-periodic motions. How many attractors exist, where exactly they are located, and what is their approximate size if indeed they really exist beyond their model specification, might have significant implications for each city. To answer these questions, if such answers are ever obtainable, requires the acquisition of far more extensive time series data than that currently available which at present allows only modest speculation about them.

For what reasons do these specific individual trajectories and their possible 'mean paths' or their strange containers and attractors exist in the phase portrait? What purpose do they serve now or have they served in the past quarter century? What historical processes are responsible for them? What do they allow by way of dynamic paths for the future? To answer these, as well as address more extensively the previous questions, a general theoretical framework must be set up.

INTERDEPENDENCIES AND ECOLOGICAL DETERMINISM

In social sciences there is a need to address and define in a dynamic context (albeit approximately) the notions of interdependency, interaction, force, and associated concepts. This is done next, in subsection 1 within the framework of what is further defined as ecological determinism. Following these definitions, the analysis focuses on some fundamental forces at work in macro socio-spatial dynamics, in subsection 2. A speculative component inherent in these forces is presented in subsection 3. The distinction between micro and macrodynamics is explored, and in a parallel mode the difference between aggregate and disaggregate social behavior is analyzed in subsection 4. Proximity and overlap among these topics makes for a rather fuzzy set of headings, and a largely artificial segmentation and sequence in these subsections.

Having dealt with the basic theoretical framework of this book, the Malthusian and other demographic models related to the main concerns at hand are looked at in subsection 5, with a critique of the basic economic growth models and some other relevant models about socio-spatial development (from Marxist sociology and cultural anthropology) being supplied in subsection 6.

1 Interactions and interdependencies

No matter which specific discipline they belong to or what their subject of investigation is, development studies address issues of growth or decline in a number of stocks. They also address issues of evolution which typically involve changes of state in the dynamics of these stocks. In particular, socio-spatial development work analyzes the growth or decline of social stocks and their possible phase transitions in space–time. Changes occur when stocks are subjected to forces. Forces operate over space and time linking spatially distributed stocks.

It is important for the analysis which follows that the reader considers a basic distinction in social-spatial dynamics: that between the spatio-temporal, socio-economic *interdependencies* among social units and the resulting spatio-temporal socio-economic *interactions* among them. Interdependencies are the forces operating among stocks in space–time, and through their ensuing interactions they produce growth or decline in stock sizes and possibly evolution in their dynamic behavior. Interdependencies are the invisible forces operating among social entities, whereas interactions are the various and largely observable (although at times unobservable, as is the case with information) flows among them.

Some socio-economic interactions among particular social stocks are not only observable in space–time but also recordable. Socio-economic interdependencies among social units on the other hand, never being directly observed, can only be indirectly deduced. Interdependencies are the basic source of speculation for theory building and social action.

Spatial economic interactions have been extensively studied in the fields of micro and macroeconomics and the allied fields of urban and regional economics and geography where flows of information, commodities, finance capital, labor, etc., have been addressed. These direct economic linkages in the form of interregional trade, transportation, and other related economic activities are neither unique nor the most important in shaping spatial development at all points in space–time. A complex ecology exists of various social economic and environmental spatio-temporal interdependencies, where many diverse linkages occur.

Interdependencies among socio-spatial stocks include, among other types, economic, social, demographic, cultural, and political elements. These interdependencies are, among other things, spatio-temporal forces of attraction (push) or repulsion (pull) among regions, nations, urban areas or other spatial units. Their effect is flow of various stocks among spatial units, in time. When aggregate net attraction at a location occurs, inflow (or invasion) of stocks results; whereas, the presence of net repulsion produces outflow of (or evacuation, abandonment by) stocks at a particular point in space–time.

Another result of spatio-temporal interdependencies is the phenomenon of spatio-temporal diffusion, not only of physical stocks, like population-built capital stock or wealth, but also of a vast variety of non-physical socio-economic variables. Examples include, without being limited to, the spatial movement of knowledge, theories, ideas, ideologies, demographic indicators (like birth- and death-rates, fertility rates, the shape of a population's age structure, etc.), cultural or social norms, various degrees of capitalist or socialist modes of economic, social, political organization, etc.[13]

Many social science disciplines study these and other forms of spatio-temporal interactions and interdependencies and attempt to connect them to many aspects of social development.[14] The core social science fields (economics, political science, demography, psychology and sociology) as well as the peripheral ones (anthropology, human geography, urban and regional analysis, history and others which draw from the central fields) constitute various disciplinarian efforts toward the recording and analysis of socio-spatial events. Analysis of events involves an examination of social (individual and collective) action, i.e., social *outcomes*, and social (individual and collective) *outputs*, i.e., artifacts, commodities, information, built capital stock, or population. Outcomes involve complex utilization of outputs.

All these fields are used to approach individual and social development in general, and urban growth and decline in settings still in existence or long extinct in particular. An observer using any one discipline alone is bound to obtain and provide only a partial, biased, potentially misguided, and often irrelevant view on social development. The main thrust of what is proposed here is to comprehensively account for all these diverse types of interdependencies and interactions, to avoid irrelevancy and minimize bias to the extent possible. Consequently, the term 'ecological' is used to indicate that all disciplinary components are meant to be included, and that the term supplies a disciplinarily aggregate theory in which a bundle of diverse (sociological, economic, political, psychological) forces and linkages[15] are incorporated. It is intended to provide a broad framework where 'elementary' components are sought which might lead toward the construction of an all-encompassing perspective.

Models or theories belonging to different disciplines do not necessarily employ different variables or parameters than those utilized in the ecological framework. They tend, however, to emphasize different social stocks and different interdependencies and interactions among them. Although they seem, to a large extent, to operate in a disciplinary isolation they all seem to have a certain overlap. This overlap provides an excellent launching pad toward identifying elementary ecological bundle-like forces. Some key indicators, which tend to be more amenable to quantification, are more often addressed in these studies than others. In particular two of these indicators, population size and per capita wealth (and their growth rates), seem to be central to these studies.

These two stocks also happen to be the two variables with the most extensive time series counts available, when compared with any other indicator.[16] Thus their choice as central components of the ecological analysis is not coincidental. Population and per capita income data figure prominently in United Nations publications, and in studies by the International Bank for Reconstruction and Development,[17] among other sources of development studies and social development policy.

A central assumption in this study, and a building block for a comprehensive ecological approach to development, is that the population size of spatial units within an environment and the per capita wealth of these units are responsible for a broad array of developmental and even evolutionary events. These two central variables are proxies and determinants, within a given environment, of a complex set of social actions. They are two dominant variables representing a composite, comprehensive bundle of outcomes which could be expressed by numerous variables in that environment.

Within the ecological framework a bundle of forces is always likely to be associated with any social event. No single event is probably exclusively due to a single factor depicted by a single variable in time, no matter how dominant that factor or variable may be in the basket of forces. This is a basic dictum in social science theory, and, coupled with the fact that such dominance might not be frequently durable, causes a great deal of difficulty in carrying out empirical, comparative, statistical or econometric research. Repetition and controlled results of experiments and fundamental assumptions in probability theory and statistical inference may be frequently violated in socio-spatial analysis. The difficulty is that such bundles themselves may not be temporally durable at a particular location, or over different locations, at a particular point in time. This issue, known as the transferability problem, is well known among socio-spatial analysts and is largely self-evident. Spatio-temporal variability may be such that no trace of the bundle may be strong enough to detect it; not enough time series data are ever left behind for observers or participants to collect, particularly from some distant point in space–time.

A celebrated example and a good test for the ecological approach is that of the literature associated with the cyclical behavior of the Maya cities. Many factors have been cited as the cause for the sudden abandonment of the Maya cities in the Yucatan Peninsula and none stands out as the dominant one, given the current evidence.[18] Similarly, the population growth or decline of modern metropolises cannot be attributed to a single force but rather to a combination of factors. The thesis is advanced that among all these variables depicting the many forces at work the most dominant and durable are most likely population and per capita wealth. In contemporary and past civilizations, the interdependence between growth or decline in population and per capita wealth is manifested through complex socio-economic and cultural events.[19]

Certain outcomes of these interdependencies have been the subject of intense investigation by historians and social scientists. Interactions associated with growth or decline in population or wealth are the various regional population migrations that have taken place over the millennia of human history. Partly, interregional migratory movements have been a response to either perceived overpopulation or to a perceived inadequacy in a population's resource base.

Within this composite bundle-type and comprehensive approach to development, population and per capita wealth play not only a dominant and lasting but also a dual role. They contain multifaceted effects of development and at the same time, through their interdependence, they entail causes of development. Thus, in these two central composite variables cause and effect coalesce. This feedback may produce a 'bandwagon effect,' whereby declines in population and/or per capita wealth feed on themselves in time, at a particular location.[20]

A very convenient, simple and efficient manner in which to model interdependencies in a comprehensive way is to employ relative rather than absolute analysis. The choice to use an absolute or relative lens is critical in terms of the insights it produces and the dynamic properties it entails.[21] When one looks at variables in relative rather than absolute values, two objectives are attained. First, a precise definition of an 'environment' is reached – something that has eluded absolute analysis in many fields, including biology. Second, one captures the stock growth elasticities, rather than absolute growth rates, when formulating their dynamics in reference to their environment. Elasticities of growth are always informative parameters.

Different environments may enter the picture depending on the total used to derive the ratios or shares, i.e., the spatial and temporal extent employed to normalize the stock size and to state the relative dynamics. Thus the dynamic behavior of the stocks may differ depending on the environment used to normalize these variables. Here, analysis of the two central variables at the urban level is carried out in terms of worldwide totals, and with developed or developing nation subtotals. As will be seen later, relative analysis may be a bit more informative than absolute development studies as it accentuates the notion of comprehensiveness.

The key element of the relative lens is that through it one can see an *indirect* force linking a specific spatial setting's stock dynamics to the dynamics of the environment over which it has been normalized. The strength of this force might be greater than the *direct* forces affecting absolute development, no matter how comprehensive the absolute analysis may be. This is a key argument and to an extent it is documented here by empirical evidence. Such indirect forces are not concrete but instead 'abstract' and 'ecological.' Their abstraction may have eluded socio-spatial analysis for too long. The subject of relative dynamics is rather new in urban development literature. It differs significantly from absolute size dynamics. At times, it paints quite a different picture to the one painted by absolute growth/decline dynamics. The two lenses also differ in the dynamic stability properties they contain. Relative dynamics are more likely to be unstable than are absolute dynamics. In relative dynamics interactions involving zero sum games and competitive exclusion are likely to be very abundant.

Undoubtedly, significant advancements have been achieved toward knowing and understanding direct and absolute linkages among social and other stocks and locations within various environments and time-periods, no matter how narrowly defined these linkages might have been by the specific disciplines employed. There are very few informative and interesting interdisciplinary dynamical approaches currently available; some are reviewed later. There is little of a comprehensive abstract ecological theory to enable one to describe and speculate on relative socio-spatial aggregate stock size dynamics. Even less is available that one would wish to have in order to reflect upon the richness of the patterns they exhibit. Next, an effort is made to bridge this gap.

2 Fundamental forces

Obviously, the study of the multiple social and economic spatio-temporal interactions and most of the interdependencies among the world's urban agglomerations are not and could not be the subject of this or any other single book. Instead, the focal point in this abstract ecological approach is the modeling of the most basic of the global interdependencies among the world's largest urban agglomerations. These interdependencies are picked up directly by an association of two central macrovariables: relative population and per capita income (product) in reference to the world's current levels in population and per capita income. Events portrayed by these interdependencies are indeed quite telling.

One can bypass modeling the complex and multifaceted interactions and numerous other interdependencies among these and other urban agglomerations, their hinterlands and their nations. Random, deterministic, or stochastic elements in these interactions and interdependencies are apparently smoothed out at the most aggregate scale of analysis. The end result is that this internally very complex dynamic ecology is quite simple and deterministic at this macroscale.

At such a macroscale determinism is not confined solely to the relative macrodynamics among these two variables. Determinism has also been found in the *absolute* population growth paths at a global scale. Such determinism has been noted by many analysts in the past for absolute population and per capita income dynamics. It was noted at the outset that according to the record during the 1950–87 period, the human population stock has been growing at an average annual growth rate of 2.08 percent.[22] Data on per capita income are not as readily available, on the other hand. Evidence seems to suggest that during the 1974–86 period the world's per capita income increased at an average annual growth rate of approximately 10 percent in terms of current US$.[23] If the per capita income is translated into constant US$ then one might see that very little has changed in real per capita income terms worldwide in the past fifteen years or so.

Looking at the available historical record, covering approximately a quarter century, one is drawn to the conclusion that absolute growth in population and the rise in per capita income have been subjected to two ecological bundle-like fundamental forces. The first, which will be referred to as the *biological force* (B), dates back to an earlier observation by Malthus: human populations, in absolute terms, tend to grow at an exponential rate if unchecked. That population stocks, in absolute terms, grow at an exponential rate may be an interesting statement from an analytical stand. It is, however, a rather confusing statement from a substantive viewpoint. In dynamics one might be able to approximate the growth rate of any stock over an arbitrary time-frame with an exponential function.

Out of this formulation, and in particular when this exponential growth rate is not simply a fixed parameter as it was in the original Malthusian population[24] model, but instead a function of either the absolute population stock level and/or another variable like per capita income, a very interesting set of conditions emerge. The growth rate is no longer fixed, but instead it changes with time. It might become positive, negative or zero at any point in space–time as a result of *endogenous* factors affecting and being affected by it, rather than being altered by *exogenous* checks as Malthus originally speculated. After Malthus, the Verhulst, Pearl–Reed and the Volterra–Lotka equations (see note 24) provided valuable extensions to the *continuous dynamics* Malthusian growth model.

Interdependent population and per capita income dynamics have been the subject of attention in modern non-linear mathematics and applied analysis in areas ranging from ecology to the socio-spatial sciences. A number of basic models in the spatial sciences have been suggested within this line of continuous and *discrete* dynamics in the past ten years or so, and certain among them produce chaotic behavior (see Dendrinos and Sonis 1990). In general, socio-spatial dynamics are prone to highly complex outcomes when choices of spatial allocations involve interdependencies or time delays. It is novel insights that make revisiting Malthus worthwhile.

That a variable exponential continuous population growth rate can be derived no matter what the spatial unit of analysis, be it an urban area, a nation, or the world as a whole, and no matter which time-frame is considered, has important dynamical implications. It is of interest that a look at the absolute dynamics of the human population from 10,000 BC to the present (see note 3 of Introduction) seems to be well replicated by a constant exponential growth rate. Malthus's original exponential population growth model seems to hold for ranges extended apparently not only over short time-periods but also millennia of human history, albeit at changing population growth rates.

Although referred to by Malthus as 'biological' this ecological force uses only its main component, i.e., the genetic propensity to reproduce, as a proxy. In reality, the B force in all societies at the micro or macroscale is a composite of many diverse factors. Economic, demographic, sociological, religious, cultural, and psychological forces affect the social (individual and collective) decision to reproduce. However, its effect is measured by a single state variable (population size), and possibly by a single and constant parameter over a relevant time-period (growth rate).

The dominance of the proxy in the bundle might possibly be very durable in space–time, particularly when compared with other bundle forces at the micro or macroscale. For example, the force might dominate such diverse factors as sociological kinship bonding, economic production or consumption processes, political and social choice rules, demographic transitions, cultural norms formation, etc. These forces, within the bundle but not likely dominant in it, are confronted with the burden that they are not readily recordable and consistently measured in space–time. Nor is there a single variable or parameter to act as their proxy thus enabling the observer or participant to derive a sharp model of their behavior in space–time. Instead, they may be composites of a number of variables and their composition may vary in space–time and possibly in a non-systematic way. In any case, these heterogeneities might severely impede one's ability to test for consistency in their behavior.

Compared with the relative clarity of and durable dynamics in global or local population counts, growth in the absolute value of the world average per capita gross product is somewhat more fuzzy. Take for instance the numbers cited earlier on per capita income, and slightly alter the time-period considered. In current terms, the scant evidence available seems to suggest that the average per capita product more than doubled in the 1974–85 period.[25] If one deflates these counts, then the 12-year period witnessed the per capita gross product increase only by approximately 6.3 percent. The count grew to a high of US$1,467 in 1980 and then continuously declined to a level of US$1,242 (in 1972 constant US$) in 1985. How accurate are these counts, however? For sure, the product and income from informal sector activities are not included in them. Much more data are thus needed before the dynamics of the absolute world average per capita gross product can be determined with some degree of confidence.

The original Malthusian model also contained a second force independent of population and dealing with growth in the available food supply. Per capita output (or income) can be considered as its proxy. There have been numerous interpretations of the combined effects of the two independent forces in the Malthusian model, since Malthus did not state them in precise analytical terms, as will be seen later in subsection 5. Malthus assumed that food per capita is likely to grow geometrically (linearly) in time, and independently of growth in population. Here, a second force – not as simple as Malthus's second force – is argued to be present. A second fundamental force, to be referred to as the *economic–demographic* (E–D) force, points to a lower exponential relative population growth rate associated with a higher relative per capita wealth (or its surrogates, current per capita income, gross national or domestic product, value added, etc.). It is apparent that differentials in population growth rates are not solely due to differentials in per capita incomes, but rather to all factors involved in the bundle of social, economic, demographic, political and cultural forces of which per capita income is their proxy. The inner structure of the E–D force must consist of a bundle of lower level factors and interdependencies. This is a matter of further theoretical speculation and certainly the subject of considerable social action by participant agents but it is not the subject of this work.

The evidence regarding the existence of the E–D force is ubiquitous at certain levels of spatial disaggregation. In 1980, 25.6 percent of the world's population lived in MDCs, and 74.4 percent in LDCs. Their estimated absolute current average per capita Gross National Product (PCGNP) was US$5,210 and US$560 respectively, while their estimated average absolute natural growth rate (NGR) was 0.6 and 2.1.[26] These differentials are much more pronounced when viewed in relative terms. No matter how much fuzziness may be found in these counts, there is at least one important message emitted from them: differences in population growth depend on differences in per capita income.

Arguments can be presented for and against the magnitude of these numbers along the lines presented in the Preface of the book. One may further argue whether particular nations at the margins of these two categories belong to the MDCs or LDCs. However, the differential between groups in both average PCGNP and NGR and the relatively small variance within groups clearly identify two different patterns of growth between the two settings. At the level of urben areas, Figure 1.1, the presence of the E–D force is hard to dispute. How the two ecological forces (B and E–D) shape the macrodynamics of the urban agglomerations specifically is demonstrated in Chapter 3 (see pp. 175–92).

One can pose a more fundamental question about the *origin* of these two fundamental forces, namely what gives rise to them? This question will be addressed in Chapter 2 (see pp. 92–5), where the notion is presented of a worldwide prevailing parity in net attractivity of locations and dynamics toward it as this might be perceived by an observer. It is further treated analytically in Chapter 3 (see pp. 192–207).

Going back to Figure 1.1, a clear *dualism* is detected between LDCs and MDCs, to the extent that their urban areas' growth is concerned. Dualism is viewed here broadly as a significant variation in behavior among two groups. This dualism seems to be the outcome of a large-scale *ecological determinism* at work at this scale of analysis. In the first two sections of Appendix 1, more details on ecological determinism are supplied. The seemingly inseparable phenomenon of dualism which accompanies it and manifests itself in many contexts, as will be discussed in Chapter 2 (see pp. 95–105), is also addressed in the final section of Appendix 1. As its base, an economic market determinism is used, although the substance of ecological determinism goes far beyond addressing economic issues alone. In the second section of Appendix 1 an introduction to a theory of individual preferences and collective (public) policy making is supplied, thus setting the stage for outlining a speculative theory of social behavior. Next, the broader lines of such a theory are drawn.

3 Speculative behavior and ecological determinism

Humans, individually or collectively, act at any point in time with some future time-period expectations in mind given past perceived experiences. The time horizon considered (both by looking into the future as well as into past) may vary from the smallest time-period (at the microtemporal scale of a minute or so) to the very long future time-frames (at the macrotemporal scale of millennia). In the case of storing nuclear waste, for example, speculative behavior extends over tens of thousands of years. Future, as well as past considerations are always to be detected in social action. Since humans cannot and do not know the future with certainty and can never attain perfect learning from the past, all such actions are genuinely speculative.

Social, individual, or collective behavior is always constrained by prevailing norms and perceived constraints as well as by opportunities be they political, social, or economic. Such norms constraints and opportunities guide the manner in which a social system evolves in space–time by replicating itself from one time-period to the next always subject to fluctuations or perturbations. Looking at perceived socio-economic norms and constraints in defining the feasibility set of possible social courses of action is only equivalent to looking at perceived opportunities in searching for social courses of action.

Through social, individual, and collective perceptions the field of psychology enters the core of ecological determinism through speculative behavior. Ecological determinism asserts that individuals and collectives alike act as agents holding and exercising *options* on events, resulting in social actions. These options cover a broad spectrum of possible events and social states. They contain a plethora of deterministic action markets where such options are traded, i.e., sold and acquired.

Market mechanisms involve an assortment of speculative demand for, supply of, and current prices for various options on social actions. Each option is associated with a social future state of affairs which has a composite value to the market participants. Interactions among all relevant social options markets governing all individual and collective activities determine the behavior of social stocks in aggregate.

A central component of these markets is that they contain either positive (self-fulfilling) or negative (self-defeating) feedback. Action by an individual agent is likely to be undertaken far beyond perfect market clearing conditions, by at least causing numerous externalities or by subjecting the market to various monopolistic imperfections. Such is the case particularly for significant social actors whose behavior has a major impact on the options market(s). Social options markets are indeed highly imperfect. Perceptions as well as social (individual and collective) alternative preferences, constraints, policies, and actions are all subject to and elements of this deterministic mechanism. In the formulation of demand, supply, prices or rationing schemes of alternative social actions, ecological factors are involved (i.e., economic, sociological, political, cultural, environmental, psychological, biological, *et al.*). (See Appendix 1 for a more detailed analysis.)

Ecological determinism operates in a dynamic framework. Speculative behavior links present social actions and events to expectations about the future which might occur from an exercising of these options. 'Speculative behavior' is used in its strict risk management definition.[27] According to the standard definition and its transfer to the context of social action, social decision making takes place under uncertainty in social events options markets. Individuals and collectives at any point in space–time express a demand for and are supplied with (i.e., they perceive as feasible) alternative courses of action or social options on alternative future states.

Acquisition of any social option by an individual or collective carries a commitment to a future series of perceived events in a prespecified space–time horizon. Adoption of any option carries also a perceived composite (not merely economic) value (what is referred to in the theory of options as the 'striking price') to the adopter as well as to the supplier of the option. This composite value identifies the net composite benefits to an individual or group in the resulting end social state which emerges from the actions of all agents.

At the particular point in space–time, and when the commitment to a specific course of action has been made by a social agent, expectations about social events in their perceived *disaggregate and aggregate* form and their associated composite values have been discounted for by the social action market participants. In time, the perceived composite current price (value, discount or premium) of these options oscillates, at times violently, as perceptions and the environment change. Elapsed time is of critical importance in the changes of the currently perceived price of these social options. Exercise of options is revealed in, among other ways, the construction of physical entities, the lasting witness of social action: the construction of buildings and settlements; the creation of artifacts and various forms of art; the build-up of armaments; the erection of monuments; the writing of books, etc; all these are examples of certain durable outputs of social action. On the other hand, the use of these physical entities, a record largely lost over the years, is only indirectly inferred from the currently available physical evidence.

Aggregate and disaggregate events involve aggregate and disaggregate outputs and outcomes of social actions. Social events, aggregate or disaggregate, may be micro (minor) or macro (major) events. Often options on actions imply options on disaggregate and aggregate social states with conflicting composite values (outcomes) to market participants. The objects of conflicting interests or social objectives (for example, the conservation of the natural environment versus its use for employment or human habitat opportunities) are the assets over which futures or options contracts are issued and exchanged in speculative social action (options) markets. Speculators in and hedgers (holders) of these assets interact in a highly complex, ecological manner.

Length of commitment to an option (i.e., to a particular course of action) varies. Lack of durability in commitment is more likely for actions with a long-term effect, as the perceived prices of holding such options contracts may fluctuate significantly in the course of their life span. Social agents have a relatively limited life expectancy relative to the durability of the option over a very durable asset like the natural environment. Their perceptions regarding the future net benefits of these options of social action to the various participants change in time.

In a traditional sense, speculators and hedgers exist in highly homogeneous commodities futures and options markets, interacting in a zero-sum game. For example, speculators in wheat, livestock, or gold futures and options markets interact in a zero-sum game with hedgers selling such options. The argument is advanced here that, equivalently, speculators and hedgers exist in highly heterogeneous and localized social events (actions) options markets, where prevailing conflicts in interest among individuals or groups make the exchange a zero-sum game, too. As social options markets are highly heterogeneous and thin, when compared with those of privately held homogeneous commodities options markets, it is much more likely that they contain significant imperfections. The study of these imperfections may be very informative (see Appendix 1, pp. 243–60).

This briefly outlined options market determinism results, from the viewpoint of an observer, in aggregate outcomes (states) where every social (individual and collective) agent's speculative behavior has been discounted. It does not contain any social action vacuums. This is the reason why models which address aggregate dynamic outcomes can only be of an observer type. One must distinguish the deterministic ecological approach as defined and adopted here, from an interdisciplinary Marxist-like deterministic view of socio-spatial stocks. Some examples of such interdisciplinary approaches are found in the field of Marxist cultural anthropology.[28] There, an abbreviated critique is supplied, structured around their lack of interesting and pertinent dynamics.

4 Micro and macrodynamics, aggregate and disaggregate analysis

At the global scale, abstract ecological determinism is attributed to durable interdependencies among the growth and decline rates in relative population and per capita gross product. The fundamental forces identified earlier, the B and E–D forces (see pp. 18–23 and 29–33), often tend to be sticky in their form and intensity in the short term, say a year. At times they remain in place intact over longer periods, possibly for decades as the evidence indicates for the case at hand. Rates of change in absolute levels in the two macrovariables are similarly sticky, although not necessarily so for all spatial units. They must be expected to remain relatively constant over longer time-frames in larger rather than smaller population size areas.

The specific interdependencies among relative population and wealth at the macrolevel of analysis are sustainable over a quarter century at least, according to the limited evidence supplied. Clearly, these interdependencies do not demonstrate ephemeral dynamic patterns to the extent that the domains of motion for the individual paths are well defined. But the individual paths resulting from such interdependencies have some freedom to float within these domains in an orderly chaotic manner. There never is pure chaos or order.

If there are severe limits to the use of these two ecological macrovariables (population and wealth) in development studies, they are due to the difficulty in obtaining relatively accurate data on the per capita product side; this is mainly because the informal sector's contribution to the gross domestic product goes largely undetected. In a relative framework, however, this might not be as severe a limitation as it would be in absolute dynamic or static analysis, since this lack of detection may be pervasive in many countries.

Although these two macrovariables seemingly play a central role in development studies and discussions by policy makers, their importance in shaping and recording large-scale and long-term development analysis has been, to say the least, unclear. Social scientists (economists, sociologists, demographers, political theorists) seem to hold widely different views as to exactly what these variables imply, as to how they are interconnected, and how they are related to other developmental variables. A brief presentation of these views and a critical appraisal are supplied in Appendix 2. The point made is that this study is not meant to be in any way a substitute for existing disciplinary studies on development; rather, it is a complement to them.

Disaggregation, or the degree of coarseness used to view the world, plays a major role in ecological determinism. Development of a specific social output, i.e., individual, firm, artifact or product, and small-scale social outcomes at a specific point in space–time, falls under the domain of local ecological *microdynamics* which deals with micro social events at particular points in space. Growth or decline of social aggregates like cities, industries, nations, regions or civilizations are examples of ecological *macrodynamics* progressively higher in geographical scale. Disaggregation is possible along *three dimensions – spatial, temporal and sectoral* – the sectoral dimension being a breakdown by social stock type. Macrodynamics exist at a geographically local scale, but microdynamics do not exist at a global scale. Micro or macroanalysis can be disaggregated by discipline, as for example micro and macroeconomics, or it can be disciplinarily aggregate and ultimately nondisciplinarian. This perspective breaks down all disciplinarian boundaries, and calls for a single (unified) social science.

Spatial, temporal, and sectoral heterogeneities are the central dimensions in the disaggregation scheme used to record, model and theorize about social outputs, outcomes, and events. Beyond being of interest to observers of social events, disaggregation along these three dimensions is of import to the extent that selective interactions and interdependencies among participant units may reinforce or dilute heterogeneity in space–time. The disaggregation scheme perceived among participants or assumed by observers may, in effect, impact the course of these participants' actions. It is not possible to dissect cause and effect between the prevailing ecological interdependencies and the spatial, temporal, and social stock disaggregation. Existence of social elites and an underclass, an example of striking social disaggregation in cities of the world during the last quarter of the twentieth century, and their prevailing ecological interdependencies cannot be viewed in a cause–effect relationship. Each is a necessary condition for the presence of the other and in combination they are both responsible for their respective dynamic behavior and possibly evolution.

Particular and distinctly different bundles of forces consisting of a bundle of variables are at work at various levels of spatial, temporal and social disaggregations. At specific points in space–time and within specific spatio-temporal horizons, to these spatially, temporally, and socially disaggregated units there is a subset of dominant forces and variables in the basket of prevailing ecological interdependencies. As one changes positions in space–time – that is, one partitions these two dimensions – the composition of the subset of dominant forces and variables also changes. These dominant forces can be perceived only to a certain extent by the participants or their observers, since social agents individually and collectively are characterized by imperfect spatio-temporal knowledge and understanding. This last is one cause among many for a prevailing fuzziness in perception and disaggregation.

Theories of social action may be viewed in a local or global scale depending upon the degree of localization in the available information. Observers and participants may derive perceptions and expectations for past or future social action at any point in space–time on the basis of their partially validated local models. The history of human societies can only be a collection of local theories in space–time. To assume a general (global) model of history, or geography, or society, at a given level of disaggregation and use it at a different (finer or more coarse) spatial, temporal, social disaggregation is not appropriate.[29] Spatio-temporal transferability of models of social action has still to be convincingly demonstrated.

To a great extent all social agents (individuals and collectives, participants and observers alike) have imperfect knowledge and understanding. They are restrained by limitations operating at various levels of sectoral or spatio-temporal disaggregation. None the less, they use their currently perceived odds on expected future outcomes of ecological interdependencies and interactions to take bets on the outcomes and act or model accordingly.[30] The various degrees of disaggregation along the three central dimensions affords them many bets, collectively affecting the systems' behavior along these dimensions. This is a major reason why it is not valid to assert the transferability of models.

Furthermore, for any particular level of social spatial or temporal disaggregation there may be a subset in the bundle of forces and variables which dominate spatio-temporally social dynamic behavior at a point in space–time. Conversely, for any dominant subset within the bundle of forces, there may be a particular social spatial temporal disaggregation in which this bundle is dominant at this or other point(s) in space–time. For the case at hand, the dominant subset in the bundle of forces at a macroscale of urban dynamics among the world's largest metropolitan agglomerations consists of the two modified Malthusian fundamental ecological forces, the biological force (B) and the economic-demographic (E–D) force. The B force is also dominant at the micro (very disaggregate) level, whereas the effect of E–D at a very fine level of analysis is not clear. Both forces' dominant macrovariables are population and wealth. Whatever the origin of the two fundamental forces, it must also be the origin of the socio-spatial disaggregation one might observe. The vast variety and diversity of social stocks must be due to the same source giving rise to the two fundamental forces (see Chapter 2, pp. 58–62).

When an arbitrary disaggregation is made for mere observation of a socio-spatial system or for taking an action in that system, it does not necessarily follow that a dominant subset of forces and variables may exist in the bundle at that level. Nor does it follow that a model of interdependencies attempting to describe and explain these forces can be arrived at routinely. It could be the case that the system at hand is under a set of forces which won't allow any to emerge as dominant; or that dominance is changing fast within the space–time horizon considered and thus not allowing enough time or space for it to be recorded. At this level of disaggregation the system may be perceived as demonstrating almost random behaviour.

Conversely, for any arbitrarily chosen subset of dominant forces and variables in the bundle one may not be able to locate any social spatial or temporal disaggregation obeying this bundle. Judicious search might be needed to solve these assignment problems. An added difficulty might be that in all likelihood the problems do not possess a global and unique optimum. It is safe to assume that due to imperfect knowledge and ecological complexity, these assignments always result in speculative and generally suboptimal solutions, at least to the extent that modeling of their behavior by observers is concerned. In case of participants, the winners in social action markets must have guessed better than the losing agents on these forces.

Many socio-spatial variables might be used to record the effects of interdependencies at a point in space–time depending on the specific focus of the analysis. The various variables used by a modeler to record these effects may remain the same, moving from the micro, through the meso, to the macroscale. Two such variables are population size and per capita income. It is likely that by switching levels of analysis new variables may need to be introduced, or old variables might have to be dropped.

More importantly, linking those state variables which might

commute among different levels of disaggregation are functional forms which do not necessarily remain the same as one moves from one level of social spatial and temporal coarseness to the next. As the analyst shifts from one disaggregation level to another, the subset of dominant variables and forces in the bundle underlying social interdependencies in space–time is altered. Shifts in dominance among bundle forces (i.e., an intra- and inter-bundle movement of forces) affect the form of the model used to describe social behavior in space–time. This may involve changes in the model's parameters, form specification, and, of course, variables.

Some examples may illustrate these points, which stand in contrast to prevailing practices in economics and other spatial sciences, where once a model is stated it is at times automatically assumed that it can be spatially and/or sectorally disaggregated at will. Were this premise to hold one could take the argument to absurd lengths by considering the micro (economic and otherwise) behavior of a consumer (or producer), inflate it by 5 billion and claim that one has a model of the behavior of the globe's population! Input–output analysis is an example of such largely arbitrary disaggregation, as is the literature on various land-use models in urban and regional analysis.

At the most *sectorally* (social) disaggregate level possible, that of an individual consumer or producer or market participant, psychological forces mostly responsible for perceptions are prominent in the bundle of dominant forces guiding individual behavior. Motivation to act in self-interest is one of them; perceived limitations is another. This perception is partly accounted for by the use of a utility function in microeconomic analysis which is maximized subject to budgetary and other constraints. Individuals differ from one another even if they belong to the same cultural milieu; they differ when they belong to different socio-temporal environments as they express this self-interest and perceived constraints in a different manner. Either different elements enter one's utility function, depending on their location in space–time, or their utility function may have an entirely different form. Physical as well as non-physical components may be arguments in these utility functions. Equivalent arguments apply to their perceived constraints.

Individuals belong to socio-spatial groups (for instance, their family, firm, social clubs, political party, religious group, city, nation, etc.). They also belong to a particular time-period or era. Being part of such spatio-temporal aggregates, they have particular preferences for and are restrained by such groups and prevailing tastes of their era. The inner structure of their individual preferences and constraints may not be merely psychological, but also group-induced and time-related.

Beyond the individual preferences and constraints, socio-spatial groups also have collective preferences and constraints at any point in space–time. These collective preferences and constraints may be much simpler than the individual ones. The multiplicity of individuals (and collectives), in their complex spatio-temporal interactions, render collective dynamical behavior simple also.

Given these observations, one is led to the conclusion that there is a fallacy in the economic literature to the extent that an arbitrary socio-spatial and temporal disaggregation of economic models is concerned. When an observer employs the same utility function and constraints used to depict a 'representative' individual's behavior in an effort to identify collective economic behavior, then the observer is bound to miss the interactions between individual and collective preferences and constraints. Further, by doing so the observer misses the point that at the individual level the objectives and constraints are different than those at the group level. One might argue that what is depicted by a 'representative' agent is indeed aggregate behavior. Then, however, the claims of individual or disaggregate behavior must be dropped. Finally, by looking at a 'representative' individual's objectives and constraints, the observer misses the complex (non-linear dynamic) interactions among the individual members of any socio-spatial group in shaping individual preferences and constraints.

An altogether different collective preference function and perceived constraints must apply to the collective, with different decision-making rules. Self-interest and perceived constraints of a group are not definable by averaging over individuals given the (at times) sharp differences within groups; or by defining a 'typical' individual's preferences and constraints at a sectoral and spatial disaggregation and then blowing up this average by simply multiplying it by the total number of individuals in the group. But that is what the fields of microeconomics routinely contain. Economists often reject models from the natural sciences on the basis that individuals are not molecules (i.e., entities with highly homogeneous behavior), but then they turn around and model individuals precisely under such or similar homogeneity conditions. Let us consider as an example the process of capital investment.

At a very *spatially* disaggregative level, say that of a small region within a nation, the construction of a roadway may be perceived as a significant event in the development of that region. To model the possible effects that such road construction might have for the region, socio-spatial analysts derive models. Clearly, capital stock accumulation is one possible variable among many to describe the event of construction, and numerous other variables have been employed to estimate its effect. But whether these variables are the operative ones in effectively describing the ecological macrodynamics at the local, regional or national levels is an entirely different question. Clearly, road construction at the national level (in contemporary US, Ancient Rome, or in the Mayan world) is not the most important variable for economic growth. What is suggested here is this: first, that road construction is not a central variable in ecological macrodynamics; second, that even the broader capital accumulation process (through savings and investment) is not a central variable in ecological determinism – at least it is not as central as population and per capita income, in spite of the preeminence that this process has enjoyed in neoclassical as well as in Marxist analysis; third, the connection between population growth and capital accumulation is not at all clear or as easily observable and recorded as that of population and per capita income, and in all likelihood it is slaved by the ecological determinism depicted by the dynamics of the two central macrovariables; and fourth, that the model one must use to simulate the microevent at the local scale must be very different than that employed to look at the road construction event at the national or global scale.

Capital accumulation (or depreciation) and consumption must obey dynamics of a different speed than those of population and per capita income. The build-up of capital stock, houses, roads, plants, etc., must be responsive to speed of immigration and outmigration of population and to the magnitude of change in per capita income. On the other hand, capital depreciation and abandonment must be lagging the spatial movement of people and changes in their level of wealth.

An important characteristic in micro or macrodynamics is the speed of movement in a model's state variables, parameters and functional form. These changes might underlie some form of temporal restructuring in interdependencies. They will be addressed briefly in turn. First, there is the relatively fast change in the state variables picking up change in socio-spatial-temporal behavior. Second, there is the relatively slow change in the parameter values connecting the state variables identifying change in the model's coefficients (parameters). Third, there is possibly the extremely slow or abrupt change in the interdependencies giving rise to the model's form.

The source for these three types of change is different in each case. In the first case, the motion is due to the internal dynamics of the model, whereas in the second case the velocity of motion affecting the value of parameters might be due to slow and continuous or discontinous changes in the environment or the local spatio-temporal conditions. Finally, the third type of change might be due to changes in perceptions or drastic and sudden alteration in the environment. Put differently, in the third type of change involving perceptions, changes in a model's form may be due to alterations in the composition of the subset in the ecological bundle containing the dominant forces or variables applicable at a specific social spatial and temporal disaggregation at a particular point in space–time.

According to a general principle in dynamical analysis, sometimes referred to as the 'slaving principle,'[31] slower-moving variables slave or determine the behavior of faster-moving components. There is another dimension, however, to record the effect of the slaving principle. Equivalent to the temporal slaving principle there might be a spatial slaving one, according to which a larger spatial unit might determine the behavior of a smaller constituent one.

More broadly, micro socio-spatial-temporal system variables may be changing faster than macro socio-spatial-temporal system variables. Thus, macro social-spatial-temporal systems may be construed as slaving the behavior of micro socio-spatial-temporal systems. In this sense, one may refer to dominant macrovariables, within the dominant bundle of ecological forces, slaving microvariables but not vice versa. Macrovariables obey longer time constants than microvariables. These constants may govern and sustain long-term stability or instability of socio-spatial-temporal systems.

This observation ought not to rule out some degree of disaggregate and aggregate socio-spatial system interactions. Microsystems may affect macrosystems, although not to the extent that macrosystem behavior affects micro socio-spatial-temporal dynamics. All this can be aptly captured by non-linear methods. Stochastic average microlevel behavior can be explicitly modeled as being affected by, and in turn affect, deterministic macrolevel dynamics.[32]

The fact that macrolevel dynamics are likely to be deterministic and non-random makes aggregate modeling possible. One can bypass the complexity of ecological interactions among the constituent elements of a socio-spatial system and go directly into modeling the aggregate dominant interdependencies as recorded and perceived. A complete accounting of the individual and collective behavior of all social agents, at all locations on the globe, at all time-periods, is obviously impossible and nonsensical to pursue if the aim is to say something meaningful about the system's macrostructure.

Aggregate analysis does not lack critics. In economics, for instance, where mathematical models of an aggregate (macroeconomic) and disaggregate (microeconomic) nature have been constructed, the arguments over their respective effectiveness, empirical validation, relevance, internal consistency and sensitivity to policy making (i.e., their use), are heated. Some economists argue that aggregate outcomes can resemble anything, implying that disaggregate outcomes cannot. Notwithstanding aggregate neoclassical economic growth theory and Keynesian macroeconomics of both fiscal and monetary type, the bias in economics and related spatial fields has been for behavioral, micro-models. The central criticism of aggregate models is that they tend to be descriptive and not policy sensitive (considered as a disadvantage), and that they lack a behavioral foundation. However, disaggregative models claiming to be behaviorally motivated suffer from the same ills they attribute to aggregate models.

Heterogeneities and associated lack of data severely restrain efforts to formally (mathematically) state behavioral micromodels and at the same time demonstrate their validity through empirical evidence. Because of this shortcoming, economic epistemology has attempted to substitute empirical evidence with formal logic and deduction.[33] The back-bones of microeconomics, i.e., the theory of consumer behavior, the theory of production, and the theory of markets (together with other aspects of microeconomics including welfare theory), are not truly disaggregate microbehavior models. They do not model the behavior of specific individ-uals, firms, governments, or entrepreneurs. They are merely claiming to represent a statistically average (non-existent in reality) rational agent, even when some stochastic element is introduced to account for variations in preferences (as in random utility theory) or acton.[34]

In social sciences, linkages between micro and macrobehavior are complex. Heterogeneous microagent behavior can produce a homo-geneous macroresult. On the other hand, homogeneous microbehavior may result in heterogeneous macro-outcomes. The vast diversity of people in the streets of London or New York City produce highly repetitive homogeneous travel patterns daily, weekly and monthly, for example. On the other hand, producers and consumers as well as entrepreneurs in highly homogeneous (even traditional) cultures, produce vastly different actions, products, and artifacts in these societies. One need only look at art and architecture to realize this. Particular types of built form are not uniquely linked to a specific function (structure), as a particular function is not uniquely associated with a specific form. Variability, hetero-geneity, and lack of non-linear interactions are among the greatest weaknesses of neoclassical economics.

Conventional behavioral microeconomics (theory of consumer behavior, theory of the firm) deals primarily with one-to-one correspondences between motives and constraints on the one hand and aggregate market actions on the other. However, in reality, one may be hard-pressed to find such one-to-one correspondence. More likely, a multiple-to-one correspondence may be prevalent.

Modern dynamical theories enables one to qualitatively address subjects of multiple-to-one correspondence between micro and macrobehavior. The possibility for chaotic and fractal properties in the dynamics of systems defined by a simple deterministic set of equations may allow for multiple level, qualitatively similar disaggregated behaviors to be accommodated.[35] It also reveals certain dynamic qualities of behavior not captured by conventional static analysis. For example, it allows the analyst to incorporate behavioral traits which tend to obey different time-scales. Slow and fast movements can be looked at through non-linear dynamics.

Habitual or traditional behavior, manifested within long time horizons, presumably follows a slowly changing dynamic motion. It 'slaves' social (individual and collective) behavioral patterns exhibited over short time horizons, that can be recorded by daily, monthly or yearly dynamic cycles. The daily trip to work is an example. Traditional behavior incorporates socio-spatially fast movement. Within this fast and slow motion framework, periodic motion, calm, oscillatory and dynamically turbulent behavior can also be studied.[36] More on this issue is supplied in Chapter 2, pp. 133–41.

Macro (collective) behavior identifies the final outcome(s) of interactions among all individual agents involved in an interactive game. Individual agents respond to and act on other agents' currently anticipated actions and to currently expected or perceived macro-outcomes. Consequently, the expected outcome of macrobehavior becomes part of the current interaction. Models of microbehavior exclusively relying on individual reaction to other agents' expected actions are not fully capable of depicting the whole interaction, or its end effect. They run the risk of falling into an infinite regression of anticipated action and counter-action.

Neoclassical microeconomics addresses the connectance between individual (disaggregate) and collective (aggregate) behavior through the incorporation of externalities into the analysis. For example, in the urban transportation economic literature the issue of road congestion externality is modeled so that individuals incur, under unregulated pure market conditions, only their marginal private cost of movement. At the same time, a social (internalized or not) marginal cost of road use is also computed.[37] However, the requirements usually imposed upon the various formulations that unique, long-run static (market equilibrium, or social welfare optimum) configurations must emerge limit the variety of effects one can obtain from this externality.[38]

As individuals react to macro-outcomes in a speculative manner, pure dynamic macromodels in the social sciences (i.e., models which do not incorporate some element for individual choice however that may be defined) are also incomplete and bound to be inaccurate. Model builders can only strive toward the most 'detached' aggregate dynamic model, without ever being able to reach it. Modelers of macrobehavior are of the 'observer' type, whereas effective modelers of the micro-behavior are of the 'participant' type. To the extent that any aggregate model placing a claim on the future's shape has an effect (no matter how significant or inconsequential) on current action – through anticipating individual or collective behavior – it cannot be of pure observer type. This topic is further addressed in Chapter 4.

Besides these substantive arguments in the modeling of aggregate and disaggregate social systems, there are also some pertinent model efficiency issues to consider when a choice is to be made between more versus less disaggregative analysis. Due to severe lack of data on individual behavior, the high cost of collecting such data sets, their very poor quality, and the sheer number of agents involved, disaggregate behavioral models are very costly to construct and not always cost-effective to pursue. More importantly, they supply a very biased view of the socio-spatial system they tend to model as they only concentrate on particular aspects of it. A far more effective means by which observers can analyze socio-spatial systems is to construct aggregate descriptive type models. The disaggregate story may best be told by the journalist, novelist, film maker and, at the final analysis, by the very agent of action.

Going back to the subject of interdependencies and ecological determinism, one notes the importance of development in this discussion. Development or evolution is the substantive issue addressed by ecological determinism. The topic of micro or macrodevelopment is diverse and multifaceted in its inner structure. It contains rich ecological interdependencies and interactions no matter what the spatio-temporal social event under investigation or the level of analysis considered. A vast number and variety of social events have occurred and are currently taking place in different regions and their subregions, each attracting some attention from observers with various degrees of interest. A huge diversity in development patterns has been and is currently being recorded; these patterns range from events of recent vintage to events which occurred in time-periods going back to remote centuries of human evolution and social development.

There is a continuous process for selecting some key to these events in order to address and study periods further in time and at various locations in space. From this soup of developmental topics only two variables and the event they trace over a limited time-period are discussed here at any length. This limited but strategically chosen evidence bears on broader issues of socio-spatial development and evolution. In a very brief survey of the available literature, the manner in which the stocks of population and wealth and their interdependent dynamics have been treated in a variety of contexts will be supplied next.

5 The Malthusian and other demographic models

In presenting the various arguments that follow, one must bear in mind that current population (x) and wealth (w) can be both viewed either as stocks characterized by negative or positive rates of accumulation ($\dot{x}$, $\dot{w} \gtrless 0$); or as current flows on already accumulated stocks ($\dot{X} = x$, $\dot{W} = w$). In the case of wealth, income (y) may be further viewed as the first time derivative of wealth ($y = \dot{w}$). Current product can be considered as a proxy to current wealth (or income), or as an addition to the available stock of wealth (or income). If one considers per capita product or income as change in wealth and thus a flow variable, then one simply transfers the argument from first to second time derivatives.

Like many other stocks, population and wealth are characterized by heterogeneity, durability and generational overlap, some mix of locational fixity and spatial mobility, and to an extent divisibility. A host of theories and models have been produced to study population and wealth interdependencies and interactions and their relationships to numerous aspects of development. In reviewing the various theories next, the reader is reminded of the notes at the beginning of the book regarding data, social events, their interpretations (theoretical models) and their markets. It is in the light of these comments that one may find the necessarily conflicting, at times contradictory, and always fuzzy structure of the available multiple theories and the policies they tend to support or reject, of interest.

At center stage in these theories is the cause and effect relationship between the two central macrovariables of socio-spatial ecology: population and wealth. Cause and effect have been extensively debated, with widely differing views as to their implications in existing development related literature being aired.[39] In accordance with accepted practices in economic theory and demography, one finds in existing economic development and demographic studies conflicting perspectives being proposed to explain the dynamics among the two central variables in absolute terms. Expensive and massive family planning programs in very many LDCs and certain MDCs, together with numerous public policies involving population growth, have been commissioned. They have been undertaken over the last thirty years or so, being motivated by the various arguments and associated fears of over- or under-population as well as by means of how best to address poverty. Extensive social experimentation and heated debate has followed these views.

There are basically two sides to the population–wealth relationship. On the side arguing that 'population growth affects wealth,' the prevailing opinion is that high population growth rates hinder rises in per capita wealth.[40] Yet others argue that for certain LDCs, high population growth rates actually stimulate growth in per capita wealth.[41] Finally, others hold that there is no direct causal effect between rates of population growth and levels of either per capita income (or wealth) or its growth rates.[42] Thus, all possible angles along this side of causality have been argued.[43]

Along the other side of causality, the 'per capita wealth affects the population growth rate' argument, certain analysts hold the opinion that high per capita birth-rates (and low death-rates, thus high natural-growth rates – the net international migration rates being perceptively low relative to these two) are due to high per capita wealth.[44] Again, others argue that there is no relationship between per capita income (or wealth) and birth-rates.[45]

An economist, Todaro (1982), even argues on both sides: at one point he argues that empirical evidence shows no definable relationship between per capita wealth levels and birth-rates.[46] He also argues, drawing from microeconomic theory of consumer behavior, that one can expect higher income households to have greater demand for children. This result is obtained when a utility model of individual behavior is formulated in which, it is argued, children are considered not to be inferior commodities when included in the preference function. As in standard microeconomic tradition, all depends upon the form of the utility function, which here has as its arguments various economic commodities and the per capita number of children.[47] Almost any outcome can be reproduced as a result of making appropriate assumptions on the form of the utility function in the microeconomic model.

The only position not extensively argued is that per capita income (not necessarily high levels or growth) negatively affects the population growth rate, which happens to be the one with the overwhelming evidence to support. Probably this position has not been widely adopted because of its perceived potential implications regarding large-scale (relative but necessarily absolute) income redistribution among MDCs and LDCs. Although policy makers of LDCs are well cognizant of the opinion that 'development is the best contraceptive,' and hints of such relationship exist in certain international agencies policy reports,[48] this single fact has not entered formal economic analysis yet! Some economists even argue that per capita income is the wrong variable to posit as a determinant of population growth rates.[49] However, that view is not expressed in a formal analytical framework and it is unclear what empirical evidence can be cited in support.

Frustrated by the failure of social, demographic and economic policies based on the various available arguments, as well as disappointed by their own inability to effect worldwide redistribution of income to a significant extent, many policy makers of LDCs have adopted the position that per capita wealth should not be relevant in the efforts toward development. Instead, improving a nebulous 'quality of life' index is proclaimed to be important. Alternatively, since presumably nothing significant could be done about poverty in LDCs at present, policy makers decided to look instead into the future for radical changes in wealth by appropriately investing in what are currently perceived as potentially highly profitable enterprises in the long term.

This 'strategic' perspective, which calls for more or less abandoning the currently poor to their present predicament and the holding of high expectations for the future (in a temporal filtering process), apparently seems to be widespread among a diverse group of LDCs' policy makers. Meanwhile, since this position became popular in the late 1970s, the 1980–84 period saw the MDCs' fraction of PCGNP to the world's average jump from 3.0 to 3.3 and that of the LDCs' drop from 0.32 to 0.27. Put differently the MDCs' share of the world's gross product increased from 76.2 to 79.9 percent.[50]

In the argumentation regarding causality among the two central variables an interesting construct, and one severely criticized by neoclassical economists, is Malthus's notion of a 'trap' in the developmental process.[51] The Malthusian trap model, a variant of the original Malthus thesis on population growth,[52] is schematically presented in Figure 1.2. There are two differential equations describing (absolute) population and income per capita growth, each as a function of per capita income: the straight line (which can have any slope, demonstrated in three of the cases shown) assumes unimodal behavior in per capita income growth, whereas the population growth is assumed to obtain a maximum at some level of per capital income.

Following this specification, the dynamics are straightforward: the lower level in per capita income is always the stable equilibrium; the upper level (no matter the slope of the income growth ray) is always unstable. The dynamics show, according to this model, that if a nation starts out to the left of E_2, the area indicated by the arrow pointing to the left, it is bound toward E_1; whereas, if it starts at the right of E_2 then unstable growth is possible. If the income growth rate equation is non-linear with respect to the level of economic wealth (income) per

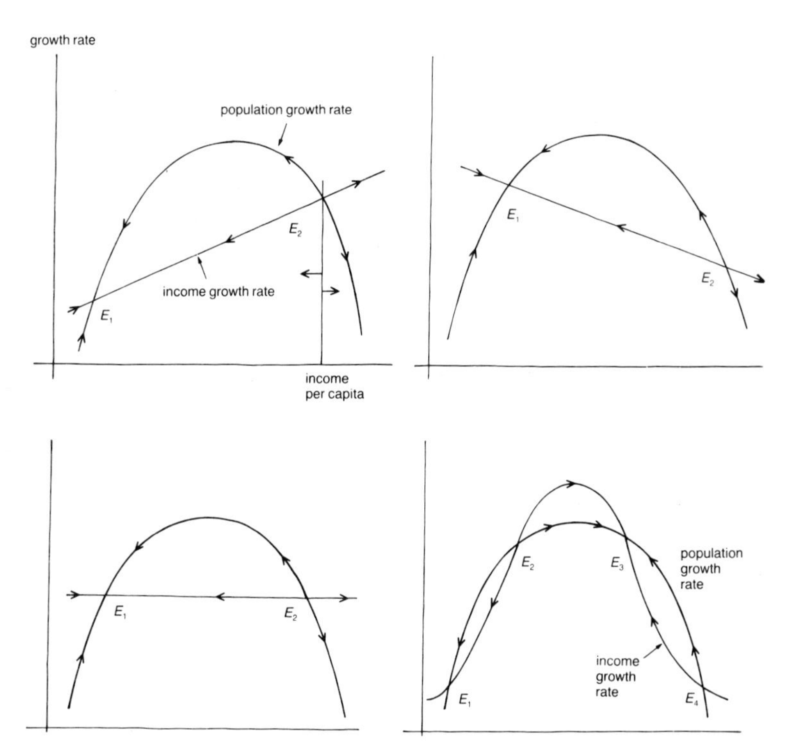

Figure 1.2 Various versions of the Malthusian trap dynamic model. E_1 is always a stable dynamic equilibrium, whereas E_2 is unstable. LDCs are trapped in E_1

capita, as the lower-right case in Figure 1.2, then a number of (stable) equilibria are feasible. According to this hypothesis, LDCs are trapped at the low income per capita (inferior) dynamic equilibrium.

There is, however, a basic flaw to this model. The trapped equilibrium always corresponds to either high or low population *and* income growth rates. According to the model of Figure 1.2, MDCs must be characterized by high growth rates in population and income. Whereas, LDCs must experience low income and population growth rates. The above statements, as they emerge from the original model by Malthus and its various neo-Malthusian versions, fail to pass the test of empirical evidence recorded over the past 25-year period. They fail to depict the E–D force: population growth is checked effectively only by relatively high levels of current per capita product (income).

How do the various checks on population *and* income growth rates exactly work? A comprehensive view of the *population* checks requires that the original list of Malthusian controls (famine, war, disease, and natural catastrophes) must be amended to incorporate at least the following (complementary or supplementary) ones: infanticide, homicide, suicide, human sacrifice, accidents, man-induced disasters, and various forms of family planning including abortion and sterilization. Thus, one obtains a spatio-temporally full view of checks that have been brought to bear upon population growth throughout history. Malthusian analysis is mute on *income* growth checks, checks which economists have been debating over the past two centuries. Within an ecological determinism perspective however, and with the possible exception of natural disasters ('possible' because human locational choices are involved), all population growth checks can be associated with rates of change in or levels of income (or wealth). In turn, any income growth checks can be directly linked to population levels and/or their growth rates from an ecologically deterministic viewpoint.

An alternative economic–demographic and social approach to development has been the theory of 'demographic transition' based largely on the original work by W. Thompson.[53] It has been extended by F. Notestein[54] among others. In summary, the theory posits the existence of an interdependence between fertility and mortality: it views high fertility rates as a reaction to high mortality rates. Lower mortality rates cause lower fertility rates, with a period of transition between the two phases.

At the first stage (represented by the primitive societies) birth- and death-rates are high and production per capita is very low and growing very slowly. During the second (intermediate or 'transitional') stage, and due to improved health standards and higher per capita production, birth-rates are still high but death-rates decline drastically. In the third and final stage, economic and social gains lower the death-rate and this in turn leads to a lower birth-rate. A steady state emerges, where birth- and death-rates are equal, as is the case during the first stage. Then a society is said to have reached its steady state. Thus, demographic transition is viewed as an agent working toward the stabilization of population levels.

Not much by way of numerical evidence has been supplied in support of this theory. The analytical aspects of this model, from a non-linear dynamics perspective, are also poor. It is quite unclear as to how they operate within a consistent, i.e., valid for all three periods, dynamic framework. Precisely what endogenous dynamic specifications produce these three phases of development, and how transitions among these three periods occur, is not evident from the theory either. F. Notestein labeled these three phrases as 'high growth potential,' 'transitional growth,' and 'incipient decline,' respectively.[55]

Marxist and neo-Marxist demographic constructs also exist, but their contribution to the theory of these ecological macrodynamics is at most marginal since they basically assume that no connection exists between the two.[56] A number of other demographic models, which are exclusively preoccupied with 'steady states,'[57] have also been suggested. Demographic models positing steady state dynamic equilibria in population growth rates are, however, conceptually congruent with the Malthusian type model presented in Figure 1.2. They therefore suffer from the same shortcomings as listed earlier, and in addition they do not satisfactorily address the problem of dynamic equilibria.

6 Economic growth models and Marxist development theory

Economists have rejected demographic, Malthusian and neo-Malthusian, models on the grounds that they do not incorporate economic growth inducing variables. In Appendix 2, a closer and brief look is made into the existing theoretical (growth theory based) macro-economics (see pp. 274–85); topics include neoclassical economic growth theory, theory of technological change, and theory of international trade. Also considered are the sociological (mainly those of the Marxian political economy type) constructs addressing the developmental process (see pp. 285–7). It is clear that the two state variables – current population size and wealth (or income) – play a key role despite their downgraded importance in the latest applied development analyses.

The theories surveyed are divided into two major categories: the purely economic theories of development and dependence theory which includes economic and non-economic (mainly sociological and political science) notions. Using ecological determinism, a unifying framework is shown in the third and fourth sections of Appendix 2, where an integration of Keynesian, Marxist and *laissez-faire* theories is attempted using tools from mathematical ecology. Dynamic stability issues in the private–public sector interaction predominate the analysis. Finally, in the fifth section of Appendix 2, two constructs from cultural anthropology are also briefly mentioned.

What is apparent from reviewing all these diverse theories is the fallacy of disciplinarian analysis in social sciences. Particularly evident is the irrelevancy of economic analysis (see Appendix 2, pp. 271–4), in spite of the fact that economics is the most methodologically advanced social science. Its prevailing theoreticians and practitioners are much better socially rewarded than the equivalent individuals from all other social sciences, in spite of the fact that there is considerable excess supply of many of their models or theories. Reasons for this apparent paradox are many, but they will not be explored here. All these theories adopt a very limited analytical and largely ideological perspective, with their conclusions being constrained by the limitations of their assumptions.

It is also evident that there are severe methodological shortcomings shared by all of these theoretical constructs, due largely to their lack of incorporating complex non-linear interactive effects into the models of social behavior they outline, although this limitation is currently eased in economic analysis and the spatial sciences with the inroads that have recently been made into chaos theory in these fields. Going beyond making marginal contributions to specific problems in these sciences, these non-linear interaction principles herald a change in the fundamental premises of social science.

CONCLUSIONS

This section set the conceptual stage for the findings of the book to be presented. It defined urban macrodynamics as the simultaneous growth or decline paths of the world's largest metropolitan agglomerations in their current share of the world's total population and their per capita gross product to the world's currently prevailing average. Empirical evidence from a 25-year period was, to an extent, used to document these urban macrodynamics. Although drawn from a relatively short time-period, the evidence seems to reveal patterns of aggregate development and evolution expected to hold over much longer time-spans. It is evident that the future of these urban areas depends on their starting (and current) position.

Two comprehensive composite modified Malthusian forces were detected in these dynamics. These fundamental forces are a biological (B) force consisting of absolute population growth, and an economic–demographic (E–D) force identifying current relative per capita income as the single most important variable negatively affecting current relative population growth rates. They were considered to be the constituent elements of a spatial-temporal macro ecological determinism operating at this scale of analysis. Issues of speculative behavior in social (individual and collective) actions were discussed as falling under the principle of ecological determinism. With this approach, interdependencies and interactions were presented as being multifaceted rather than specialized in terms of economic, political, environmental, cultural or any other specific type. Ecological interdependencies and interactions provide the grounds for a unified (and single) social science.

Existing development-related theories, as expanded in Appendix 2, were shown to say little and provide contradictory statements with regard to the macrodynamic events analyzed here. Disciplinarian neoclassical economic growth theory, international trade theory, technological innovation, dependence theory, and some constructs from cultural anthropology were found to supply little guidance, in the light of the empirical evidence and theoretical framework provided here. It was argued that the issue of public policy can be addressed within the framework of the ecological determinism principle proposed to hold at that level of analysis, and this was expanded in Appendix 1. Public intervention was viewed as severely constrained, strongly obeying the prevailing global ecological interdependencies found to govern social systems in space–time.

2

NATIONS, CITIES, INDUSTRIES, AND THEIR CONNECTANCE

INTRODUCTION

Like nations, urban areas interact in space–time. These interactions render national boundaries and the political notion of sovereignty obsolete to a great extent. Inter-spatial flows of commodities labor capital and information, among other stocks, cross national borders under various degrees of easiness. *De facto* limits rather than *de jure* boundaries become largely effective. Extent and strength of spatial interaction contributes considerably to the dynamic instability of national economies, mainly through growth or decline of the major urban areas they contain. Often interdependencies and interactions are attributed to national economies or industries in aggregate, but vary rarely to cities as this intermediate spatial level is frequently overlooked or bypassed in the analysis. In this chapter an effort is made to bridge this gap. A variety of significant comparative advantages are directly allotted to specific locations in space where major cities are found, fueling spatial interaction and through them a host of spatial disparities. A striking example of such spatial differentials is the significant variation in population densities over space.

At the tail end of the twentieth century one witnesses in most of the cities of the world, particularly the giant cities of the globe, a collage of urban landscapes. In certain instances parts of these patches can be traced back several centuries, and they currently coexist with urban land uses and activities which herald the beginning of the third millennium AD. A vast diversity of social and economic activity is found to permeate these urban landscapes, where at times extreme poverty and backwardness seem to live in close spatial proximity to vast wealth and, by current standards, technologically very advanced production and consumption hubs. Heterogeneity within and among the urban scenes of many nations in all corners of the world is well captured by basic principles of dynamic instability. Urban quilts in space–time stand to offer powerful and convincing testimony to the explanatory and descriptive capacity of socio-spatial dynamical analysis.

In this chapter, one of the main causes for instability in socio-spatial stock dynamics is expanded. A central source for this instability is the degree and extent of inter-spatial interaction. The analysis addresses foreign trade *as a representative* component of a variety of inter-regional flows. By focusing on international trade, highly selective relatively weak inter-sectoral and inter-locational linkages leading to instability are detected. The end result of this instability is worldwide dominance at any point in space–time of a few industrial sectors, nations, and very large urban agglomerations. Spatial hierarchies are also viewed as the outcome of these linkages. Instability and chaotic spatial dynamics can be obtained by assuming that all nations constitute a set of coupled oscillators. A simpler and potentially more insightful path is traced by looking at instability from a mathematical ecology standpoint.

Some additional empirical evidence is given in this chapter to demonstrate certain selective arguments. Two economies provide the testing ground and laboratory for making observations central to the arguments presented: the US and Mexico's international trade flows, their continuously changing industrial base, and the evolution of their major urban areas' economies. Direct trade between nations is found to be relatively weak in absolute terms when compared with domestic markets in determining sectoral and urban growth of nations. As exchange rates are central components in international trade, the analysis widens further in scope by incorporating international currency flows. Inter-spatial movement of stocks, including commodities, labor capital and information, are all linked in causing dynamic instability in the absolute and relative population and per capita product dynamics of the world's largest urban agglomerations within a global environment of interaction. A variety of socio-spatial dualisms emerge.

In Chapter 1, two fundamental forces were discussed, the biological (B) and economic–demographic (E–D) force. Where do these fundamental forces originate? And where do interdependencies and interactions come from? It is argued in this chapter that, for an observer, two notions underlie these forces – that of relative parity and spatio-temporal arbitrage producing flows of spatially liquid stocks among the world's urban agglomerations. A number of interactions caused by this relative parity are demonstrated with reference to certain US metropolitan areas and Mexico City. How socio-spatial dualism can appear under a relative parity in net attraction among locations principle is discussed in Chapter 3, where the link is identified as that of dynamical instability.

Relative spatial macrodynamics are then addressed, together with the subject of cycles. The argument is put forward that the relative lens is quite helpful in capturing dynamic composite multiple spatial interdependencies and interactions. A mathematical ecology framework for relative spatial dynamics is then advanced, with a few examples to demonstrate its empirical foundation. Cases include the relative macrodynamic paths of major US urban areas in reference to the US economy, and the US aggregate relative macrodynamic path in reference to the world's economy. The exposition sets the stage for the ecological relative dynamics of the globe's largest urban agglomerations, in reference to the world economy, to be presented in Chapter 3.

A diversity of subjects are displayed in this chapter. Although they may not be covered at any significant depth, this not being a key objective here, they are indicative of the broadness in scope of the nonlinearity theme and its ensuing multiple dualisms. From looking at frozen shrimp production in Mexico, to the ecological macrodynamics of the New York Metropolitan area in reference to the US, to the fluctuations in the price of oil in international markets and the fortunes of the Austin Metropolitan Area, to the currencies of developing nations, the reader is led through a diverse landscape. A common thread – that of instability – runs through all these features. Indeed, instability is a major dynamical theme of this chapter and an important nucleus of the notions found in this book.

INSTABILITY FROM INTERNATIONAL EXCHANGES

In international commodity markets, in 1970, a barrel of Middle Eastern light crude oil was exchanged at about US$4, and the price of gold stood fixed at US$35 per ounce. Approximately 12.5 Mexican pesos were exchanged in international trade for one US dollar, while the latter was commanding 357 Japanese yen. By 1982, a similar barrel of crude oil was traded at about US$30, the price of gold was fluctuating around US$500, and about 100 pesos were exchanged in international currency markets for one US dollar. When a first draft of this book was completed in the first quarter of 1987, crude oil was trading at US$17 per barrel, the price of gold was around US$400 and 1,200 pesos were traded in formal currency markets for one US dollar. The US

Department of Commerce estimated the legal immigration from Mexico in the 1960–70 period to have been approximately 44,300 per annum. During the 1980–84 period this count stood at 109,600 per annum.[1]

These seemingly diverse dynamics apparently indicate some instability in certain international markets, possibly turbulent events taking place in the relatively short time-period of one decade. As will be argued later, all these events are linked and in these linkages one can detect certain broadly applicable forces at work connecting social stocks in space–time.

Nations, regions, urban areas and their industries continuously interact in space–time. Their spatio-temporal interactions are multifaceted and dynamically varied in strength. Some of these interconnections are manifested through the flow of heterogeneous *commodities*, i.e., flow of natural resources, agricultural and manufacturing semi-finished or finished products, or services, or spatial spillovers from their extraction, trade, production and consumption externalities. Other interactions involve the flow of heterogeneous *capital*, i.e., flow of money stocks, financial and other assets.

Spatial linkages are further manifested through the international migration of heterogeneous *population* (labor), where in this case the attributes of heterogeneity include demographic, economic, social, ethnic, cultural, educational and other characteristics of the migrating populations. Finally, there is international flow of heterogeneous stock. It includes scientific knowledge and technological innovation, social, cultural, religious and other norms, ideas and ideologies, news in various media, etc. Since spatio-temporal flows of commodities, capital, and information follow and serve human actions it is spatio-temporal reward differentials to population which must be found underneath these flows.

All these components, to be referred to here as the 'four elementary components of spatial interaction,' cross national, regional, and urban boundaries. Their continuous flow gives a sense of quantum fuzziness in their *de facto* borders, no matter how spatio-temporally fixed or precise or constant their *de jure* aspect is. Among the four spatially flowing factors, the easiest to count may be population movement, while information flows are the most difficult to record. Issues ranging from an appropriate unit of measurement, to the existence of formal and informal flows make an exact recording of inter-regional movements rather elusive. Economic variables in these four elementary components of spatial exchange enjoy a significant advantage over all other types of variables: they are partly recordable. This is not a strong enough reason, however, to consider the economic forces as dominant in the complex inter-spatial and inter-temporal ecology in the community of nations. They can be used, none the less, as examples to demonstrate some important points in this dynamic ecosystem.

Exchanges and flows produced by inter-spatial interdependencies (or forces) result in micro and macro spatial systems undergoing dynamic change (development) or structural instability and phase transitions (evolution) in space–time. Inter-spatial interaction in the form of flows results in relatively fast changes in the size of various socio-spatial stocks in space–time. Relatively slow changes in spatial interaction factors (parameters) affecting these flows may result in sharp shifts in the economies, cultures, social groups, and institutions of interacting social systems. They can also significantly impact the collective hierarchy of nations, their urban areas, and their industrial sectors. Shaped by the short- and long-term spatial interdependencies among the world community of nations, regions, and urban areas, these spatial units' micro and macrodynamics are played out.

An examination of international trade allows one to sharply focus on the economic (foreign trade related) interactions among nations. Available empirical evidence allows one to document the absolute dynamics of and the effects from direct exchange among national economies. A major conclusion drawn from this examination is that direct economic national links, as evidenced mainly through international trade, are weak. They constitute only a fraction of the overall economic activity of large and medium-size nations. Domestic markets are overwhelmingly more important for relative or absolute economic growth when examination of growth is confined to a national economy.

Although these trade linkages may be weak, they are gauges of significant structural conditions and dynamics of nations within the global community. They are associated with unstable dynamics of significant magnitude and duration. Multifaceted international linkages are instrumental in the restructuring of a nation's economy. The US and Mexican economies' performances during the 1960–83 time-period provide the testing ground as they are discussed in some detail next to back the arguments.

A central thesis in this study is that the international flows of the four elementary components in inter-spatial exchange are ecologically (i.e., in a multifaceted manner) interconnected. In particular, the role of capital flows, as it relates to the other flows, is of special interest in this discussion. It is hypothesized that flows obey a perceived effort toward achieving an inter-spatial relative parity in net attraction among various locations. This effort is subject to the macro spatio-temporal ecological determinism as discussed in Chapter 1.

Absolute spatial economic growth and international trade in commodities are correspondingly driven and dominated, at any time-period, by a relatively few leading industrial sectors and selected places of origin and/or destination for products. International movements in population, information, and capital (particularly certain currencies) also seem to be highly skewed toward specific origins and destinations. The impact of this selectivity upon the dynamic instability in the economic base of interacting nations (large, medium, and small) is pronounced.

All four elementary components of exchange are not freely traded in international markets. Political boundaries and other restrictions selectively interfere with their international exchange. These restrictions have major consequences for the ecological dynamic instability of the prevailing spatial and sectoral hierarchies.

Direct economic exchange among nations predominantly occurs through direct flows among their major urban centers. Direct links are also found to be spatially and sectorally very selective. Selectivity is partly responsible for their unstable urban dynamics. Urban centers' macrodynamics are mainly tied to the domestic hinterlands and markets they serve. Dependence upon international markets is, however, stronger for them than for their nation's economy. It is even more so for their specific industries, particularly their basic (export-oriented) industries and firms, and their import-sensitive domestic-oriented sectors.

Data from the major US and Mexican urban agglomerations demonstrate that in general, economic development of very large urban settings is tied to flows of spatially liquid stocks. Changes in socio-economic, production, consumption, market and government processes, as well as in the mode of property ownership, introduction of new products, technological innovation diffusion, and industrial base shifts, are also shown to be critical in the socio-economic development process no matter the nation under study. Such changes occur in space–time not in a haphazard manner. On the production side, for example, transitions involving shifts in labor occupation from agriculture and trade sectors to manufacturing, from manufacturing to services, and from rural to urban settings, punctuate development at specific points in time. At times these transitions are fast enough to involve the same generation of workers. They may also be violent enough to transform the socio-economic structure of a nation in a relatively short time-span.

Such events are not confined to individual nations, national subregions, or metropolitan areas. They are not restricted to a single time-period, but instead they are spread worldwide; although they may not occur concurrently in space, they seem to follow some rough time schedule. Broad societal shifts diffuse in space–time, having originated in one (or a few) source(s). Transitions and large-scale societal transformations are not restricted to economic variables. Being multifaceted, they must be viewed from a broader, ecological perspective. Transformations not only in production, consumption, governance, and market processes encompass social and cultural changes. They are captured by and recorded in the accumulation or breakup of various socio-spatial stocks, in a proxy-type manner. Relative macrodynamics are far more appropriate to depict these ecological events than absolute, disciplinarian, disaggregate analysis.

TWO NATIONAL ECONOMIES; THEIR INTERNAL GROWTH AND FOREIGN TRADE

From the exposition hereon, the reader should not infer an attempt toward deriving another theory of international trade. The effort is rather to identify the unstable dynamic effects that foreign trade in commodities has upon spatial economic entities within a theory of inter-spatial, inter-temporal ecological interaction. What follows is a more detailed presentation of two nations, in terms of absolute economic growth and direct foreign linkages, at a particular phase of national and worldwide economic development during the third quarter of the twentieth century. One is a large nation belonging during that time-period to the MDCs: the United States of America. The other is a medium-size nation belonging at this time-period to the LDCs: the United States of Mexico.

The reasons for analyzing these two neighboring nations in more detail are twofold. First, to show that foreign trade is spatially very selective, and that in strict economic terms it plays a small part in the overall absolute economic activity of the two economies notwithstanding its political, social, and broadly cultural importance. Spatio-temporal instability is clearly in evidence within this context, as arguments about dynamical instability are presented, drawing from the example of spatial commodities flow. Second, to demonstrate certain basic similarities and dissimilarities in the current and past industrial base of these two nations and their evolving structure, and then to pose some central questions about socio-spatial evolution. These points are used to support the relative parity principle within the broader perspective of ecological determinism. An attempt is made to link these two national economies to the phase of their absolute and relative development, and to the dynamics of their major urban centers. Particular attention is drawn to the role that the manufacturing sector and its evolution played in these linkages.

The purpose of this brief exposition is *not* for it to act as a substitute for the many and far more detailed studies of these two national economies. But rather to highlight some important, up to now rather neglected, elements to be used in structuring the arguments regarding indirect ecological-type linkages to be further elaborated in Chapter 3.

The US economy and its foreign trade

In 1962 the US population numbered 186.7 million or about 6 percent of the world's total. By 1987 the absolute population level had risen to 243.8 million, but the share had declined to approximately 4.85 percent. In 1974 the US current per capita gross national product stood at US$6,640 or 4.88 times the estimated world's average per capita gross product. By 1985, US PCGNP increased to US$16,400 – slightly more than 5.69 times the world's average – whereas the GNP of the US was approximately 28 percent of the world's gross product – a significant share of the world's economic activity and by far the most dominant among all other nation's product over the 1974–85 time horizon.[2]

Data on the world's average per capita income for 1960 are not available. However, were one to use the available information on the per capita gross domestic product for the world's market economies (closely related to the world's average for the period data are available), then the United States' ratio in 1958 would be slightly higher than five times, whereas its then absolute level was US$2,602.[3]

Clearly, from the above snapshots there are two different images which emerge concerning the US economy: in an approximately 20-year period, the US economy experienced significant growth in absolute population and per capita product terms. Even when per capita product is measured in constant US$, the US per capita income grew from US$11,786 in 1960 to US$14,941 in 1985 measured in 1982 constant dollars.[4] However, looking at these numbers in relative terms and in reference to worldwide totals, one sees little PCGNP growth and sharp population declines in relative shares. These large-scale dynamics were occurring while certain significant changes were underway in the industrial base and composition of the US and world economies.

The analysis will not enter into the underlying factors for growth or decline of these particular sectors at any length. Every industrial category, and every firm therein, is affected in the short as well as the long term by a very complex set of circumstances. Each industry and firm has a history of its specialization or diversification process, of its technological progress or lack thereof, of the spatial and temporal expansion or contraction of demand for its output, of the conditions surrounding the supply of its input factors in production, of changes that have occurred in both input and output factor prices in space–time, of the conditions surrounding the transport sector impacting upon its shipments, of the spatial and temporal competition among firms, in its industry, and in the ensuing fluctuations in earnings and profit margins. To record each industry's development one must study, in addition, the dominant firm within each industry and the history of the firm's management of its labor, financial and other capital assets.

This is not the forum to address such complexities for each firm or industrial category, or to record in detail the various inter-industry (input–output type) linkages. It will be noted, however, that detailed conditions depend upon the level of disaggregation in the SIC code considered, as well as the spatio-temporal horizon under study. For the purposes at hand, one can bypass the specifics, because some aggregate outcomes are quite telling.

In Table 2.1, the US GNP at the first digit Standard Industrial Classification (SIC) Code is supplied for 1960 and 1980. The 20-year period saw very few changes in its industrial valuation mix at this level of disaggregation. At the same time, in terms of employment, the manufacturing share declined from 31 percent in 1960 to 22.4 percent in 1980. One must attribute this change to the capital for labor substitution which must have occurred at a large scale in the manufacturing sector of the US economy over the 20-year period. This technological development, which had a significant impact upon the major urban areas of the US, was not confined to the US economy. It also characterized a large number of national economies in all continents.

Easing in US manufacturing employment was accompanied by a shift toward higher shares of employment in finance, insurance and real estate, trade and particularly domestic market-oriented services. Viewed at the third digit SIC level, one discerns two points. First, there is a highly selective mechanism in that a few leading sectors dominated the US industrial mix. Second, there was a pronounced change in the ranking of the US major manufacturing categories over that time-period

Table 2.1 US GNP by industry at first digit SIC, 1960 and 1980 (in constant (1972) US$)

	1960 $bn	%	*1980 $bn*	%
Agriculture, forestry, fisheries	32.1	4.4	39.1	2.7
Mining	13.5	1.8	21.6	1.5
Contract construction	46.1	6.3	53.3	3.7
Manufacturing	171.8	23.5	351.2	24.2
Transportation, communications, public utilities	57.4	7.8	140.8	9.7
Trade	117.5	16.1	243.3	16.7
Finance, insurance, real estate	102.7	14.0	237.9	16.4
Services	83.5	11.4	189.0	13.0
Government	107.7	14.7	177.3	12.2
Total	732.3	100.0	1,453.5	100.0

Source: US Department of Commerce, Bureau of the Census, *Statistical Abstract of the U.S.* (SAUS), Table 693, 1982–83.

(see Table 2.2). The US economy's manufacturing sector in 1980 was dominated by a relatively few products: motor vehicles and equipment, aircraft and parts, industrial chemicals, and communication equipment manufacturing.

The rapid rise of electronic components and equipment in 1980 (see Table 2.3) points to a major shift in US industrial production. In the 1950s the steel mill products dominated the US manufacturing sector, later succeeded by the motor vehicles, parts and equipment (still the dominant sector by 1983), with electronic components and equipment emerging in the early 1980s as the future expanding sector. Factors influencing these structural changes in the US economic base were ecological and were globally driven as a result of the US being a large economy open to worldwide interdependencies.

The analysis now turns to addressing the issues of national growth and international trade, and to demonstrate the selective process and dominance of particular (small in number) sectors and locations (origins and destinations) in a nation's foreign trade. The US total foreign trade amounted to US$458.6 billion in 1983, with imports at US$258.1 billion exceeding exports at US$200.5 billion.[5] Combined, their valuation was 13.85 percent of the US GNP for 1983 (totalling US$3,310 billion). The net imports valuation of US$57.6 billion was

Table 2.2 US manufactures, current value added, third digit SIC: ranking of selected industries (with value added more than US$15bn in 1980 and US$2.7bn in 1960)

		$bn
1980		
1	Motor vehicles and equipment	30.5
2	Aircrafts and parts	27.6
3	Communication equipment	22.6
4	Basic steel products	22.3
5	Petroleum refining	22.2
6	Office and computing machines	18.2
7	Construction and related material	17.2
8	Electronic components, accessories	16.9
9	Fabricated structural metal products	15.6
1960		
1	Motor vehicles and equipment	10.1
2	Steel rolling, finishing	7.7
3	Aircrafts and parts	6.6
4	Basic chemicals	5.1
5	Dairy products	3.2
6	Beverages	3.2
7	Structure metal products	3.0
8	Newspapers	2.9
9	Petroleum refining	2.8
10	Communications equipment	2.8
11	Metalworking machinery	2.7

Sources: 1980: *SAUS*, Table 1382, 1982–83.
1960: *SAUS*, Table 1095, 1963.

only 1.74 percent of the GNP for that year. This pattern, although fluctuating over the study period, remained relatively stable: in 1973 the US foreign trade amounted to about US$140 billion, freight on board (f.o.b.), approximately one-tenth of that year's GNP. It reached 20 percent in 1981 (US$490.4 billion out of US$2.5 trillion), the year that also saw for the first time in contemporary US history the valuation of imported merchandise starting to exceed the valuation of exports.

By the end of 1986, the US international trade deficit was approximately US$152.7 billion or about 3.63 percent of the US GNP.[6] Worldwide, the US foreign trade was about 12 percent of the total in 1973 – a share which remained the same for 1981. This share is much below its GNP share of the world's product. Thus, international trade is

Table 2.3 US foreign exports and imports by major category, 1967, 1970 and 1981 (current US$bn in parentheses)

	1967 ($bn)	*1970 ($bn)*	*1981 ($bn)*
	100 (30.7)	*100 (42.5)*	*100 (236.3)*
Exports:			
Foods, feeds, beverages	16	14	16
Industrial supplies and materials	33	32	30
Capital goods, except automotive	32	34	35
Automotive vehicles, parts	9	9	8
Consumer goods (non-food)	7	7	7
Imports:	*100 (26.9)*	*100 (39.9)*	*100 (264.1)*
Foods, feeds, beverages	17	15	7
Industrial supplies and materials	45	38	52
Capital goods, except automotive	9	10	13
Automotive vehicles, parts	9	14	11
Consumer goods (non-food)	16	19	15

Source: *SAUS*, Table 1486, 1982–83.

a relatively small component of a large economy's overall activity level. Similar is the case for other large economies, as well. For example, in the case of the People's Republic of China, the value of imports and exports were Rmb 73 billion with China's gross output value at Rmb 749 billion in 1980, i.e., about 10 percent as well.[7]

In summary, the economic activity destined for domestic markets far outweighed foreign trade oriented, both in net and total production and in distribution quantities. No matter how much the trade deficit was considered as a political issue in US economic growth, it was rather slim relative to the size of the US economy in the study period.

From 1967 to 1981 (see Table 2.3) the US foreign trade, when studied at a very aggregate level, had not changed much. Industrial supplies (mostly petroleum) and materials, and capital goods (except automotive) still dominated US exports, and industrial supplies in 1981 far outnumbered in valuation all other imports (individually) as they did in 1967. At a finer level, the fourth digit SIC, one obtains a more clear view of the US trade pattern then: the US major export categories were aircraft, electronic computing equipment and parts, corn, wheat, construction machinery and motor vehicle parts and accessories (see Table 2.4). From this table it can be seen that US imports were dominated by crude petroleum, motor vehicles and bodies, petroleum

Table 2.4 US international trade, fourth digit SIC, 1981, ranked according to valuation (more than US$3bn)

	$bn	%
Exports:		
Aircraft	10.3	(42)[1]
Electronic computing equipment, parts	8.5	(28)
Corn	8.0	(40)
Wheat	7.9	(77)
Construction machinery	7.3	(47)
Motor vehicle parts, accessories	7.3	(17)
Soybeans	6.2	(48)
Motor vehicles and bodies	6.2	(10)
Industrial organic chemicals	4.6	(14)
Oil field machinery	4.3	(42)
Aircraft equipment	4.3	(43)
Petroleum refining products	3.5	(2)
Imports:		
Crude petroleum	63.8	(50)[2]
Motor vehicles and bodies	23.5	(27)
Petroleum refining products	16.4	(7)
Blast furnace and steel mill products	11.0	(15)
Non-ferrous smelting and refining products	5.9	(62)
Radio and TV receiving sets	5.6	(50)
Natural gas	4.1	(10)
Tree nuts	3.5	(29)
Solid state semiconductor devices	3.5	(25)
Paper mill products	3.4	(14)

Source: US Department of Commerce, Bureau of the Census, *Commodity Exports and Imports as Related to Output (1981, 80)*: Table 2.1 (exports), Table 2.2 (imports).

Notes: [1] All figures in parentheses in exports section indicate percent of total domestic output.
[2] All figures in parentheses in imports section indicate percent of the total supply.

refining products, and steel mill products. Again the dominance of a very few commodities is evident.

Contrasting this to 1960 foreign trade patterns, at the detailed level of the fourth digit SIC, a change in dominance among exported commodities emerges in US foreign trade patterns. In 1960 exports of certain agricultural products (cotton and grains) were the main foreign trade merchandise, followed by aircraft engines, and construction and mining equipment (see Table 2.5). At the import scene, the table shows

Table 2.5 US foreign trade, third digit SIC, 1960, ranked according to valuation (more than US$0.3bn)

	$bn	%
Exports:		
Agriculture		
Cotton	1.0	(41)[1]
Grains	1.8	(19)
Tobacco	0.4	(32)
Minerals		
Coal	0.3	(17)
Manufactures		
Chemicals	0.3	(18)
Refined copper	0.3	(14)
Construction and mining equipment	0.8	(33)
Aircraft, engines	0.8	(20)
Aircraft, parts	0.5	(13)
Imports:		
Agriculture		
Raw coffee	1.0	(100)[2]
Crude rubber	0.3	(100)
Minerals		
Iron ores	0.3	(31)
Manufactures		
Sugar products	0.5	(73)
Pulp mill products	0.3	(23)
Paper mill products	0.7	(19)
Copper, refined	0.4	–
Platinum and tin	0.3	(36)

Source: *SAUS* (1963): Table 1207 (exports), Table 1208 (imports).

Notes: [1] All figures in parentheses in exports section indicate percent of total domestic output.
[2] All figures in parentheses in imports section indicate percent of total supply.

that agricultural (raw coffee) and paper mill products dominated import trade. The type of commodities dominating trade change in time, with very few always dominating it.

The dominance by a few commodities is also accompanied by dominance in origins and destinations in foreign trade. Looking at the nature and location of foreign imports and exports (see Table 2.6) in

Table 2.6 US foreign trade in 1959 in major categories of merchandise, by location (in current US$bn)

	Total[1]	*Canada*	*LA*	*WE*	*FE*	*Other*
Exports:						
Total	15.9	3.7	3.5	4.5	2.6	1.5
Grains	1.4	–	–	0.6	0.3	0.2
Foodstuffs	0.9	0.2	0.1	0.3	0.1	–
Fats, oils, seeds	0.7	–	–	0.3	0.1	–
Cotton	0.4	–	–	0.2	0.1	–
Tobacco	0.4	–	–	0.2	–	–
Machinery	3.7	1.1	1.0	0.7	0.5	0.3
Automobiles	1.1	0.4	0.4	–	–	0.1
Chemicals	1.5	0.3	0.4	0.1	–	0.1
Textiles	0.6	0.1	0.1	0.1	–	0.1
Petroleum products	0.5	–	0.1	0.1	0.1	–
Imports:						
Total	15.6	3.3	3.6	4.5	2.6	1.6
Coffee	1.1	–	1.0	–	–	0.1
Meat	0.5	–	0.1	–	0.1	–
Sugar	0.5	–	0.3	–	0.1	–
Foodstuffs	1.4	0.2	0.4	0.3	0.2	0.1
Non-ferrous metals	1.3	0.3	0.3	0.3	0.1	0.2
Paper	1.0	1.0	–	0.1	–	–
Petroleum	1.5	–	0.8	–	–	0.5
Textiles	0.8	–	–	0.3	0.4	–
Machinery	0.6	0.1	–	0.3	0.1	–
Automobiles	0.8	–	–	0.8	–	–
Iron and steel mill products	0.6	–	–	0.5	0.1	–

Source: *SAUS*, Table 1217, 1962.

Note: [1] Only items with a total more than US$0.4bn are listed.

1959 most of the US foreign trade was grain exports to Western Europe and machinery to Western Europe, Canada and the Far East (mostly Japan) which contributed in building their industrial bases. Imports mainly included coffee from Latin America (mostly Brazil); paper from Canada; petroleum products from Latin America and the Middle East; automobiles from Western Europe; and iron and steel mill products from Western Europe.

By 1980, exports were dominated by machinery to Western Europe, Latin America, the Far East and Canada; crude materials (except fuel) to the Far East and Western Europe, and transport equipment to Canada (see Table 2.7). On the import side the table shows that machinery from the Far East, mineral fuels from Latin America, and transport equipment from the Far East dominated the import trade markets in valuation.

A more detailed breakdown of US trade links with these particular regions and nations which contained the then world's largest urban agglomerations (see Table 2.8) reveals that from 1958 to 1981 no drastic changes had occurred in their share of US total foreign trade (with the exception of Iran) over the study period. In 1958 exports to

Table 2.7 US foreign trade in 1980 in major categories of merchandise, by location (in current US$bn)

	Total	*Canada*	*LA*	*WE*	*FE*
Exports:					
Total	216.7	34.1	35.4	66.3	47.8
Agricultural	41.3	1.9	5.7	12.0	13.5
Non-agricultural	175.3	32.3	29.7	54.3	34.3
Food, live animals	27.7	1.5	4.5	6.6	8.3
Crude materials (except fuel)	23.8	1.8	1.8	9.2	9.6
Mineral fuels	8.0	1.8	0.8	2.7	2.1
Chemicals	20.7	2.1	4.9	6.4	5.1
Manufactured goods	22.3	3.8	4.4	6.9	3.8
Machinery	55.8	9.5	10.6	16.5	9.7
Transport equipment	28.8	8.5	4.6	6.2	4.4
Instruments	5.3	0.7	0.6	2.2	1.0
Imports:					
Total	244.9	41.5	30.0	46.6	61.6
Agricultural	17.4	1.1	7.1	2.8	2.6
Non-agricultural	227.4	40.4	22.7	43.6	58.9
Food, live animals	15.8	1.4	7.4	1.6	1.8
Beverages and tobacco	2.8	0.4	0.2	2.0	0.1
Crude materials (except fuel)	10.5	5.2	1.1	0.7	1.4
Mineral fuels	82.9	7.1	12.8	4.6	5.6
Chemicals	8.6	2.5	0.5	4.0	1.0
Manufactured goods	32.2	8.2	2.4	7.8	10.1
Machinery	31.9	4.0	2.3	10.2	15.0
Transport equipment	28.6	8.3	0.4	7.4	12.3
Clothing	6.4	–	0.6	0.5	5.2
Footwear	2.8	–	0.3	0.9	1.5

Source: *SAUS*, Table 1487, 1982–83.

Table 2.8 US trade with selected countries in 1958, 1970 and 1981 (in current US$m)

	1958[1]		*1970*[2]		*1981*[2]	
	Exports[3]	*Imports*[4]	*Exports*[3]	*Imports*[4]	*Exports*[3]	*Imports*[4]
Argentina	249	130	441	172	2,192	1,124
Brazil	535	565	840	670	3,798	4,475
Mexico	893	454	1,704	1,219	17,789	13,765
Peru	171	123	214	341	1,486	1,224
France	438	308	1,483	942	7,341	5,851
UK	852	864	2,536	2,194	12,439	12,835
USSR	3	17	119	72	2,431	348
Egypt	53	18	77	23	2,159	397
Iran[5]	105	42	326	67	300	64
India	312	190	572	298	1,748	1,202
Indonesia	63	169	266	182	1,302	6,022
Korea	215	2	643	370	5,116	5,227
Japan	845	666	4,652	5,875	21,023	37,612
China (P.R.)	–	–	2	1	3,613	1,897
US total[6]	17,910	12,792	43,224	39,452	233,759	261,305

Sources: [1] *SAUS*, Table 1218, 1962.
[2] *SAUS*, Table 1488, 1982–83.
[3] Freight on board (f.o.b.).
[4] Cost, insurance, freight (c.i.f.).
[5] During the 1976 period trade with Iran reached: US$3,685 million (exports), and US$2,877 million (imports) levels.
[6] With all nations.

Mexico, Japan and the United Kingdom, and imports from the UK, dominated US foreign trade. By 1981 imports to and exports from Japan and Mexico dominated US international trade.

No matter how spatially extensive the particular connection of the US is with other nations the links are weak in absolute size: relative to US GNP the valuation of US foreign trade with these nations (imports plus exports) was relatively small. It is estimated, by the US Department of Commerce, that in 1984 every billion of foreign exports to all nations the US traded with represented 25,000 jobs in the US. Even at this average level, 6.5 million jobs were involved, or less than 6 percent as of 1984 of all workers in the US labor force.

Although these links may have been weak from the US standpoint, one might argue that they were much more significant in their valuation for the other side, particularly the LDCs. This viewpoint, however, is watered down by the spatial distribution of the US foreign trade: very few destinations overwhelmed all others in terms of total volume, and these regions and destinations were not small nations.

Mexico's economy and its foreign trade

In 1960 the United States of Mexico's share of the world's population was about 1.2 percent. By 1987 this share had jumped to 1.6 percent. In 1974 the per capita gross national product of Mexico stood at US$1,000: 73.5 percent of the world's average. In 1985 Mexico's GNP was estimated at US$166 billion accounting for 1.2 percent of the world's gross product (same as in 1960) and 72.2 percent of the world's average per capita GNP.[8] Thus, in a snapshot (of approximately a quarter century) Mexico's population grew considerably in both absolute and relative terms. Per capita gross domestic product slightly dipped in reference to the world's average. Mexico, of course, experienced a rapid increase in the petroleum sector during the past quarter century.

Was this rapid growth of Mexico's population a dynamically stable pattern? Or is the once heralded 'Mexican miracle' of the 1970s evidence of dynamic instability, where spurts of population growth are succeeded by (at times sharp) declines for nations with strong ties to selective industries (like petroleum?). Some more elaboration on this issue, to the extent that Mexico City is concerned in the macrodynamics of Figure 1.1, will be supplied in Chapter 3, together with a theoretical perspective under which the above two questions might be answered.

Mexico's total foreign trade amounted to US$36.9 billion in 1982 with exports exceeding imports: US$20.9 to US$15.9 billion.[9] Combined, their valuation was 24.3 percent of Mexico's GDP for that year, 10 percentage points higher than the corresponding share for the US in 1983. Net income from foreign trade amounted to only 2.8 percent of GDP. As in the case for the US, Mexico's foreign trade was far below the portion of economic activity destined for domestic markets.

The Mexican economy underwent a significant change in the 1910–60, 1960–78 and post-1978 periods. In Table 2.9 Mexico's GDP by industry at the first digit SIC is shown for the years 1910, 1960 and 1980. Comparing the industrial shares of 1980 to the corresponding ones for the US, Mexico had a higher share for agriculture, mining, manufacturing and construction. Whereas, it had a much lower share in

Table 2.9 Mexico's gross domestic product by industry, first digit SIC share, 1910, 1960 and 1980

	1910[1]	*1960*[1]	*1980*[2]
Agriculture and allied sectors	35.7	15.2	7.4
Mining (including petroleum)	9.2	4.6	5.5
Manufacturing	15.8	17.9	34.6
Construction	0.9	4.7	8.5
Transport, communications and public utilities	3.0	3.5	6.7
Trade (including restaurants, hotels)	20.4	29.2	17.2
Services	13.1	19.6	20.0
Government	1.9	5.3	

Sources: [1] Mexico: *Informacion sobre Aspectos Geograficos, Sociales y Economicos: Aspectos Economicos, Volumen III*, Mexico, 1983, Secretaria de Programacion y Presupnesto (SPP).
[2] *Agenda Estadistica 1983*, SPP; Table 4.1.9.

GDP for government and services. During the 1960–80 period the Mexican economy experienced a significant increase in the manufacturing share, a very sharp change over a 20-year period compared with the preceding almost insignificant change during a 50-year period (1910–60).

The very high share in construction and high level in mining, in conjunction with the explosive manufacturing growth, were the result of a sharp rise in exploitation of the oil and natural gas reserves both inland and along the Gulf Coast. At the same time an extraordinary growth in the Mexican population occurred during the 1960 to 1984 period (from 37.2 to 77.7 million). It could be associated, in accordance with traditional economic and sociological theses, with the rise in manufacturing and the concomitant decline in the agricultural share of GDP (from 15.2 to 7.4, Table 2.9). However, in view of the evidence discussed earlier, whether this was the main reason for such growth is subject to dispute when Mexico's growth is examined in relation to the world's population and its growth rate and in particular to global socio-economic events occurring during this time-period.

The employment composition for Mexico at the first digit SIC (see Table 2.10) reveals a rather low labor productivity in agriculture and trade and relatively high manufacturing productivity. In the agricultural and manufacturing sectors the labor productivities increased sharply

Table 2.10 Mexico's employment share by industry at the first digit SIC, 1960 and 1979

	1960	*1979*
Agriculture and allied sectors	54.1	28.9
Mining (including petroleum)	1.3	1.0
Manufacturing	13.8	19.5
Construction	3.6	6.4
Transport, communication, public utilities	3.5	4.0
Trade	9.5	13.8
Services (including government)	13.5	25.9

Sources: Mexico: *Informacion sobre Aspectos Geograficos, Sociales y Economicos: Aspectos Sociales, Volumen II*, Mexico, 1982, SPP; Tables 2.4, 6.

during the 1960–80 period from their 1910–60 levels. Although the share of value of manufacturing output doubled in the 1960–80 period, its employment share increased by 41 percent only. Clearly, Mexican manufacturing was becoming more capital intensive.

Taking a closer look at the manufacturing sector, the kind of capital intensity taking place in Mexico becomes clear. In Table 2.11, the value of output for the top four categories is shown at the third digit SIC, whereas Table 2.12 indicates the top five categories in terms of employment. In summary, Mexico assembled automobiles and was involved in light manufacturing output and employment. There was very little production of capital goods, and labor productivity was low in light manufacturing as in beverages and textiles.

The foreign trade of Mexico was quite specialized, with exports particularly so, whereas its imports were more diversified (see Table 2.13) but fewer than those of the US. Mexico's exports were overwhelmingly petroleum and natural gas related; whereas, its imports were mostly industrial machinery and equipment as Mexico attempted to build on its industrial base. In slightly more detail, at the fourth digit SIC, the major categories (at least US$400 million) of ranked exports (with their value in US$ billion in parenthesis) were: oil (9.4), metal products (1.0), natural gas (0.5), derivatives of oil (0.4), coffee (0.4), and frozen shrimp (0.4).

On the other hand, Mexico's imports were: assembly materials for automobiles (1.0), corn (0.6), metal sheets (0.6), pipes (0.5), automobile parts and equipment (0.4), machinery for processing metal (0.4), and

Table 2.11 Value of output in manufacturing for the four major categories, third digit SIC, 1982–83

Category	*Pesos (bn)*
Automobiles	164.2
Iron and steel mill products	166.0
Beer	63.4
Oils, margarine, vegetable products	56.8

Sources: *Agenda Estadistica 1983*, SPP: Tables 4.2.12, 3.

Table 2.12 Employment in manufacturing for major categories, 1982–83

Category	*No. employed ('000)*
Total	580.1
Beverages	52.7
Automobiles	48.2
Iron and steel mill products	23.5
Textiles	23.2
Cellulose and paper products	22.1

Sources: *Agenda Estadistica 1983*, SPP: Tables 4.2.12, 3.

agricultural machinery (0.4). That assembly materials for automobiles played such a dominant role in Mexico's imports in this time-period represents a broader event: certain LDCs were in this time-period an assembly line for products destined for MDCs.

The spatial distribution of Mexico's foreign trade for selected countries was much more specialized than its product. In Table 2.14 Mexico's foreign trade with selected countries for 1982 is shown. It is quite clear that Mexico's foreign trade was very closely tied to the US market: it exported oil to the US (accounting for approximately 53.6 percent of all valued export for Mexico) whereas it imported industrial output (particularly steel mill products and electrical equipment) from the US (which accounted for 56.6 percent of all Mexico's value of imports in 1982).

All this activity in commodity or final product trade was occurring

Table 2.13 Mexico's foreign trade, 1982 (in US$bn)

Exports (f.o.b.).	
Total	20.9
Agricultural	1.1
Mining	16.6
Manufacturing	3.1
Imports (c.i.f.):	
Total	15.0
Agriculture	1.0
Manufacturing	13.0
Chemicals	1.3
Iron and steel mill products	2.0
Industrial machinery, equipment	3.9
Professional, scientific equipment	1.1
Transport and communication equipment	1.8
Electrical, electronic equipment	1.1

Source: *Agenda Estadistica 1983*, SPP: Table 4.3.

Table 2.14 Mexico's foreign trade with selected nations, 1982 (in US$bn)

	Exports	*Imports*	*Balance*
Argentina	0.05	0.13	−0.08
Brazil	0.70	0.35	0.35
China	0.09	0.06	0.03
Egypt	0.02	–[1]	0.02
France	0.90	0.35	0.55
India	0.03	0.02	0.01
Indonesia	0.03	0.02	0.01
Iran	0.10	–[1]	0.10
Japan	1.50	0.90	0.60
Korea, S.	0.21	0.03	0.18
UK	0.90	0.28	0.62
USA	11.20	9.00	2.20
USSR	0.08	0.01	0.07
Spain	1.80	–	–
Israel	0.70	–	–
Canada	0.60	–	–
W. Germany	–	0.90	–

Source: *Agenda Estadistica 1983*, SPP: Tables 4.3.4, 5.

Note: [1] Level below 0.005

while the recorded Mexican annual population migration into the US grew approximately by 146 percent. At the same time, the exchange rate for the Mexican peso in international currency markets decreased by 9,600 percent. These dynamically unstable patterns in the United States of Mexico were associated with highly selective, specialized trade with very few outside sources, mostly in the US, and with selective growth in a few industries mostly centered around petroleum. It is within this context that one must view the political, economic, social and other forces behind the present formation of major trading blocs globally, including the North American bloc of the US, Canada and Mexico.

As evidenced by the Mexican case, very closely associated with the flow of raw materials and finished products is the flow of population. LDCs export raw materials and people to selected MDCs, whereas MDCs export finished products to selected LDCs. There are economical, political, cultural and other barriers at the origin and destination of these flows impeding their free movement. Trading blocs, *de facto* trading zone boundaries, and the formation of a few cores is the picture emerging from the above brief look into trade. But these trading entities are not only fuzzy, they also tend to be non-durable.

International labor migration and its interaction with international capital flows partly created by the large-scale redistribution of manufacturing production worldwide bring attention to turbulent, dynamically unstable, patterns in international flows.[10] Potential for turbulent motion (i.e., dynamically highly unstable) arises as a result of the multiplicity of interaction in socio-spatial dynamics. A particular type of instability is due to labor migration: migrants flow into high wage rate regions whereas capital (or job openings) flows into low wage locations, thus creating conditions for unstable flows. This fundamental incongruence in spatial dynamics can bring about violent events, involving high-frequency non-periodic spatial cyclical movements in all components of spatial exchange.

Foreign trade and economic stability

A national economy's production and international trade patterns impact the nation's economic dynamical stability. Although the empirical evidence supplied covers only two nations within a specific time-period, the US and Mexico, the implications seem to transcend specific spatio-temporal horizons. What one observes at this point in time must be typical of what has occurred at all prior and will occur at all future time-periods, unless one assumes that our current position is a privileged one – something that this analysis rejects. Next, the stability properties of such production and trade features are discussed. The empirical and theoretical arguments to base the conclusions are encapsulated in the following two points:

1 The production of the US and Mexico in the study period demonstrates a dominance by a few, but changing, industrial and agricultural sectors. Foreign trade was extensive in space and in sectors, but it was dominated by a few locations and products. Its absolute size was relatively small, when compared with the magnitude of the respective national gross domestic product. Foreign trade is a representative economic linkage exhibiting patterns of dynamic instability; it is further representative of many socio-spatial linkages demonstrating dynamical instability.
2 Theoretical evidence regarding the broader interaction of populations in time indicates that there are critical thresholds underlying the presence of dynamical stability. Models from mathematical ecology, employing the dynamic Volterra–Lotka equations of interacting populations within a community, produce results of particular interest in this case. They identify two components in the growth or decline paths of individual stock abundances, an intrinsic growth or decline factor (self-growth parameter) and a factor which is due to comprehensive interaction with other spatially distributed stocks (an interaction parameter). As the interaction term becomes diluted, it loses its importance in the growth- or decline-producing force which is the result of the type and magnitude of the eigenvalue associated with the stock in question in the dynamics. The sign and size of the intrinsic growth coefficient become the dominant components in the real part of the stock's eigenvalue, which identifies the stability properties of the dynamics.

As the interactive species (stocks and/or locations) increase in number, and/or their strength of interaction is lessened, and/or their interactive patterns become more random, the species' dynamics experience an increase in the likelihood of becoming unstable. These conditions contribute to the diminution of the interaction effect upon the growth or decline paths of individual stocks. As the intrinsic growth or decline rate coefficients become the sole determinants of the stock dynamics, exponential growth or decline patterns emerge. Namely, populations suddenly tend to demonstrate patterns of explosive growth or extinction. Thresholds in the number of interacting species, strength of interaction, and randomness in interaction are routinely found, through numerical simulations, to be very low and associated with a relatively few stocks or locations (around ten) in the case of absolute dynamics.[11]

Combining the empirical and theoretical evidence, and by slightly broadening the frame of reference, one arrives at the following conclusion: if foreign trade is used as a proxy to a more composite set of international interactions then a nation's aggregate growth picked up by appropriate variables like population or any other stock (for example per capita wealth) must exhibit unstable dynamics in reference to the global ecosystem.

The above conclusion is obtained because two out of the three requirements for dynamic instability presented in 2 above are met. Only the requirement that the interactions must be extensive rather than highly selective is not met by international trade patterns, to guarantee dynamic instability. Their strength of interaction was found to be weak particularly for large and medium-size nations. Certainly the number of interacting nations is relatively large. Thus, foreign trade and, broadly speaking, international linkages must be considered as destabilizing for most economies. There must be destabilizing for both relatively small nations, even in the short term, and for large nations in the long term.

This instability is detected when one uses the relative lens to record population and per capita income changes. When, however, the absolute population and wealth lens is used to look at overall economic growth for a number of medium and large nations (for example, the Western European nations) these instability effects are not always detected in space–time. The Federal Republic of Germany and the German Democratic Republic, what is Germany since 1990, have, for example, reached a point in their absolute population dynamics where population decline is now registered. Other nations in Europe also are thought to be close to their 'steady state' absolute population size.[12] The argument for selecting a relative lens now becomes more apparent: the absolute lens does not pick up hidden instability.

But the evidence also suggests that there are a few very large nations dominating the global socio-economic system. Thus, a key requirement for presence of dynamic ecological stability is met: namely, that a few species must dominate the ecology. That is exactly what one sees: a very few nations and industrial sectors dominating, in size or wealth, the global ecosystem. Thus, one must presume that the global economy and its subparts must be stable, and stably interconnected. On the other hand, one could also witness the rise and fall of the economic fortunes of nations and industries through the centuries. Seemingly, a paradox emerges.

A closer look into the unstable dynamics, however, resolves the paradox. The tail end of a dynamically unstable ecology is that an initially more balanced population and/or wealth distribution gives way to a scheme where a few locations or stocks *become* dominant, the rest having experienced decline throughout the ecological association with some quickly becoming extinct. That there are so many nations and industries ecologically interacting is by itself cause for instability in the current global ecosystem. That a few tend to dominate (in trade and otherwise) is the effect of instability. That inter-national, inter-sectoral interactions are highly selective is the cause for and/or the effect of dynamic instability. Dominance by a few nations and/or industries is the cause for and/or effect of dynamic instability. The proposition thus that the interaction dynamics are unstable does not violate any of the three observations.

Why each nation has a few trade partners can be analyzed under this perspective in two ways. It can be viewed as the end result of only very few nations having survived and evolved from the past into dominating international linkages, with the nations in question having started within a much more extensive or no foreign trade interconnectance. This end result must be due to (a bundle of) ecological forces. It can also be viewed in the light of an effort by nations and/or industries at times to be part of a stable hierarchy. To do so, they must maintain a highly selective interaction pattern, continuously reinforcing spatially or sectorally specialized linkages and forming trading blocs. However, to be stable these inter-national inter-sectorial linkages might require such selectivity which could be ecologically impossible to sustain. Economically, socially, politically, technologically, morally or otherwise the required degree of selectivity may become infeasible or obsolete beyond a critical level and/or strength of interaction in space–time. At that time old trading blocs break down and the process leading toward new blocs sets in.

One can also posit similar arguments on national and sectoral selectivity and maintenance of dynamically unstable urban hierarchies. Equivalently, one might describe in combination the dynamics of national urban (i.e., spatial) and sectoral hierarchies. In particular, these arguments are very apt in describing the qualitative dynamic properties of inter-national urban hierarchies.

Foreign trade and the effect of distance

Having recorded the foreign trade volume for the US and Mexico by place of origin and destination with selected nations for various years, one can analyze the role that distance played in these international trade flows. Factors associated with geological and topographic features, particularly natural resources distribution in space–time, are paramount in comparative advantages for the purpose of trade. Whether social, political, and economic variables not directly associated with spatial advantages or propinquity are more dominant in shaping trade flows between nations can be directly examined by regressing trade volumes versus a variety of variables.

Analysis of the US and Mexican inter-national trade flows was carried out to detect any spatial regularities using the nations of Tables 2.8 and 2.14 as a sample.[13] The evidence seems to suggest that physical distance does not play a significant role in trade flows between nations when compared with other important variables such as per capita GDP, or population level with the trading country. This is of particular interest in the case of Mexico, since more than half of its foreign trade is with the US, a neighboring nation.[14]

As spatial impedance is seemingly ruled out as an important factor in explaining trade patterns globally, one might propose that initial endowments (natural resources) may be the key components in the comparative advantages enjoyed locally in reference to worldwide prevailing conditions. However, a brief and casual look at the trading flows of the recently industrialized nations of the Far East and Japan might convince the analyst that this is a questionable hypothesis. Ecological rather than merely topographical (distance related) factors alone, and/or simple initial endowments seem to impact inter-national trade, a supporting argument for the main thesis of this book.

UNSTABLE URBAN MACRODYNAMICS

Some general observations

The study of urban development and evolution is inextricably linked to national development patterns. Although this realization has been acknowledged from early on in almost all urban development research,[15] it is rarely recognized in national development literature which is overwhelmingly non-spatial. In particular, the theory of economic growth fails to recognize the contribution of individual urban agglomerations to aggregate national development.

Major efforts by theoretical and applied economists in the study of national economic growth are testimony to this omission.[16] Matters are not much different in economic cycles literature.[17] This may be a severe theoretical shortcoming because national economic cycles are tied to cyclically behaving industrial sectors. In turn, major firms in these sectors may be located at certain urban areas in a nation. Thus, national economic and industrial cycles may be directly associated with growth and decline in these urban economies.[18]

It can safely be asserted that overall national development patterns affect the growth or decline of individual urban areas. In turn, urban growth or decline affects overall national development, particularly the development of urban agglomerations at the top of a nation's urban hierarchy. Nations with dominant primate cities are prime examples.[19] The argument that urban growth affects national and regional development is well known to urban and regional analysts and is apparently self-evident. What is not well known or documented is the specifics of this interaction. How many of these urban areas affect and are affected by national development, for how long, how much, at what phase of development, and under what time lags? Both quantitative and qualitative evidence is scarce.

As indicated earlier (see pp. 78–81), and drawing from international trade, recent qualitative findings from the theory of non-linear dynamical analysis seem to provide some new and powerful insights into this interaction. Some form of dynamic stability could underlie the interdependence between a nation's aggregate economy and its major urban centers. These interdependencies cause ecological interaction through such flows as income, capital, employment, population, pollution, etc. Qualitative conclusions from theoretical mathematical ecology seem to suggest that within a community of interacting locations the qualitative dynamics must also exhibit unstable patterns.[20]

Were the urban hierarchy of a nation to be stable, i.e., free from extinctions or primacy, then it can be assumed that interdependencies among urban areas of the nation are highly selective and strong. However, only in a few MDCs do urban growth patterns come anywhere near being dynamically stable. In the vast majority of nations, particularly LDCs, primate cities exist and relatively high exponential growth or decline rate paths are widespread. Intra-nationally urban interactions are extensive, weak, and not selective. The underlying interdependencies apparently produce unstable urban growth patterns within the international urban hierarchy.

This conclusion may sound counterintuitive, in full view of primate cities being the major sinks for population, capital, wealth, and other economic stocks. To capture the event, however, one must see this as the dynamic outcome of an evolving system, which might have started earlier under interdependencies that gradually led the system into such explosive and unstable patterns of growth. We are able to speculate on the kind of linkages which might have existed prior to these dynamics being triggered. At early and critical time-periods, the urban system must have been in an extensive, weak, and almost random interactive pattern. Such a state of affairs cannot have lasted long, and must have quickly settled into what we currently observe. This important stability versus complexity argument was used earlier to view international and inter-sectoral flows.

It may be useful to look at interaction among urban settings within a national economy under some sort of efficiency criterion. The minimum (necessary) interactions must occur at any given stage of development. Instability may result when non-efficient interaction occurs, or when the efficient interaction scheme is too extensive. From a development policy and implementation (action) perspective this viewpoint might be informative. Intervention must be selective and strong if stability is sought. Otherwise it might destabilize the system, or it might be ineffective, or both.

Whether a nation's urban system or urban hierarchy is stable or not may impact the dynamic stability of the nation's aggregate economy – particularly when the nation's economy is very much tied to a dominant primate core urban region where the bulk of a nation's economic activity is amassed. Most of the economic activity of national economies occurs in their main metropolitan areas (their definition will be addressed later) where the bulk of industrial production takes place, with the obvious exception of farming and mining (on land and/or in the sea) of natural resources. Even though production of output in the last two sectors may occur in the hinterland, the income received is normally recorded (at least for census purposes) in cities – particularly when the headquarters of mining firms or seats of governments in the case of public ownership, reside in cities. That is the place where the bulk of the income received by agriculture and mining firms and workers is expanded and distributed.

Modeling urban and national growth

Few studies address interdependencies in economic growth among urban and national economies.[21] Two recent ones by Williamson *et al.*

adopt a strictly economic standpoint in viewing such interactions.[22] They study the ways by which a national economy affects the nation's aggregate urban sector rather than its specific urban areas. From these two urban development models the Kelley and Williamson model will be briefly reviewed. The issues to be brought up in relation to this model are shared by both, and by many other urban models. Thus, the criticism leveled against it transcends its specific formulation and applies to conventional urban econometric models with similar statistical features.

Kelley and Williamson present an economic 'dynamic general equilibrium' model of a national economy in an attempt to explain growth and migration in Third World cities. They state an eight-sector national model, which includes agriculture, housing (three types), services (three types) and manufacturing. The three kinds of housing involve the urban non-slum, slum, and rural kind; the service sectors include urban capital-intensive, urban labor-intensive (informal sector type) and rural labor-intensive categories.

Aggregate constant elasticities of substitution Cobb–Douglas production functions and prices and labor demanded equations are used for the formal agricultural and informal sectors, with appropriate market distortions added to account for various national policies. The dynamics are introduced through capital-accumulation and skill-accumulation equations. Households are assumed to have consumption patterns according to an 'extended linear expenditure system' which involves 'subsistence' needs, and it is based in standard microeconomic analysis of consumer behavior.

The model's dimensionality is formidable: it employs more than 150 parameters, more than 50 exogenous variables, and more than 120 endogenous variables for every national economy. From what one can infer from the documentation provided by Kelley and Williamson, less than one-fourth of the variables are directly observable and only for three time-periods (1960, 1965, 1970), and not for all nations. At least twenty key parameters in the model are assumed by the authors to have specific values without any reference to data sets.

The authors then use this specification (calibrated for a 10-year period) to derive predictions for 1980, 1990 and 2000 (over a 30-year period). Further, they use this model to test some 'policies' related specifically to urban wages and slum settlement. They conclude through an elaborate argument that rising wages in the formal sector chokes off its employment, and in turn this event gluts the informal sector labor market, having as a result falling wages there with concomitant declines in migration flows into the urban areas.[23] One may question how the authors reconcile this to observed rises in wages in the formal sector in Third World cities and rises in migration during the 1960–70 period.

Although the Kelley-Williamson model simulates a national economy as a whole, three endogenous variables are the focus of their study: urban share of certain exogenous variables, their urban growth rates, and immigration rates. They conclude that Third World urbanization, past, present and future is most sensitive to nine variables: prices and productivity, changes in manufacturing and agriculture, labor force growth, capital accumulation in manufacturing, accumulation of rural and urban housing stocks for the poor, and skills.[24]

The authors seem to be aware of the severe limitations of their absolute growth direct linkages model. Throughout the text the question appears on whether this is 'fact or fiction?' The question mark on the book's title (*What Drives Third World City Growth?*) indicates that the central question was apparently not answered in their analysis. Many substantive and technical issues are not adequately addressed by this model and theory of urban to nation linkages. On the substantive size one would question whether national and urban economic interdependencies are of an equilibrium rather than disequilibrium type. As a result, one would question the dynamic form of the Becker–Mills–Williamson model, which is really static with dynamic adjustments of an exponential type in the capital and skills sectors. This is a general criticism of most neoclassical growth and their subspecies models.

Dynamic flows of capital, labor, skills, and other input factors in production occur in accordance with driving differentials between a specific urban area and its average background expected marginal rewards. Whereas, jobs supplied by particular industries flow into areas where the per unit output expected profit is the highest over a time horizon, jobs demanded flow into areas where the expected wages are the highest. This may create an imbalance, whereby capital investment may flow into low wage regions while population may be migrating into high wage regions.[25] On the analytical side, the dynamically stable patterns obtained by the Becker–Mills–Williamson model fly in the face of ubiquitous dynamic disequilibria and the unstable dynamic patterns observed in the urban sector of Third World nations, and in the interaction of these urban areas with their national economies.

A positive externality of this model is to document what is fundamentally disturbing with large-scale economic and econometric models. There are too few data to even qualitatively test them. They contain too many degrees of freedom which allow various kinds of outcomes to be reproduced and rationalized at will. The most critical of these models' shortcomings, however, may be their failure to depict broader socio-economic forces affecting urban areas worldwide and their inability to address the manner in which a (large) urban area affects its national economy.[26] Current urban development literature, at this scale of analysis, is seriously deficient on conceptual, methodological and empirical counts.

Interdependent urban and national growth

J. Jacobs recently advanced the thesis that, in reality, only city-economies ought to and in certain instances do exist at times with an agricultural and/or urban or sub-urban hinterland.[27] Although her arguments may not be feasible to test directly with available data, they do provide some useful guidance for formulating a few testable hypotheses. However, they must be transformed, to a great extent, along lines presented in later argumentation before they can be used as working hypotheses.

According to Jacobs, nations are merely artificial and at times temporal political jurisdictions impeding the efficient functioning of city-economies. Their main purpose is to redistribute wealth from more productive subregions to less productive ones, presumably for social equity purposes. Thus, national governing bodies act as agents for 'transaction of decline,' as she calls these transfers, of otherwise potentially prosperous urban economies. She does not point out that occasionally these entities act also as brokers for growth, as well as agents of decline transfers and regulators of externalities and other spatial market imperfections.

This hypothesis is not very distant from conventional spatial literature, which views the role of governments as balancing the leading (fast) growth regions (or sectors) of a national economy with the lagging (slow) growth (or declining) regions (or sectors).[28] By intervening, through a variety of policy instruments, governments distort the pure spatial and sectoral market functions. The total level of output produced, nationwide and for all sectors, is lowered for purposes of a more even spatial and sectoral distribution. This line of argument underlies the 'balanced growth' versus 'maximum output' (the so called 'social equity' versus 'efficiency') debate in the literature.

Were one to accept the call for city-economies, one must take the argument one step further and argue that only industrial production units do or should exist. These producers are defined over different firms, as well as over different locations. Nations *and* cities must then be considered to be arbitrary political jurisdictions impeding the efficient economic performance of firms. Interference in production occurs through regulations firms have to observe in prices and taxes they incur. These interferences are presumably motivated by attempts to internalize production externalities occurring locally, and redistribute income locally and nationally. They also come in the form of tariffs and quotas motivated by attempts to account for various economic or political externalities occurring in international markets.

Even further, the argument can be pushed to the point that each consumer, like each producer, is impeded by similar considerations. For each individual there is a hinterland, and thus a *de facto* border identifying the individual's locational (consumption, production, trade) niche. There are clearly ecological reasons why this extensive de-nationalization and increased territoriality did not and will not occur. To spatial economists these reasons are bunched together under the encompassing term 'agglomeration economies.' Clearly, there are other even stronger forces in setting national boundaries, at times resulting in increased isolation. They are of an ecological type, too.

At this point, however, a novel element for spatial economists is brought into the discussion: it has to do with currencies, a subject to be addressed more fully later (see pp. 95–105). A basic interference in efficient production, the argument goes, is an indirect tax (or subsidy) cities and their firms incur when they transact with various national currencies for purchasing input factors or natural resources and in selling their final output. As the currency rates fluctuate in international currency markets, in the short term nations, cities, and firms incur a tax or subsidy attributed to the relative increase or decline in the strength of their national currency. In spite of its novelty to spatial economics, this is not a controversial argument. Traditional non-spatial economics has employed currency fluctuations in national development models since currencies began to be traded freely in 1971 in view of the wide fluctuations demonstrated in international currency markets, particularly during the 1980s.

Survival of firms, industries, urban areas, regions, and even nations might be at stake under these fluctuations in currency exchanges. This is an interesting element in spatial and sectoral dynamics because it links spatial macrodynamic stability to relatively very fast moving parameters. Since firms are relatively durable entities, and so are cities, they change relatively slowly. Neither their built capital stock nor labor are very mobile compared to the ephemerally behaving currency exchange rates impacting fast changing input and output prices.

This is an exceptional case where a very fast moving variable, in this case price, affects a slow moving one – for example the quantity of the built capital stock. These effects will be explored in more detail in the next section, which records changes in the US dollar as well as changes in the price of oil as having affected particular US regions and some of their urban agglomerations. Tying city growth and decline to relative price fluctuations, and by following earlier suggestions by F. Hayek on 'the denationalization of money,'[29] the argument goes that, at the currency front, cities ought to be allowed to use their own currencies.

This is a novel perspective, its impact being difficult to assess analytically or empirically. Jacobs makes an effort, citing qualitative evidence from different settings, to back certain of her arguments. The difficulty in accepting this hypothesis should not take away from the thrust of the observations which have major urban agglomerations incurring the brunt of international currency fluctuations. How important these fluctuations are in the short run (say, a decade or so), the medium term (quarter century), and the long run (one century or so), on the growth or decline in the population and wealth stocks of large urban agglomerations, is not entirely clear.

Interaction between currency exchange rates is partly based on perceptions regarding the future demand for the various urban/regional/national output(s) in international markets; so rates reflect general economic conditions surrounding the production and consumption sectors of the spatial units. In effect, rates do not by themselves dictate the economic fortunes of spatial entities; they are largely determined by perceptions (admittedly fast-changing and fluctuating) regarding the future prospects of these spatial entities' production and consumption sectors.

Firms and industries are the ultimate medium for development. Consequently, the evolution of urban areas and nations directly and largely depends on the changes in conditions related to output, profits, employment, capital investment, technology, etc., that the industries and firms they contain undergo over time. There is, however, lack of enough time-series data on such variables at the firm or industry level for almost any nation to be able to trace such industry-related spatial evolutionary patterns. Thus, one is forced to carry out the dynamic analysis at qualitative and relatively aggregate spatial, urban or nationwide, and sectoral levels.

This is not as damaging, however, as it might initially sound. Urban and national economies, in spite of their overall diversified industrial base, have been in the past and currently are basically driven by a very few industries located in a few of their intranational regions at any time-period. Whether their primary function was and is administration, production, trade or any combination, destined for domestic or foreign markets, the dynamics of a very few economic activities and a few regions at each time-period determine largely broader spatial and sectoral economic development. Again, we come across the 'very few' condition.

Primate cities and nations

As is the case with most of the material in this book, the observations made are based on current spatial urban and national growth and interaction patterns; they could, however, be applicable to cities and dominions of the past. So what is outlined next could supply insights into the functioning of nations and urban settings of the past at many locations on the globe.

Not all major nations and urban agglomerations are engaged in the production and trade of all the major commodities and merchandise traded and produced worldwide; sectoral specialization occurs in space–time due to prevailing comparative advantages. Most of the very populous (primate and few) nations and urban settings are very diversified when compared with less populous countries and urban areas. Large nations' foreign export-oriented production and total international trade activity is significantly higher, in both volume and valuation, relative to the more numerous, smaller in size nations and urban areas.

Foreign trade volume, however, is relatively small when compared with that destined for domestic markets. As a result, primate nations and cities grow or decline in absolute terms mostly because their domestic markets grow or decline in absolute terms. There are a few large city-states, i.e., nations of very small areal size, which do not obey this rule because of their special spatial status. City-states like Singapore and pre-1997 Hong Kong, the special economic districts of South-East China, and Taiwan,[30] among others, are not open economies with a national hinterland vast in population and area. For them foreign trade is much more important than domestic markets: their hinterland is international.

Both large nations and large urban areas perform a dual function: they serve and are served by broadly defined extensive national and international hinterlands, including rural and urban areas. Relatively high levels of concentrated economic, political, social, cultural and other stocks and activities are found there. Being the major centers of production and distribution of commodities in both raw and finished form, they are the major origins and destinations of information and financial capital flows. Primate cities in particular are locations where intranationally the demand for and supply of labor and population is relatively high, along with the demand for high wages. They occupy large quantities of the national space, and are distinguished by significant levels of locational comparative advantages, enjoying relatively high accessibility to foreign and domestic markets. Their central cores demonstrate very high population, capital, and information densities. Primate cities are the depositories of a vast variety of human activity. They are the *very few* locations where much of the world's and most of their nation's economic, social, political, and cultural assets are to be found; that is, where the most significant gains (including additions to) and losses (depletion of) these stocks occur. These few nations and urban areas are most frequently (but not exclusively) the source of technological innovation in production and consumption for their international and their domestic hinterlands, and the sink of innovation as it propagates out of international nodes of origin. They are areas for which residents exhibit strong locational preferences by being willing to incur a premium over other nationally or internationally competing locations.

These nations and urban areas form 'hyper-concentrations' of economic activities. Primate cities in particular are at the top of a highly skewed national urban hierarchy. They are referred to in the literature as the 'dense core' regions of national economies.[31] These very large and very few concentrations exhibit the bandwagon effect in their population and capital growth. As they grow in these and other stocks they increasingly become magnets for economic concentration, apparently under conditions which have the currently perceived overall benefits of further concentration far outweighing the costs of agglomeration.[32]

Large nations and their large urban agglomerations, to the extent that one can detect from the available record in market economies, are locations where negative effects of agglomeration are the most pronounced. Pollution, congestion, and poverty gradients have their steepest slopes there. They are the sites where political and economic failures (including market imperfections and corruption), are at their highest levels. Sociological and cultural externalities like consumption and production waste, political violence and social deviance, are very pronounced too.

They are the locations where housing rents and political power attain their peak, as do land prices and the price of information. The most expensive of commodities are also traded there. Naturally, all these gradients are inter-nationally, intra-nationally, and inter-sectorally linked so that primate cities are connected through the strongest worldwide flows of the four elementary components of spatial interaction. Levels of stocks encountered and flows of stocks observed among these cities are not haphazard, random, or unrelated.

Flows and spatial arbitrage

Below, the first reference to the notion of relative parity in net attraction among locations is made; the full analytical exposition of this notion is given in Chapter 3. Here, an introduction to the qualitative aspects of it are given.

In Chapter 1, two fundamental socio-spatial forces were identified, the B and E–D force. These forces were the outcome of spatio-temporal interdependencies resulting in spatial flows of mobile factors. Now, the most basic question is posed: where do these interdependencies, forces and ensuing flows originate?

To an observer, there are spatial differentials in attractivity for various social and economic factors, and an agent which must govern the response to such differences. The agent which regulates the inter-spatial levels of socio-economic variables within a national and international frame of reference, and simultaneously generates inter-spatial flows, is 'spatio-temporal arbitrage.' This agent of development or evolution represents an effort toward and a sustaining of worldwide, inter-national, inter-sectoral 'relative norms' (or targets) in each urban factor. Collectively these targets are components of (arguments in) a perceived aggregate background composite *relative parity* in net attraction (benefit, gain) for each urban area. Prevailing levels in these relative norms and in relative parity are due to current conditions in the production–consumption market and governmental processes world-wide, nationwide, and urban-areawide.

Although the effort is *toward* the attainment of this relative parity in net attraction among all locations at any point in time, it does not necessarily follow that such parity in attraction ever materializes. Among other differences, this is the major departure of this work from traditional static equilibrium-bound spatial models of the past. In fact, the dynamic behavior in this effort toward relative parity in net attraction may entail turbulent and chaotic dynamics far away from dynamic equilibrium conditions.

Population size and per capita product are two among the many variables in the bundle of variables composing the relative parity concept. Relative parity, and its associated 'factor relative norms,' acts as a valve. If at some location one sector's component happens to be greater than its norm-target level would dictate in the relative parity in net attraction measure, then part of this sector (through spatio-temporal arbitrage) will outmigrate to other spatial units. Net attractivity of the location will be affected in turn.

In the case of a negatively perceived factor (like pollution), a negative difference will result in a level of attractivity higher than the location's par. An observer will expect an inflow of economic activity with a concomitant fall in the location's attraction. In case of a positive difference, the location's attractivity will be lower than its par, and economic activity will outmigrate while net attractivity there will rise. On the other hand, if the component's level is below the 'factor's relative norm' dictated by the parity in net gain level currently prevailing, then inflow in that component will occur. Equivalently, the net attractivity will be changed at that location.

Aggregate average relative parity in net attraction is possible through 'spatial liquidity' in its component factors. Ecological (i.e., diverse) transaction costs limiting spatial factor liquidity are themselves elements in the relative parity bundle. They are part of the 'costs' in the net attraction indicator, which consists of beneficial (positive) and costly (negative) factors.

Consider, again, the issue of environmental pollution as an example. If some of the world's largest and most polluted cities, like Mexico City, Los Angeles, Seoul, Cairo, São Paulo, etc., were to have a level of pollution below their par within their relative spatial parity bundle, then capital investment would flow into them from other regions within their respective countries, or from other nations. It would tend to equalize its actual level with its inter-nationally, and intra-nationally 'expected' or targeted pollution level under the relative parity principle. Thus, pollution stock might be thought of as having been 'imported' into these cities from other urban areas on the globe, together with capital for instance. Adjustments in actual factor levels within urban areas occur in bundles and are not confined to one factor at any point in time. In combination, to an observer, these factors alter their levels in a quest for overall composite net attraction parity.

Being at par in a factor does not necessarily mean, as in the previous example, that pollution levels in Seoul, Mexico City, Cairo, São Paulo and other relatively high pollution cities must be equal at any point in time; or that, for example, the current air pollution level of Shanghai must match that of San Francisco (a relatively low in air pollution city). On the contrary, their corresponding at par (relative norms) pollution levels must vary, apparently remaining at relatively high levels in the cases of Seoul, Mexico City, Cairo, and São Paulo. Variations in air pollution levels reflect the effects of other factors in their composite relative parity levels among the globe's large urban agglomerations. A technical presentation of the concepts of relative norms (at par), and composite relative parity is supplied in Chapter 3 (see pp. 192–209).

Aggregate relative parity in net attraction, for an observer, identifies a weighted measure of a bundle of positive and negative factors equal over space to a background prevailing level at any point in time, if reached. Dynamically, however, the globally prevailing (background) net attraction level changes as environmental, economic, and other (that is ecological) conditions are altered. As a result of these changes, there are internally as well as externally induced readjustments in each component factor in each urban area. The relative norm level for each factor at each large (and small) urban agglomeration is reset.

Each factor being at par, or at its relative norm level, can also be construed as this factor having reached its 'carrying capacity' within a specific urban area at a particular point in time.[33] Carrying capacity fluctuations always occur in international, national, and urban ecologies and as a result the various urban factors are constantly in a state of flux. Their carrying capacities are being continuously either under or overshot, becoming moving targets. Consequently, continuous in-migration or outmigration of stocks occurs in these urban areas.

Trade in commodities is one form of direct economic spatial arbitrage, where driving differentials in relative parity associated with production or consumption are diminished. Another form that this arbitrage takes is the net spatio-temporal trade in capital, information, labor opportunities (or unemployment), poverty, etc., where driving differentials are smoothed out by these flows.

A spatio-temporal arbitrage agent fills vacuums of worldwide equivalent levels of stock sizes, partly through the inter-spatial/temporal movement of the four elementary components of interaction: population, capital, commodities, and information. Filling in these vacuums, or the releasing of overfills, occurs mostly through intra-national (endogenous) adjustment of resources rather than inter-national (exogenous) movement of stocks. Endogenous growth or decline is simply faster in adjusting to these differentials than exogenous inmigration or outmigration of stocks is. This issue is further addressed in Chapter 3 (see pp. 195–201).

One may ask how is this aggregate relative parity in net attraction and associated carrying capacities being computed? Further, how does one (other than the observer) know when a carrying capacity has been reached? To answer the first question one must adopt an 'observer'-type dynamic model and trace the observed relative dynamics in reference to the relevant environment along the lines of the ecological models suggested in urban mathematical ecology.[34] The second question requires that 'participant'-type agents, individuals and collectives be explicitly recognized. It is answered by assuming that these participants find out whether and when such carrying capacities have been reached by trial and error, through broadly defined social speculative behavior.

Individually and collectively, participant agents act on expectations regarding these and other variable levels pertinent to the individual agent and their expected net rewards to them. As a result of timely responses and speculative social (individual and collective) action, some agents succeed in filling factor vacuums and thus they gain. Other agents fail in their attempts to add to already saturated factor levels, and they lose resources having made the wrong guess. It must be remembered that aspects of gain and loss are multifaceted in this ecological perspective. Further, one must distinguish between an individual or collective (participant) action and an observer's expectations regarding aggregate behavior.

In summary, this type of interdependency among nations and among the world's largest urban agglomerations underlies the forces and linkages modeled in this book. As is to be discussed later in this chapter (see pp. 141–57) and demonstrated analytically in Chapter 3, relative urban macrodynamics are the end effects of these interdependencies. Urban areas have a particular size and wealth at a particular point in time because relative parity is dictating a set of kinetic conditions producing these levels within a worldwide environment.

CURRENCIES AND DUALISMS

It should be noted at the outset that the discussion which follows is not intended to make a contribution to the theory of monetary economics or the theory of finance. Its objectives are to address a phenomenon not usually found in formal theories of monetary instruments – namely, dualism in currencies – and to point out the possible effects of this dualism on global urban dynamics.

Among all four elementary components in spatial interaction, labor may be the most homogenous stock to flow among nations and the most restricted in its inter-national and at times intra-national mobility. Inter-national labor flows may be substituted by domestic flows. It is more difficult to carry out such substitutions for the other three elements, particularly capital. As will be seen later, foreign capital from certain MDCs and local currency from LDCs are not perfect substitutes. Although not as restricted as labor flows, capital flows encounter significant barriers in international currency markets. All four elements, their inter-national movements and their effects on spatial (national, urban) dynamics, are intertwined.

Dualism, traditionally viewed as the sharply uneven distribution of a stock in space[35] or its equivalent the very uneven allocation of wealth among social groups, is indeed a general phenomenon observed in the dynamics of socio-spatial stocks, including the stocks of population, commodities, capital and information. All four elementary component flows affect each other's dualism. In this subsection, the focus will be on capital flows, and in particular on the monetary aspects (i.e., money-related issues) of this flow. For the exposition that follows, it is of little consequence whether a distinction is made between commodities and capital or whether any specific definition of capital is adopted. If capital is viewed as another commodity, then the inferences drawn next hold good. In the formation of a dualism among national monetary stocks, one might detect general processes producing dualism in the other three stocks as well.

Consequently, how, through trade, currency flows affect urban and national economies and ecologies will be addressed. Currency exchange instabilities and dualism in currencies (formal or hard, and informal or soft currencies) are related to the phenomenon of dualism found to be ubiquitous in the dynamics of LDCs and MDCs, as well as in the dynamics of different socio-economic groups within these nations and their large urban areas. The arguments to be presented lend support to the unstable dynamics hypothesis driving international interdependencies and interactions at both the national and urban scale.

Capital, finance capital that is, is an indirect factor in production assuming that built capital stock is the direct factor; it is traded in international capital markets as currency stock. Other factors in production do the same thing, too. However, capital in the form of money performs another function not performed by other input factors in production. it is used as the *numéraire* to weight relative prices of labor, commodities, output, and other factors in production and consumption.

In economic analysis, currencies as money are peculiar stocks. Micro and macroeconomic theory considers money to be neither an initial consumer (or producer) endowment, nor a produced consumer output. It does not enter the consumer (or producer) utility (or production) function as an argument.[36] In the Walrasian general (static) equilibrium model of all consumers and producers of an economy, the system is at market long-run equilibrium when all sectoral markets are cleared – which implies that all excess demand conditions for factors and products are zero – and when all sectors' profits are zero.[37] Excess demand is defined in Walras's equilibrium as the difference between the quantity of a factor a consumer decides to consume less his initial stock, where his excess demand for a final product is the quantity he consumes as a result of allocating part of his income in purchasing it.

In this static and highly restrictive model solutions for any *absolute* set of prices do not exist. Money enters the system of simultaneous linear equations to provide a *numéraire* quantity so that a set of *relative* prices can provide the opportunity for obtaining a solution. The Mayas, who never discovered money, used cocoa beans as their *numéraire* commodity instead. Other nations in various time-periods have used consumable commodities as the *numéraire* commodity, including tobacco in colonial Virginia in North America. The Mayas, as well as all these other nations, may have had a highly unstable monetary system in the long run, although it might have been stable for short time-periods.

For the purpose at hand, more detail of this model (including J.B. Say's law) is not needed. What is of interest is the main conclusion emerging from this general equilibrium construct according to which, 'consumers (or producers) will never desire to increase or decrease their money stock, i.e., they will never desire to exchange money for commodities (factors or final products) or commodities for money.'[38] This conclusion is clear testimony to the restrictiveness of the Walrasian theoretical model, as at times consumers and producers have preference for holding money stocks.

There are analytical difficulties as well which further restrict the validity of this general long run market equilibrium construct. Definitional difficulties aside, prices may be non-linearly related to quantities of commodities exchanged and consumed. As a result, conditions for uniqueness of a solution are further restricted. But the major limitation of this model is its static structure. Although economists are aware of the dynamic nature of the iterative process in attempting to locate the static and unique solution of Walras's model, many of them do not seem to be aware of the complicated dynamics that may ensue, including instability and turbulence.[39]

If conditions are so complicated in the case of one economy, where the local currency can be thought of as the *numéraire*, they are far more complex in a multiple economies framework. Each nation of course issues its own money. Intra- and inter-nationally, not only are input factors in production (labor, land, and fixed capital assets), natural resources (primary commodities), and final products (commodities for consumption) exchanged, but also currencies are traded in international currency markets.

In the Walrasian general equilibrium model, money is the *numéraire* quantity closing the static system of incomes, prices, profits, quantities of factors and commodities, so that it has the possibility for a (unique or multiple) solution(s). For international markets, however, there is no *numéraire* quantity to close the system, statically or dynamically. As a result, it has infinite unstable solutions. At times, nations resort to artificial composite currencies, consisting of bundles from national currencies, to close the system.

Prior to 1971, following the 1944 Bretton Woods accord, the quantity of a precious metal – gold – was used to set the price of a nation's currency. Till then, money was simply a promissory note stating the gold equivalent quantity it represented. To an extent, flow of money was accompanied by flow of gold. The amount of gold in reserve dictated the quantity of money a nation could print. Following the decision to abandon the gold standard and to float the US dollar in international currency markets, it is no longer clear (a) what determines the exchange rate among currencies, where certainly the real interest rates (the nominal rate of interest minus the inflation rate) of various economies and their available money supply must be determining factors, and (b) what regulates the quantity of money a government prints and supplies domestically and internationally, at any point in time. As usual, a multifaceted bundle of causes must affect such monetary policies ranging from economic to political, to social. Moreover, a highly non-linear interdependency must govern the stock and flow sizes and their relative prices giving rise to potentially violent and turbulent dynamics in these indices.

Often, as is the case with nations with 'soft' currencies, the quantity of 'hard' currency in reserve is used as a barometer of the amount of local currency supplied by the local central bank. In this case, the 'hard' currency is used as the *numéraire* to close the system. The end result is that, inter-nationally, selective preferences for trade are reinforced by high selectivity for particular currencies. This has contributed to the creation of, and is reinforcing, a dualism in the quality of various world currencies.

A worldwide impact has been that there are now formal and informal currencies. Formal currencies are the global 'hard' currencies, those which are recognized by most banks and are exchanged worldwide. Informal currencies are those which are locally exchanged. The former are relatively durable in time, having maintained relatively stable (although at times oscillatory) exchange rates for years, decades and at times even centuries. The latter are frequently changing unimodally, meaning they are continuously depreciating relative to the former, and in a rather fast manner.

Almost all LDCs, and a great many countries classified as MDCs, issue informal currency. A hybrid between the two is the currency of the recently industrialized countries (the new Taiwan dollar, the Hong Kong dollar, the South Korean won, and the Singapore dollar among others); these currencies are quite strong with reference to the formal world currencies, although they are not exchanged in international currency markets. There seems to be a strong connectance between the type of currency a country has and the level of development the country enjoys.

In fact, informal local currency turns out to be, partly because of the restrictions it faces in foreign exchanges, the currency of the middle classes of these informal currency issuing nations. Local elites of LDCs, and those MDCs with informal currencies, hold primarily financial assets in formal currency nations. The underclass mostly deals in a barter exchange system.

There are far fewer formal than informal currencies, as there are far fewer developed than underdeveloped countries. It can be argued, as it most often is, that internal policies of nations (linked to both 'restrictive' or 'irresponsible' monetary and fiscal governmental actions, degree of development of capital markets and institutions, extent of corruption, etc.) are among the very many internal and external variables directly affecting local soft currencies. It is obvious that such factors influence the course of a currency as to whether it becomes soft or hard. The exact extent to which they are influential in doing so is open to question still.

It is argued here that no matter how multiple and complex (austere or responsible, lax or irresponsible, or whatever characterization one might wish to attribute to such policies for a particular nation at a particular time-period), such internal policies and local conditions might be simply irrelevant in explaining the total number of hard and soft currencies worldwide. No matter what the local actions and what the worldwide and locally prevailing conditions, these factors may be irrelevant to the aggregate dualism in quality of the currencies. Merely because there are so many currencies interacting worldwide creates conditions which inevitably drive the great majority of the nations' currencies to be or become soft. Dualism in currencies is simply the result of dynamical instability prevailing when many stocks interact in space–time. Exactly where and when soft or hard currencies appear might be linked to the presence of dualism in other stocks at the particular points in space–time already alluded to in this book.

Another instability, due to such worldwide linkages among multiple currencies, is that characterizing currency exchange rates. Exchange rates are central components in the interaction coefficients linking the growth rates among the various currencies' stock size and their quality (i.e., whether they are of the soft or hard variety). Mathematical ecology stock interaction models can be used to study such instabilities which are of considerable interest to approach.

To a great extent, the *very few* formal currencies are in order of stock size: the US dollar, the British pound sterling, the West German mark, the Swiss franc, the Japanese yen, and the French franc partly ranked according to currently held size in foreign reserves.[40] Shifts in dominance among currencies in this currency hierarchy are associated with potentially violent events. The economic depression of the late 1920s was the era that the pound sterling gave way to the US dollar at the top of the currency hierarchy.[41] Shifts in the currency hierarchy and dynamic changes in the national, sectoral and urban hierarchies are ecologically linked.[42]

Formal currencies flow into the LDCs and the low per capita gross domestic product MDCs through a variety of modes. Besides trade, policies are implemented for worldwide marginal redistribution of income by way of subsidies from the high per capita product MDCs to other nations. At times this mode of subsidy is referred to as, or takes the form of, foreign 'grants in aid' or 'loans.' Many social economic and political purposes are served at both the origin and destination by these capital flows, some of which are inter-national while others are intra-national. At the international front, the commercial banks' objectives as well as the objectives of the IMF and the World Bank, both sources of such foreign loans, are manifold, central among them being an explicit inter-national economic goal of stimulating the recipient economy. It also makes it possible for the LDCs, and neighboring in per capita income MDCs in receipt of these subsidies, to purchase with formal currency durable goods from the high per capita income MDCs (often, at least up till 1990, military hardware).

At the recipient front, these subsidies to informal currency nations serve to carry out local economic, political and social objectives. At the same time, they also provide formal currency to the local elites. They stabilize the exchange rate, which among other things allows for an orderly transfer of such informal currency stocks by these elites outside these nations as they participate in informal speculative markets of the local currency. Finally, they serve to control, within limits, local apparent inflation.

LDCs obtain formal currency stock through international trade. They do so by trading their natural resources, or their quasi-finished products. Governments in the LDCs strictly control the exchange rates in formal markets, as well as the entry or exit of local money into or from the nation. Of course, they similarly control the accumulation or dispersion of the foreign formal currency reserves. Ecological-type forces govern these controls.

In a local, informal currency economy a number of key variable growth rates are linked: rates of growth in the supply of the local informal currency; the extraction, depletion and export of the nation's natural resources; the net outmigration of population (labor); the net inflow of formal currency; and the net inflow of information. As a result, the relative and absolute population size and respective growth rates of the underclass, the middle class, the elites are interrelated.

It was previously pointed out, and has been widely recognized, that an interdependency exists between exchange rates, natural resource exploitation, and economic development. They are all linked to the growth and decline of particular key industries in a national and urban economy. One could also explore much less discussed and more difficult to identify interdependencies like those among currency exchange rates, national real interest rates, and the other state variables mentioned here, although data limitations may hinder the search.

No matter what the detailed linkages, one can safely conclude from formal dynamical analysis findings that these growth rates are subject to dynamic instability properties found in model ecological systems. Turbulent regimes which might involve explosive growth or decline, including extinction or competitive exclusion, in many stocks are some of the dynamically violent events one must expect in many locations. Dualism is a phenomenon resulting from such dynamic instabilities.

Currency rate dynamics simply reflect the broad implications of these fundamental qualitative statements. In the past few years, cataclysmic socio-political events ignited during the tail-end of the 1980s and occurring worldwide but particularly in Eastern Europe and in the Communist Bloc, supply convincing testimony to these hypotheses.

One might argue that what happens to the informal currencies of a local economy is of little direct consequence to that nation's various socio-economic groups, particularly its underclass or even its middle class. Keeping intra-nationally and inter-sectorally relative prices unchanged, the fact that the national currency might be worth less now in international currency markets than a year ago is of little importance to local consumers. Certainly, it is of importance to producers dependent upon international trade, although they can adjust to them by employing various hedging strategies. The fate of an informal currency nation's legal tender is of little significance to its elites, a possible reason why it has been tolerated in so many countries and for so long. The impression that a nation's currency is of little import to local consumers may have been one of the reasons why such effects were never considered in urban, regional, and at times even national, economic models of development.

Although some of the currency related factors may be short term and weak in directly affecting urban and regional development, in the long term and indirectly they may determine intra-nationally urban and regional growth paths. Historically, depreciation of informal currencies has had an affect on the growth of a barter system in commodity exchange. Another effect has been the rampant apparent and real inflation, and all the concomitant consumer versus producer conflicts it brings to bear upon local informal currency economies. Real inflation might be measured by the change in the amount needed to purchase local goods and services with formal currency, in time, whereas apparent inflation might be gauged by the same change when measured in terms of local currency.

Both currency devaluation and inflation can be looked at as capital depreciation processes similar to those widely studied in built capital stock sectors – for example, housing or transportation. Depreciation seems to be a universal principle applicable to all stocks, including labor, information, and commodities. Key factors for finance capital (money or currency) depreciation relate to the per capita rate of growth in output and information (labor productivity), relative to the per capita rate of growth in the supply of currency. Clearly, if the rate of growth in output and in information generated per capita does not keep pace with the rate of supply of money, the local currency will be devalued and inflation will occur.

Without investing in it, upgrading and maintenance of the finance capital (currency) quality level is not feasible. Similarly, upgrading and maintenance of the built capital stock cannot occur unless a constant effort is made to invest in it no matter how speculative that investment might be. In the case of finance capital upkeep (maintaining that is the relative strength of a nation's currency), investment takes the form of relative increase in both relative quality and quantity of output or information produced by the local economy, i.e., increase in the relative valuation of the stock's marginal product in terms of both (non-physical) information and (physical) commodities output.

Similar to the currency flows, and strongly linked to them, are information flows. Examples of such flows include the spreading of technological innovation, scientific knowledge and information, ideas, or ideologies. Sources of technologically innovative information tend to be, by and large, nations with formal currency. One may argue that they have formal currency precisely because they are the source of such innovation. Indeed, the *numéraire* quantity in an economic system, that is the reserve from which the nation's currency draws its worth, may be (primarily within a bundle of items that no doubt constitute such an ecological reserve) the stock of current (and not depreciated, outdated, or obsolete) technological and scientific knowledge, and the manner in which it is used to carry out the various socio-economic functions (including production, consumption and exchange) within the specific socio-spatial context. Another key component in this argument is the amount of redundancy (the number of carriers of such current information stock) found in a spatial economy. Most of the time, a higher level of redundancy is preferable to a lower level, up to the point of diminishing marginal returns.

To the extent that the governments of nations with informal currencies control both their local currency and foreign reserves through their central banks' monetary policies, they are also forced to control the inflow and outflow of information. They do so as economic agents speculating on the effects that such information flows might have upon the flow of the other three elements, in or out of their local economies. Macrodynamics of the socio-spatial systems they control might critically depend on such restrictions at times.[43]

The dualism observed between the underclass and the elites, in population and in the informal and formal currencies (finance capital), is not restricted as a phenomenon to only these two socio-economic sectors. It extends to commodities and information as well. Existence of dualism in commodities, particularly finished products, is exhibited by the sharp differences in quality level of these products. In economies where this dualism in quality is pronounced the low-quality products are used mainly for domestic consumption. They are mostly purchased by the underclasses and the middle classes of these nations, whereas the high quality products are channeled in international trade to attract foreign currency or they are acquired by the elites.[44] Finally, duality in information is also widespread.[45]

Dualism is also recorded in the distribution of power and social status. Hyper-concentrations of these socio-political factors onto a very few and small segments of a nation, and concomitant lack of access to power and status by the vast majority of these nations' populations, are also examples of dualism. These socio-political disparities are fed by and feed on each other.

Aside from dynamic instability, dualisms are also linked to barriers. It is not clear which is the cause and what is the effect. It may not matter, as the two are strongly related. Barriers are supplied obviously, because there is demand for them at the composite bundle price at which they are provided. There are multiple factors giving rise to the demand for barriers, as there are multiple factors contributing to their supply. One can further argue that there is a demand for dualism and that this is the reason for its presence, at some composite price. In this sense, they obey an ecological (which includes economic) determinism. For a fuller exposition see Appendix 1.

Barriers to intra- and inter-regional trade in commodities, capital, population, and information are also present in nations. In most of these nations, governments attempt, through such barriers, to control the growth of their largest urban agglomeration(s). Particularly strong are the barriers to the flow of population from other regions of these nations into their capitals. These barriers are not always effective, despite governmental pronouncements. For example, although the government of Indonesia reports the population of the Jakarta metropolitan region to be about 7.5 million people[46] in mid-1985, other observers estimated the actual size of the Jakarta region to have exceeded 12 million by that year.[47] Strict control over the population of capital cities has had a positive externality for urban model builders. It has supplied time series for the official population on a very frequent basis and other variables on some large cities, like Taipei.[48] These counts allow one to obtain a glimpse into the very fast movements in the population stock of large urban agglomerations.

Intra-national barriers to the flow of these four elements (commodities, capital, population, information) are often set to address problems that in fact transcend national boundaries. Social and ecological conflicts within nations cannot be effectively checked by intra-national policies – no matter how restrictive, consistent and cohesive these policies may be – when their main causes lie largely in factors determined in inter-national markets.

Large-scale rural to urban migrations, hyper-concentrations of capital into the core regions, explosive growth in the underclass, demographic time bombs associated with highly skewed age pyramids, lack of know-how in the peripheral regions of the globe and environmental degradation, are all manifestations of worldwide spatially ubiquitous and mobile socio-economic (ecological) events. Most of these problems or issues are partly due, according to participants, to political and market failures resulting from the conflicting effects of inter-spatial and inter-temporal flow barriers. According to an observer they are due to the underlying instabilities from interdependencies and interactions attributed to the spatially sought-after relative parity in net attraction among locations.

Dualisms and instability have been cumulative over the years. It is certainly unreasonable to expect individual and local governments acting under severe resource constraints to be successful in attempting to deal with these long-term, large-scale, complex and interrelated issues, in isolation and in their short time-spans.

ANATOMY AND EVOLUTION OF SOME VERY LARGE URBAN ECONOMIES

In this subsection the industrial structure of urban economies is looked at and its evolution within a quarter century is analyzed. Both the industrial base and its dynamics are linked to the presence of a relative parity principle. An attempt is made to demonstrate the manner in which the effort toward attaining an aggregate relative parity in net attraction regulates the growth and decline paths of certain giant urban agglomerations in the US and Mexico. Some attention is paid to the role of speculative behavior in these dynamics.

Through the widespread application of import substitution policies, many national economies and their major urban areas seem to have become less dependent on foreign trade (by a decrease in imports) and more diversified over time. Industrial diversification is clearly evident in the largest metropolitan areas of MDCs and those cities of LDCs for which data are available. Two major urban agglomerations of the US and the largest Mexican metropolitan regions are discussed next. A snapshot of a Chinese city's industrial base for 1980 is also given in Appendix 3 (see pp. 318–21).

From an absolute growth standpoint, it seems that national and urban economies are very specialized at the start-up phase of their development. With increasing wealth, as increased production occurs in a small number of basic output sectors and demand for these sectors' outputs from inside or outside the nation gathers momentum, an increase in interaction and demand for auxiliary production and services in national and urban economies reach a stage where these economies could diversify. This is a critical junction for a potential take off. Certain nations and urban areas are successful in diversifying their industrial base, triggering a process of internal growth. Others fail to do so and consequently they either stagnate or decline.

Since the industrial revolution and at the start-up of manufacturing production, the primary urban function has been considerably altered. Large urban areas, which till then were predominantly centers for trade administration and limited services, were transformed into manufacturing agglomerations. Examples include the seats of recent European empires such as London, Madrid, Vienna, Lisbon, Amsterdam, and Paris, or older ones such as Rome and Istanbul; examples also include the seats of Asian empires such as Tokyo, Beijing (to a limited extent, Shanghai being the urban area in China which incurred the brunt of the manufacturing transition), Bangkok, Bombay, and Seoul. The transformation is also apparent in what were initially administrative and trade-based centers in the North American continent: cities like Boston, Philadelphia, New York, Chicago, and Mexico City. New urban areas were created, too, with manufacturing production being the exclusive focus during their early stages of development. The cities of Manchester, Pittsburgh, Detroit, and São Paulo are examples of what originally were merely factory towns.

In the later stages of the production process, large-scale capital for labor substitution (process innovation) occurred due to technological innovation in the manufacturing sector and manufacturing employment declined sufficiently to offset the absorption of new labor into manufacturing due to increased demand for new manufacturing output (product innovation). Large-scale shifts from relatively high to low wage regions of manufacturing-related jobs permeated this subphase of industrial development. On the other hand, population (job seekers) kept flowing into regions of relatively high real wage rates from regions of low real wage rates. Urbanization was only one aspect of this basic spatial employment incongruity.

At the same time, these shifts were reinforced by a sharp decline in the agglomeration economies present in manufacturing production during its first state of development, which initially pulled different plants spatially close. This breakdown was partly due to multimodal improvements in the transportation sectors and involved highways, railways, waterways, airways, pipelines and telecommunications, and partly due to congestion and exhaustion of scale economies. Transport improvements drastically increased (in absolute levels, although differentially in relative terms) the accessibility of many locations in the national and international space, significantly altering the urban form and structure of cities.

The last quarter of the twentieth century has witnessed an acceleration of another structural shift in production and consumption, namely the manufacturing to services transition. This latter transition is partly a reversal of an earlier transformation (the trade services and administration to manufacturing transition at the start of the industrial revolution). Rises in real incomes, changes in preferences, introduction of new labor-intensive relatively low-wage services, and high marginal utility services in consumption brought about a rise in the service sector of spatial (national, regional, urban) economies. The manufacturing to services transition is another major force altering the form and structure of contemporary urban areas. Cases where this shift has been extensive include all the major urban agglomerations of North America and in particular Pittsburgh, San Francisco, Chicago, New York, and Los Angeles.

All of these events may be at the core of the large-scale metamorphoses that cities have undergone through the centuries and are currently undergoing. One of these changes may be the unprecedented spatial spreading of urban activity, and the decline in the average population density in cities of developed nations. Many urban regions throughout the globe which had predominantly manufacturing-based industrial structures have been, or are currently in the process of being, converted to service-based economies at various degrees. Among the many services they offer, some of these newly transformed urban areas are becoming financial centers, the source and sink of considerable capital flows. The seats of stock exchanges worldwide are the primary centers of such capital flows.

Again, some new, large, urban agglomerations have evolved into centers for financial and other services having bypassed the manufacturing phase. In North America, examples include many southern and western new urban areas, like Atlanta, Dallas, Phoenix, and Tucson. A striking example is that of Miami, a services and financial center which happens not to be the seat of a stock exchange but the place of a large-scale informal capital flow activity. These cities represent examples of urban areas without any significant 'memory' of manufacturing.

All of the above economic factors in combination have resulted in a major switch in the main urban labor force employing industry over the past century, from manufacturing to services. Although services seem to be emerging as the dominant urban economic activity, there is no evidence to suggest that the other two preindustrial modes of production, predominantly trade or administration, have reappeared as the dominant employer in any existing major urban center. New cities may have emerged with trade (the new free economic zones in China are cases in point) or administration as their main planned function. Examples of the latter case are the various new national capitals: Brazil's Brazilia in the 1960s, Nigeria's Abuja still under construction in the 1980s, and even Viedma the proposed new capital of Argentina. But scant evidence seems also to suggest that services (formal or informal) quickly overtake these two planned functions, as is now clear from the Brazilia experiment where the population and employment base is a far cry from the original conception by its planners. The predominant movement seems to be for an existing manufacturing town to be converted to a mainly service economy or for a new town to appear with service as its main economic function, at specific points in space–time.

Planned or unplanned urban growth in space–time lends itself to analysis based on speculative behavior. Experience drawn from the US as well as from other nations seems to indicate that a national economy seemingly follows a spatially diversified investment strategy. The various regions (or larger divisions) and their urban centers within the current boundaries of the US for example, were developed at different points in time, with different speeds or type of development at any given time-period. Earlier in the regional and urban history of the US, the Eastern and Southern regions were developed with the Mid-Western and Western frontiers following suit. Similarly at the urban front, the older Eastern cities gave way to the development of Mid-Western, South-Western and Western urban agglomerations.

Collective speculation in its strict definition may help to explain why towns are transformed or new towns appear throughout history, either at the agriculture, trade, and administration to manufacturing shift or during the manufacturing to service transition. Certain towns are converted from old to new functions, while at the same time a number of new towns emerge with altogether new functions as their main focus. Closely related is the current phenomenon in North America, the mushrooming of small-scale, highly specialized, homogeneous (although of a vast inter-settlement variety), middle to upper-middle income exurban communities appearing with a lightning speed at the spatial fringes of existing urban agglomerations, enjoying easy access to freeway capacity. Speculative development schemes with very high expected rates of return and short economic lifespans, these spatial entities clearly demonstrate the strong linkage between capital formation and spatial expansion.

A nation might be viewed from an angle whereby its various consuming, producing, and governing units collectively act *as if* the nation approaches regional development from a speculative position. At certain time-periods certain regions develop, while in others development is still on hold. When technologically, economically, or politically (i.e., ecologically) it becomes feasible and/or desirable (profitable) enough relative to existing composite returns from currently developed regions, then development switches from the mature to the underdeveloped and developable regions. Exploration and exploitation development processes of space underneath, on, or above the surface of the ground or sea level are similar in their underlying mechanisms.

Individual nations or their spatial units may employ specific private and public sector development instruments and policies to attain development objectives.[49] To study these individual cases is not enough, however, to obtain a broader picture of the development process globally: one may want to detect collective speculative development behavior in the international scene as well. Nations, giant urban areas, or whole regions of the globe are developed in stages, not according to a ground plan or anything even remotely resembling a blueprint, notwithstanding colonialism or highly centralized socialist planning.

One might approach development globally as an observer. Development may not follow a blueprint, but rather it proceeds *as if* it is put on hold at certain urban areas, regions, or nations. While the always limited current development resources (through a complex conflict resolution process) are channeled into the subset of the world's regions currently being perceived as the more (collectively and ecologically) profitable, other regions await their turn. Developing all the world's regions at once over a long time horizon *collectively* and *ecologically* may not be the most profitable strategy to pursue.

Regions, nations, urban areas, industrial sectors and other socio-economic spatio-temporal stocks may be perceived as competing currently for development in space–time. They compete for a position or niche in their respective spatio-sectoral hierarchies. A specific position in the spatio-temporal matrix for development is accompanied by an interconnectance scheme with other positions which lasts over specific space–time horizons. In the development game, where conflicts due to varied interests are abundant, socio-economic spatio-temporal stocks compete for dominance and priority in the spatial extent of their interactions. Beyond space, they also compete for the corresponding time-period it will take for the interactions to play themselves out, i.e., for socio-economic development or evolution at a particular point in space–time: they bid to use a platform at a specific time to play out their roles.

That is, they compete for both space and time. What actually occurs is simply the collective conflict resolution scheme of spatio-temporal ecological interconnectance (development) prevailing in space–time. Resources available for any spatial unit and its agents (individual and collective, i.e., governments) to compete in this speculative game are limited and exhaustible; very rarely are these resources replenishable. Inefficient speculation in this complex ecological interdependence (development, evolution) game depletes the resources fast and removes these inefficient players from the game.

For an observer, the 'as if' principle included in the principle of global relative parity is a restraint on or stimulant to speculative behavior on development, including speculative socio-spatial hoarding and arbitrage. A look into a relatively recent and confined time horizon and to a limited geographic space reveals informative insights into the dynamics of spatial units.

Current inter-urban variability is pronounced within any nation; dynamic instability in it is manifested in the dominance and extinction of numerous urban settings in space–time. There is plenty of diversity and dynamic instability at the intra-urban level as well. Cities in the US are notable examples of such instability. The old, decayed and abandoned manufacturing-built capital stock of old inner North American cities rests as a testimony to a phase of urban production now rendered mostly obsolete by lack of intra-industry technological innovation, by a sharp rise in real relative input factor prices including manufacturing labor wage rates and in land rents, and to a shift in input factors and output markets. Profitable opportunities for new construction in alternative intra- or inter-nationally competitive locations, attributed to shifts in comparative advantages, made the maintenance of such capital stock prohibitive.

Spatially, in North America the distribution of new and decayed capital stock (a spatial dualism in built capital stock) is highly non-random. Specifically in the US, in the newer urban agglomerations of the Southern and Western states, not much decayed capital stock is found at the beginning of the 1990s. Whereas the older ones in its Northern and North Central regions, its Rust Belt, are containers of the obsolete built capital infrastructure. From an intra-urban standpoint, the older structures are found mostly in the central cities, whereas the vast majority of the newest built capital stock is located in the suburban and ex-urban counties.[50]

The implementation of labor-saving technologies in manufacturing production, the rise in demand for services and shifts in the comparative advantages of metropolitan areas (primarily due to changes in input and output markets), were factors working together in affecting the evolution of metropolitan areas worldwide during the past quarter century or so.

Older manufacturing-based cities softened the comcomitant decline in population which followed such transitions in the production process by appropriately altering their industrial employment base. Pittsburgh, St Louis, Philadelphia, Minneapolis, and to an extent Baltimore, are examples of older, larger US metropolitan areas which underwent the manufacturing to services transition to various degrees after they saw their manufacturing-based economy plummet during the 1960s and early 1970s. Chicago, Kansas City, Cincinnati, and Oklahoma City are milder cases in point.

As a result, population and wealth cycles became abundant and recordable even over very small time-spans. Once shunned by real estate developers and speculators as decaying manufacturing urban areas of the North on the verge of fiscal bankruptcy, these older metropolitan agglomerations did to an extent bounce back in the 1970s and early 1980s. In the 1975–85 period they were in the midst of a mild construction boom centered around services and concentrated in their previously decaying central business districts.

There was overbuilding by developers of the Southern, South-Western and Western urban areas during the 1970s and 1980s, the result of (what proved *ex post facto* to be) overreaction in real estate speculation, overreaction to the rapid increases in the price of petroleum and land prices in the 1970s, and to the massive exodus of manufacturing from the Northern cities. Dallas, Denver, Colorado Springs, and

Phoenix are a few cases in point, and the thrift problems of the 1980s and 1990s are testimony to such real estate and land speculation practices. This overreaction followed speculative underreaction by land holders in the Northern cities during the 1960s and early 1970s. In the late 1970s and the beginning to mid 1980s overreaction in real estate speculation seem to reemerge in the North-East again, as seen in the North-East and Pacific bank failures of the early 1990s carrying investment portfolios with heavy exposure to real estate loans.

Numerous economic incentives attempted to and partially succeeded in bolstering the industries of old metropolitan regions like the Boston and New York areas in the 1980s. They were partly enhanced by a reversal in the movement of the oil prices, a significant increase in technological innovation in processes and products (particularly electronics and telecommunications), an increase in spatially diffused defense-related government expenditures, and a new infusion of capital into undervalued financial instruments forcing the restructuring of old firms.[51] These events were partly responsible for a limited revival of the central core of the older cities of the North and North-Central regions of the US. In these cities, as a result of new construction, one now observes a dualism in their built capital stock.[52] Advanced quality built capital stock is found in close spatial proximity to decaying inner city slums where poverty and informal sector activity flourishes.

Capital and labor restructuring processes are not unique to North America. Similar to other spatio-demographic shifts which started there, like the phenomenon of suburbanization or that of aging in the population stock, they have been observed elsewhere in the MDCs and certain LDCs. Urbanization, industrialization, modernization, technological innovation, demographic shifts, etc. are events bound to spread in space–time. With a time delay, and with different speeds dependent upon the level of relative development and institutional political, economic and cultural barriers, these phenomena are propagated in many areas on the globe. One must expect that these newest developments in capital and labor markets will spread to other regions of the globe in time, as did the older events.

In containing capital stock of different vintage, urban areas are historical mosaics where each historical period's stock competes for space and time in the urban landscape. The presence of many such claims by many different past periods makes this competition particularly stiff in the relatively limited space of European and Asian cities. It is not as competitive in North America, where the thinness of past historical periods and abundance of space eases the conflict. One can look at this competition, however, with the future capital stock in mind, too. Through the notion of speculative holding of space for future construction, it is not only the past eras' stock which is competing for space at present. Future capital currently puts claims on space–time through a variety of socio-economic (ecological) processes. If so, then what one currently observes in space contains not only the remnants of the past but also the kernels of the future.

Are these intra-national, inter-regional, inter-urban or intra-urban oscillatory movements of long-term and large-scale type, or are they simply ephemeral changes like the temporal fluctuations in prices of stocks in stock markets? Are they induced by intra-national factors or are there broader inter-national factors involved in this cyclical movement? Before focusing on the changes in the industrial base of certain major urban centers in the US and Mexico (New York, Los Angeles and Mexico City) to answer these questions, some foreign trade implications will be analyzed. How certain inter-national flows affect, among other things, intra-national factor mobility will be ascertained.

Parallel to the changes mentioned earlier in the industrial base of metropolitan areas in the US, shifts in foreign trade have affected its metropolitan sector to a much larger degree than they have affected the US economy as a whole. Particularly affected were the coastal metropolitan agglomerations – New York, Los Angeles and Houston are striking examples. In Table 2.15 the volume of merchandise traded through these and other major ports of entry and exit are shown for 1965 and 1981. New York has lost its dominant role in imports, from enjoying an overwhelming share of 30.8 percent of all valued imports it only just retains its ranking as first among custom regions in imports with a share of 17 percent.

All other urban areas picked up the lag, with the Houston metropolitan region lifted from the second lowest to second highest position (from 2.8 to a 14 percent share). New York also declined as a major exporting metropolitan area, dropping from a 26.4 percent share to 15.3 percent. Houston is again a major beneficiary rising to the position of second largest exporting metropolitan area in merchandise evaluation. Chicago has also experienced a major change, from a share of 1.8 percent to 12.1 percent.

This directly affected the north-eastern Atlantic seaboard as a center for international trade in durable manufacturing products. At the same time, the emergence of the Japanese exports – particularly in the bulky category of automobile vehicles, parts and equipment – impacted on

Table 2.15 US exports and imports: share by major customs region, 1965 and 1981 (current US$bn in parentheses)

	Exports		*Imports*	
	1965	*1981*	*1965*	*1981*
Total	100 (27.3)	100 (233.7)	100 (21.4)	100 (259.0)
Boston	0.7	0.4	3.3	9.4
New York	26.4	15.3	30.8	17.0
Baltimore	2.6	9.5	4.1	10.4
Miami	2.2	8.5	2.3	7.8
New Orleans	8.1	10.0	3.7	10.1
Houston	7.0	13.5	2.8	14.0
Los Angeles	2.9	8.4	5.6	9.4
San Francisco	4.0	13.1	3.2	10.3
Chicago	1.8	12.1	2.8	11.5

Source: *SAUS*, Table 1483, 1982–83.

Los Angeles and San Francisco. The shift from Western Europe as the dominant foreign trade market to the Far East and Latin American trade markets partly accounted for the rise in importance of the Pacific seaboard and Gulf Coast.

With the extraordinary population base of China, India, and other South and South-East Asian countries (Japan, Indonesia, the Philippines, Pakistan, Bangladesh being notable examples), and the rapid rise in per capita income in some of these and other South Asian nations (Thailand, Malaysia, etc.), the trade focus is currently being shifted from the Atlantic to the Pacific basin. US trade markets are moving from Western Europe and the North American Atlantic coast to East Asia and the North American Pacific seaboard.

As the population of Latin American nations grows their incomes rise and their economies become more industrialized and modern-service-oriented (particularly those of Mexico, Brazil, Venezuela, Argentina, Chile, Peru, and Columbia), with the Gulf Coast of North America becoming more accessible to larger markets for US agricultural and manufacturing products, as well as services to these nations. For Venezuelan and Mexican oil exports, and agricultural products from Brazil and Columbia, it acquires additional accessibility as a final destination.

New York and Los Angeles

Notwithstanding the above specific observations on port activity, the major transformations in the New York and Los Angeles metropolitan statistical areas' (MSAs) economic bases were not mainly induced by changes in the spatio-sectoral patterns of the US foreign trade. Rather, they were due to large scale shifts in spatio-sectoral economic and political conditions within the US domestic market. They were also attributed to complex socio-cultural factors within the *de facto* US border. Most of the spatio-sectoral shifts have been ongoing since the end of the Second World War, and some even before then. Suburbanization, inter-national migration, increases in family size, incomes, the average age of the population, the economic rise of the South and West, capital for labor substitution in industrial production, the rise of the service sector, heavy investment in the transportation infrastructure, and the development of electronics, were among the events which had commenced earlier and by 1960 were severely affecting the US spatio-sectoral structure.

Besides events within the *de jure* US boundaries, there were also events within the more fuzzy *de facto* extent of US influence and the global factors affecting the US economy which influenced the demographic and economic changes in its two largest urban agglomerations: New York and Los Angeles. Both were differentially affected by these developments. New York was being negatively impacted upon while Los Angeles was being positively impacted upon by these events. Shifting growth into the younger and smaller MSA in the West was more (ecologically) profitable than accumulating more capital and population (labor) into the older and larger MSA of the North-East. In the bundle of forces impacting upon the growth of these two metropolises, domestically driven forces were operating side by side with internationally prevailing conditions. Global shifts in trade patterns, technological innovation, LDC to MDC labor migration, low European and high Asian and Latin American population growth rates, are a few of these long-term, large-in-scale international conditions. In combination, they were fueling not only changes in these two urban agglomerations' economic bases, but they were also shaping major US socio-economic spatio-sectoral shifts. The Second World War was the beginning of the decline in New York City's population dominance (see Figure 2.1) with its share of US population dropping from about 6.4 percent in 1940 to almost 4 percent in 1980. Erosion of its share of income within the US economy had started much earlier with almost double the

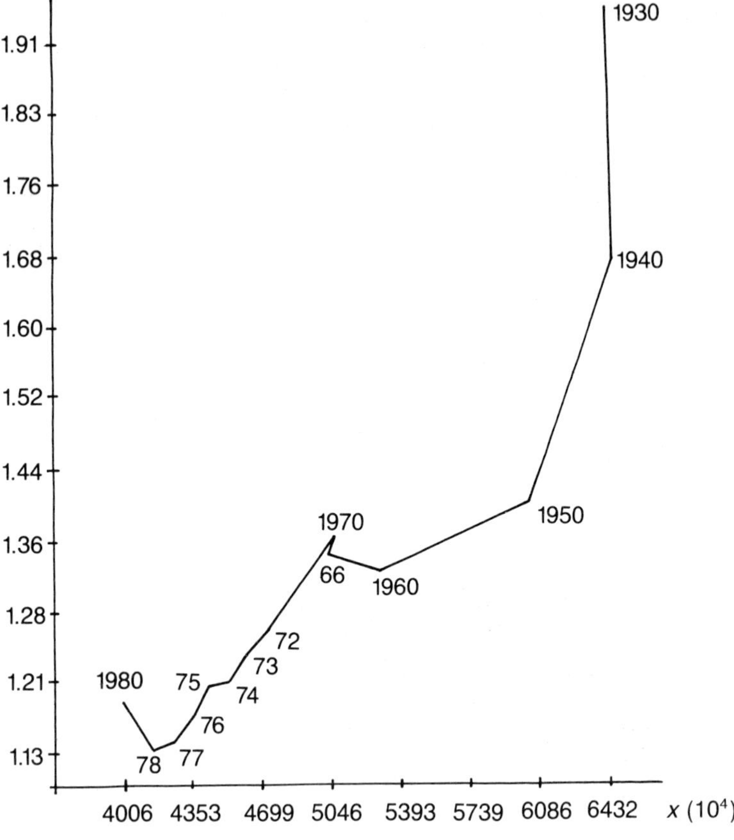

Figure 2.1 The actual path of the New York MSA's relative dynamics, 1930–80

per capita income of the US in 1930 the ratio dropped to 1.2 by 1980. One could argue, too, that the war was also the beginning of the ascent in Los Angeles' dominance.

These are merely a few examples of the very large number of events and factors, some larger in scope and of longer-term impact than others, which were shaping the location and relocation patterns of capital and population within the US space during the past quarter century. Next, a look into the industrial bases of the two US MSAs is provided. The manufacturing to services transition is documented, and a very disaggregated employment mix comparison is made between the two MSAs. Possible effects of these transitions upon their overall growth are

Table 2.16 The New York MSA: employment by industry type, first digit SIC, 1960 and 1980

Category	*1960*[1,2]		*1980*[3,4]	
	Absolute ('000)	*Share (%)*	*Absolute ('000)*	*Share (%)*
Agriculture	16.7	0.4	10.8	0.3
Mining	4.5	0.1	3.1	0.1
Construction	198.5	4.5	121.9	3.1
Manufacturing	1,126.8	25.8	709.6	18.2
Transportation	371.2	8.5	371.6	9.5
Trade	828.7	19.0	742.3	19.0
Finance, insurance and real estate	336.9	7.7	423.1	10.8
Services	543.6	12.4	1,345.2	34.4
Government	208.8	4.8	180.4	4.6
Other	265.3	6.1		
Total	4,372.6[5]		3,907.7	(100.0)

Notes: [1] US Department of Commerce, Bureau of the Census, 1960 *Census of Population: General Economic Characteristics*, Table 75.
[2] Industry group of employed persons (both sexes).
[3] US Department of Commerce, Bureau of the Census, 1980 *Census of Population: General and Economic Characteristics*, Table 122.
[4] Industry of employed persons 16 years and older.
[5] The total for all reported categories equals 3,901.0, whereas it is reported as 4,372.6. In computing the industrial share the 4,372.6 total is used.

traced. In Table 2.16, the New York MSA's employment shifts are shown for various industrial categories at the first digit SIC code for the years 1960 and 1980.

The overall 10 percent decline in employment for the MSA was unevenly distributed among different industries. At a very aggregate industrial level, the major loser was manufacturing and the major gainers were services (their employment share almost tripled) and finance, insurance and real estate. The New York MSA is a typical example of a manufacturing metropolis having been transformed to a service center. Looking at the third digit SIC code, the major losses in employment share can be traced to five sectors. They occurred in textile, food and kindred products, fabricated metals, electrical machinery, and transport equipment. Major gains in employment occurred in banking, insurance and real estate, and all major categories of services (see Table 2.17). Among the latter, notable is a doubling in the category of

Table 2.17 The New York MSA: employment for selected industrial categories, third digit SIC, 1960 and 1980

Category	*1960*		*1980*	
	Absolute ('000)	*Share (%)*	*Absolute ('000)*	*Share (%)*
Food and kindred products	99.1	2.3	39.1	1.0
Textile	301.1	6.9	174.7	4.5
Fabricated metals	48.9	1.1	26.4	0.7
Electrical machinery	88.8	2.0	50.8	1.3
Transport equipment	69.3	1.6	23.0	0.6
Business services	120.2	2.7	206.8	5.3
Entertainment	49.3	1.1	61.9	1.6
Hospitals and health related	132.0	3.0	349.3	8.9
Educational services	178.6	4.1	287.0	7.3
Social services	63.0	1.4	121.6	3.1

Sources: See Table 2.16 notes.

hospitals and health related services, social services and entertainment employment. More than one-third of New York's employment is currently in service-related industries.

The Los Angeles MSA has undergone drastic changes similar to those New York experienced during the period 1960–80. A mainly manufacturing center in 1960 with 30.6 percent of workers in manufacturing, it became a mainly service center in 1980 with services capturing 30.8 percent of the workers to 25.5 percent in manufacturing. The 36.3 percent increase in the services share represents a substantial rise from the 22.6 level in 1960. Trade ranked third throughout the 20-year period with around 20 percent of the workers (see Table 2.18).

Los Angeles' transition in industrial employment composition has not had as pronounced an effect on the total urban economy as that of the New York MSA. It is noted that despite the fact that New York is a declining (in terms of total population and employment) metropolitan area and Los Angeles is a growing one, the 1980 aggregate industrial composition of the two metropolitan agglomerations are very much alike (see Tables 2.18 and 2.19). Further, the transformation trend, from a mostly manufacturing economy to a predominantly service-oriented one, was very much in evidence in both cases. Since the manufacturing to services transition is in evidence in both urban agglomerations, the cause for relative growth in Los Angeles and relative decline in New

Table 2.18 The Los Angeles MSA: employment by industry type, first digit SIC, 1960 and 1980

Category	*1960*		*1980*	
	Absolute ('000)	*Share (%)*	*Absolute ('000)*	*Share (%)*
Agriculture	37.4	1.4	35.7	1.0
Mining	8.5	0.3	8.1	0.2
Construction	144.6	5.5	154.6	4.5
Manufacturing	800.0	30.6	884.1	25.5
Transportation	162.2	6.2	248.4	7.2
Trade	487.1	18.6	700.1	20.2
Finance, insurance and real estate	139.1	5.3	249.3	7.2
Services	591.0	22.6	1,070.1	30.8
Government	112.5	4.3	121.4	3.5
Other (not reported)	133.2	5.1		
Total	2,615.6	(100.0)	3,471.8	(100.0)

Sources: See Table 2.16 notes.

Table 2.19 The Los Angeles MSA: employment for selected industrial categories, third digit SIC, 1960 and 1980

Category	*1960*		*1980*	
	Absolute ('000)	*Share (%)*	*Absolute ('000)*	*Share (%)*
Textile	50.1	1.9	87.2	2.5
Primary metal	25.1	1.0	27.0	0.8
Fabricated metal	82.2	3.1	63.3	1.8
Machinery, except electrical	63.9	2.4	86.9	2.5
Electrical machinery	103.4	4.0	94.4	2.7
Transport equipment	188.6	7.2	184.5	5.3
Business service	62.1	2.4	144.2	4.2
Entertainment	50.7	1.9	85.2	2.6
Hospitals and health services	58.4	2.2	250.4	7.2
Education related services	127.8	4.9	240.3	6.9
Legal, engineering and other services	89.1	3.4	121.4	3.5

Sources: See Table 2.16 notes.

York must be attributable to the changes in their detailed industrial composition, and in relative importance from an international standpoint as demonstrated earlier with their ports' activity data. The Los Angeles manufacturing activity is strongly related to the defence industry, an industry which is at the forefront of technological innovation and political wishes. It is not only affected by events within the *de jure*, but also, and considerably so, by events within the *de facto* extent of US influence, and by prevailing social, economic, and political conditions.

The US Bureau of the Census projects that the Los Angeles MSA will surpass the New York MSA in the early part of the twenty-first century. Computations from an urban mathematical ecology model[52] (see Figure 2.2 also containing other US cities), seem to indicate a close size in relative population (but not per capita income) between the two settings in the long term. In any case, the results are indicative of some ongoing restructuring in the US urban hierarchy, a manifestation of the large-scale long-term socio-economic changes currently underway in the US.

Looking at the Los Angeles MSA third digit SIC code employment share in manufacturing, one detects an increase in textiles and a decrease in metal manufacturing (durable and non-durable) categories

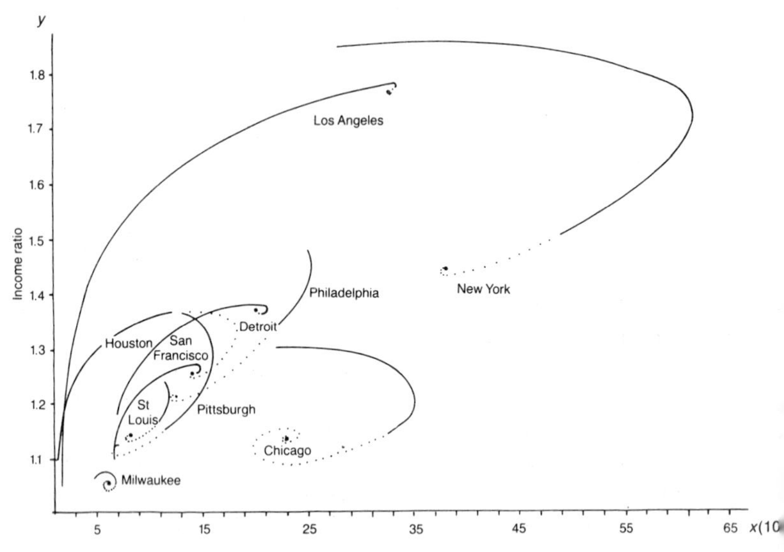

Figure 2.2 The relative macrodynamics for selected US MSAs, 1890–1980. Continuous line is simulated path; dotted line indicates projections and steady state

(with the example of non-electrical machinery, shown in Table 2.19). By far, the most noticeable change which occurred in the Los Angeles MSA, as it did for the New York MSA and indeed for the vast majority of metropolitan areas in the US during that period, was in the services-related categories of finance, insurance, real estate and selected retail trade sectors.

Table 2.19 demonstrates the doubling in the share of business services, the more than tripling of the hospitals and health related services, and the moderate increases in entertainment, education and other (legal, engineering, etc.) professional services shares. These are very similar to the New York patterns. In both cases, the changes in industrial composition of employment are attributable to worldwide changes in the production and consumption processes (labor savings technologies in manufacturing, higher demand for services) and in international flows in the four elementary components of exchange. They are also attributable to broader, nationwide, demographic changes: as the average age of the US population increases, the ethnic composition of the US population is altered, and regional shifts in the location of population and employment occur. A drive toward relative parity in net attraction between New York and Los Angeles resulted in growth patterns which favored Los Angeles during this time-period.

Before examining how these events have impacted a major urban area in another national environment, Mexico City, a look will be taken into an urban region of the US without any prior memory of heavy manufacturing. The reason for looking at this region is to identify the urban macrodynamics under conditions of continuous change in the environment affecting them significantly.

The Texas urban conglomeration and the Austin MSA

The Gulf region contains the Dallas–Houston urban corridor, not far from the Austin metropolitan area. It is of interest to look closely at this urban agglomeration. One can analyze direct and absolute linkages between population and per capita income growth in this particular MSA within this tripartite Texan urban conglomeration, and also see what effects exchange rates and prices in key commodities have had. The effects, which can be detected, were very fast moving and profusely fluctuating variables like the recent oscillations in the price of petroleum, exerting pressure in the short-term upon less fast moving ones, like population size.

During the 1960–80 period, this particular triangle of urban centers, with its technology and oil industrial sectors combination, was at the top of the fast growing league among North American urban conglomerates in absolute population and income. Austin, in particular, was emerging as a new center for electronics production and it was undergoing explosive growth in absolute and relative population and employment during the 1970–80 period. Austin, in the 1970–85 period, offers the opportunity to urban analysts to learn a few lessons on direct linkages, and on dynamics cycles fluctuations expectations and social spatio-sectoral speculation.

In 1970 Austin's total population, which includes the Hays, Travis and Williamson counties, amounted to about 360,000. By 1980 the population had reached 537,000 according to actual decennial census counts. Interesting events follow these recordings, which had the Austin area population growing during this decade by an average absolute annual growth rate of about 4.9 percent.

Annual rates of increase in the following three years of 3.3, 4.4, 5.5 percent, respectively, were estimated by the US Bureau of the Census.[53] These estimates or speculation (in its broad definition) by the Bureau of the Census, based on the decennial actual counts of 1970, 1980 and local reports, had in 1986 estimated the population of Austin to be 695,500 by mid-1985.[54]

A later estimate, however, had the 30 June 1985 absolute population of the Austin Metropolitan Statistical Area at 646,000, or 7.7 percent below the previously cited estimate.[55] The difference cannot simply be attributed to statistical error. It is too pronounced to dismiss it on this count. The P-25 and P-26 series on which these two counts were based reveal that, due to the different sources, methods, and time-periods used, expectations and actual events were diverging in their estimates of Austin's absolute population growth by a significant margin. The P-26 estimates[56] were derived analytically using a variation of the Administrative Records Method, employing Federal Income Tax data, without feedback from local sources, whereas the P-25 Report[57] contains corrections as a result of interaction with local agencies. Statistical error in overestimating county populations by the Administrative Records Method was found to be approximately 1 percent[58] from actual counts from selected counties in the US. Thus, the difference in the case of Austin must clearly be due to local changes in population growth paths from earlier expected counts by the Census.

Meanwhile, per capita income in constant 1982 US dollars declined in Texas during the 1982/3 period from US$5,523 to US$5,470 while the US average increased from US$5,395 to US$5,471.[59] Austin's current per capita income increased in the 1982/3 fiscal year from US$11,337 to US$12,148.[60] It mildly increased in constant dollars.

The background to these changes were fluctuations in the price of petroleum and in exchange rates, as well as in the growth of the electronics sector during the second half of the 1970s. In early 1980, the price of oil reached its highest level in reference to the US dollar, the currency it trades internationally, then standing at US$34 per barrel for the Western Texas Intermediate (currently, light sweet) Crude, the benchmark in the US petroleum markets. The turn of the decade saw the value of oil starting to decline and then proceeding almost to collapse, reaching approximately US$8 per barrel by June of 1986.

At the same time, the value of the US dollar went from a temporal high of 275 Japanese yen to one US dollar in late 1984, to about 145 in March of 1987. For Japan, the near collapse in the price of oil, since tied to the US dollar, meant a considerable windfall as the price Japan faced for oil dropped from about 9,000 yen to about 1,350 yen per barrel.

Significant cost saving advantages experienced by the Japanese manufacturers in energy costs, among other factors, allowed them to predatory price electronic products in international markets, including semiconductors. Thus, the pressure, not only in the oil industry but also in its closely related technology sector, was felt by the Texan urban conglomeration. The region's relative fortunes took a nosedive, as did Texas and its neighboring oil producing states of Oklahoma and Louisiana, and the state of Alaska. In relative terms, Louisiana saw its per capita income as a percent of the US average, in current dollars, go from 89 to 83 percent, while Oklahoma's dipped from 94 to 87.[61]

A number of conclusions emerge out of this event. They relate both to absolute versus relative dynamics, to the interconnections between fast and slow moving variables, and the absolute and relative effects they have on spatial systems of different scale or level of disaggregation. Expansion of capital investment in oil exploration and in oil-servicing manufacturing, as well as in the electronics-related sector which is strongly linked to the oil industry, was steadily underway in the 1970s. Employment and population were growing in these sectors and so were the areas in which they were located. These slow moving variables, capital accumulation and labor growth, were riding on 'rational expectations' of a continuous rise in oil prices, taking their lead from what had occurred in a period of one decade.

Expectations of growth were also built into the models projecting future growth in these industries and regions. Large banks (for example the Bank of America under A.W. Clausen in the 1970s) made large-scale commitments based upon such 'rational expectations,' in oil-related ventures, real estate development, etc. The fortunes of the Bank of America nosedived in the mid-1980s and, under Clausen, picked up again in the early 1990s. The broader effects that fluctuations in the price of oil and associated expectations by real estate speculators have had in this time-period upon savings and loan institutions and banks, not only in the south-west but US-wide, are well known.

The 1970s were years when the value of tangible assets from mining, agricultural and forestry products (especially petroleum, metals and timber), and land and real estate in particular, was rapidly increasing and thus fueling unprecedented rises in inflation and interest rates. During that period speculation was focused in certain tangible assets and an innovative mode of financing referred to as 'junk' (high yield, high risk) bonds appeared. In tandem with these events, the fortunes of various microregional and urban economies, mostly in the southern and western parts of the US and in the newest segments of existing urban agglomerations, rose sharply. They also had a rough landing in the last years of the decade.

This significant level of capital investment was not carried out in a theoretical vacuum. Speculation was extensively using economic efficiency criteria from macroeconomics and financial economics as its base to place its bets. Economic growth models of the neoclassical type were used to place these resources at stake. These models were concluding that the price of oil, as a depletable and non-replenishable resource, is and ought to increase as the compounded rate of interest[62] which by early 1980 had risen to about 21 percent in the US, with inflation hovering at about 17 percent. From early 1980 to the beginning of the fourth quarter of 1985, the price of oil had steadily declined from US$34 to about US$28 per barrel. Interest rates fell to 7.5 percent.

How did these movements affect the population of Austin and Texas and per capita income so far? From the evidence supplied, one detects that these fluctuations resulted in the decelerating of the absolute growth rate of population at the urban level and in the reversing of the absolute income path for only one year (the 1982–83 period) at a state-wide level. One might conclude that the absolute size behavior of the macrosystem contains either time lags or buffers which cushion it from volatility in ephemerally behaving variables, like prices and exchange rates. It must be underlined that these conclusions apply to specific absolute size levels of the slow-moving stocks and for a certain width of fluctuations in the fast moving variables (prices).

Then in a six-month period, from January to June 1986, the price of oil plummeted to below US$8 per barrel, clearly an unsustainable level. So much so, that by the end of the year it stood again at about US$17 per barrel. At this point, data are not available to detect the spatial impact of these latest changes in the Austin metropolitan area. Future research, undoubtedly, will fill this demand. Since this window of price fluctuations in the oil markets and fluctuations in currency exchanges may not remain open for long, this event provides a unique opportunity to gather evidence on its effects upon spatial economies. This concludes the analysis of the US urban scene and the speculative nature of growth in metropolitan development. Next, a look into Mexico City's industrial composition and macrodynamics is taken, with particular attention being focused on the industrial and demographic shifts as diffused through space–time from the US into Mexico.

Mexico City

Similar, although not identical, events to those outlined earlier for US metropolitan areas can be detected within the urban sector of Mexico. Changes in the economic base proper of the Mexican economy, in combination with foreign-trade-related factors affect the evolution of Mexico's primate urban agglomeration, the Mexico City Metropolitan Region. However, due to the fact that Mexico was at a different phase of development than the US economy during the study period, these factors had differential effects, in terms of strength, upon Mexico City than they did on its US counterparts.

Mexico City's primacy, in 1970, was much more pronounced in terms of employment than it was in terms of population.[63] This was particularly so in manufacturing (55.3 percent of the largest 37 cities' total employment in this sector), public utilities (53.0 percent), services (52.9 percent) and government (61.2 percent), with an average for all non-agricultural sectors of 51.2 percent of employment in the 37 cities.[64] For the same year, the share of Mexico City's population to that of the largest 37 cities was 44.8 percent.[65]

Looking at the employment by industrial type for the years 1960 and 1970, for the Federal District (a part of the Mexico City Metropolitan Region, accounting in 1975 for approximately 70.6 percent of its total metropolitan population [66]) one sees the following in Table 2.20: during the 1960–70 period the manufacuring employment share declined slightly and that of services increased to make it the primary employment category in the Federal District. It is noted that government employment is included in 1960's 33.7 share of service employment.

Table 2.20 The Federal District of Mexico: employment share by industrial category, first digit SIC 1960, 1970 and 1980

	1960[1]	*1970*[2]	*1980*[3]
Total employment (1,000)	(1,752)	(2,166)	(2,294)
Agriculture	2.7	2.0	8.8
Mining	0.7	0.9	14.6
Petroleum related		0.6	
Other extraction		0.3	
Manufacturing	30.4	30.2	17.7
Construction	6.8	5.5	14.0
Utilities (gas, electricity)	0.9	0.6	3.2
Commerce (trade)	17.5	13.7	5.9
Transportation	5.8	4.4	1.6
Services	33.7[4]	32.0[5]	9.7[6]
Finance, insurance, real estate			7.1
Government		6.9	
Not specified	1.7	3.3	17.3

Sources: [1] *VIII Censo General de Poblacion, 1960*, Distrito Federal, 1963, Table 21 (p.186).
[2] *Anuario Estadistico de Los Estados Unidos Mexicanos*, 1980, SPP, Table 3.1.5 (p.394).
[3] *X Censo General de Poblacion y Vivienda, 1980*, Resumen General Abreviado, 1984, Table 9.

Notes: [4] The count for services includes government employment as well.
[5] The count includes finance, insurance, real estate employment.
[6] The count includes government employment.

The reason for the manufacturing share not to have decreased more drastically was that as the Mexican economy was still in the industrial expansion phase, demand for labor in manufacturing was outpacing (slightly then) the capital-for-labor substitution.

A drastic change seems to have been recorded in the 1980 census regarding the industrial employment composition of the Federal District. As shown in Table 2.20, the 'not specified' category has increased significantly over the 1970 and 1960 count. One tends to associate this category with the size of the Federal District's informal sector, in view of the fact that of the 3.3 million active workers in the labor force only 2.3 million are recorded, i.e., about 70 percent of the labor force – an expected share.[67] From these employment shares in the industrial composition of the Federal District one can detect a stability in the mining and manufacturing share (in combination remaining at about 30 percent of the employed), a steep rise in construction, and a decline in services-related employment.

Mexico City is presented because, beyond merely identifying what is currently underway in a large LDC urban agglomeration, it vividly demonstrates a transition. For the first time since the last century, urban centers are under two forces, the remnants of two transformations in the industrial base of cities: the remnant of the force due to the industrial revolution, causing a transformation in employment from trade and government to manufacturing (the 'industrialization' phase); and at the same time the growth of the service and information sectors, caused by a transformation of manufacturing-based employment to services. One force acts as an impetus, in combination with comparative advantages in labor wage rates worldwide, for manufacturing workers to increase; the other in combination with the capital-for-labor substitution aspects of labor-saving technological innovations, acts as a deterrent for such increase in manufacturing employment. Mexico City is an example of an urban area where two forces are playing themselves out, currently. It is a case where the broader forces in the worldwide relative parity are at work.

Mexico City is a hyperconcentration of economic activity as well. The very strong agglomeration effects in the population, employment, capital, political power, etc. concentrations of the 1960–80 period are, however, now beginning to be offset by the deglomeration forces operating in the vast central highland urban agglomeration of Mexico, in absolute terms, and in reference to the rest of Mexico's economic space. As a direct result of trade patterns, so much dominated by petroleum (exports) and machinery (imports) with the US, the Northern and Southern Gulf regions of Mexico have been the field of two counter agglomeration forces currently affecting population and employment allocations within Mexico. The Gulf Coast and particularly the areas around Monterrey in the northern, Tampico in the central, and Vera Cruz in the southern coastal regions have attracted comparative advantages away from the central highland region (mainly Mexico City).

This pull for decentralization is taking place in combination with a certain push for decentralization surrounding the Mexico City Metropolitan Region. The last is mostly due to the high negative externalities of agglomeration (namely, particularly high levels of congestion, air, water and soil pollution, and very high population density), in Mexico City compared with worldwide standards. Above all, the push factors are due to the low quality and quantity of public infrastructure in Mexico City and to constraints resulting from limited availability of water in the central plateau.

Many analysts currently expect the population size of the Mexico City Metropolitan Region to reach 30 million, either by the turn of the century or eventually.[68] The Mexican census bureau projected in 1980 that by the year 2000 the region would reach about 23.5 million.[69] Certainly, events during the 1960–80 period partly justify such absolute population projections. There are municipalities in the region that appeared suddenly and in a quarter century experienced explosive growth. For example, Ciudad Netzahualcoyotl did not exist in 1960, but by 1970 it had a population of 580,436 people.[70] Are these growth rates sustainable? If past experience is any guide, they are not. Is the 30 million threshold attainable? It is highly unlikely, given other experiences, in other regions of the world, at different time-periods.

The Mexican census bureau, by 1985, was already revising their earlier forecasts downward.[71] Under the revised forecasts the Federal District's population is expected to reach 11.51 million (or an 11.07 percent share of the Mexican population) by 2000 and 12.2 (9.9. share) by 2010. The State of Mexico is anticipated to contain 17.64 million (16.97 share) by 2000 and 23.89 (19.39 share) by 2010. In 1980, the Federal District had 9.2 million (13.2 share) and the State of Mexico 7.88 million (11.32 share). In combination, in 1985 the Mexican census bureau projected that by 2010 their total would be 36 million or 26.3 percent of Mexico's population. It was once thought that New York would reach the 20 million mark, but having reached 17 million in the late 1960s it has declined in absolute population during the 1970–85 period.

In relation to the US scene, another factor subtracting from the agglomeration strength in Mexico City (and in all of Mexico) has been the growth of the Southern and South-Western regions of the US, particularly Southern California (Los Angeles and its neighboring MSAS), Texas (Dallas, Houston, Austin, San Antonio), Arizona (Phoenix, Tucson), and New Mexico (Albuquerque). Population migration in the US is currently estimated at slightly above one-half of 1 percent of Mexico's total population per annum.

Migration of population into the US (not only from Mexico but also from other Central and South American nations) has created an interesting, from a spatial standpoint, phenomenon along the US–Mexican border: the birth of twin cities on both sides of the border in the 1970s. The Mexican side of these urban areas counterbalances the agglomeration forces toward Mexico City and other large Mexican cities to an extent – particularly, since US capital flows into the Northern Mexican States (particularly the state of Sonora) for manufacturing (particularly auto assembly manufacturing by the Ford Motor Company).

From an ecological perspective, the interaction between these twin border cities can be classified as 'cooperative,' i.e., the growth in one positively affects growth in the other. It could be of interest to study in more detail these border spontaneous spatial concentrations as intermediate steps within large-scale, long-term migration phases. They may be examples of locational arbitrage behavior, in a movement from the large urban agglomerations of Mexico to the large urban areas of the Western and South-Western US.

They certainly have exhibited typical arbitrage patterns. These cities have appeared rather fast, and rather recently, in the post-1970 period. Most of their economic activity is transitory, being tied to population migration movements from south of the border to the north – the two-way nomadic commercial activity – and to currency trade. Their rate of growth is not unrelated to the value of the US dollar relative to the Mexican peso, and to the population growth rate of Mexico.

Relative to the other cities in the Northern States of Mexico, the Mexican part of these twin cities enjoy a per capita (formal) income level much above average. Relative to the other Southern US urban areas, the American portion of these cities has on average a per capita (formal) income below the US urban areas' average, but higher than their Mexican counterparts. They are locations where US–Mexican income differentials tend to even out. Overall, however, income and other differences between the US and Mexican sectors of these twin cities are still quite pronounced. They are clearly detected by areal photography of their land use patterns.

Are these 'border' towns (although some are not exactly on the border), sinks of employment opportunities slowing down the inter-regional labor migration rate? Or are they indeed sources of employment migration into the most prosperous region from the less prosperous one, acting as a one-sided stable spatial equilibrium (i.e., attracting population from the less-developed and sending population to the more-developed region)? More empirical work is needed to answer this question.

Similar, but altogether smaller, border cities, can be found in many regions of the world. Another example in the 1980s, where similar economic conditions prevail, but on a much lesser scale, is the northern (coastal) section of the Colombian–Venezuelan border. These towns are similarly dependent upon currency exchange between the relatively stronger Venezuelan bolivar and other weaker South American currencies. One can make the argument that the Guangzhou special economic district is another case of locational arbitrage between Hong Kong and the rest of the Pacific Rim on the one hand, and the People's Republic of China on the other.

Border towns present an interesting laboratory to test the locational arbitrage part of the relative parity hypothesis. To the extent that significant disparities exist between two neighboring regions, and transportation costs being favorable enough, their common border becomes the beneficiary. These border towns are the intermediate spatial markets, which regulate the flow between the two regions found in a state of spatio-temporal disequilibrium. Whether spatial arbitrage, within the principle of relative parity, does produce a process leading to dynamic equilibrium can also be tested. As will be argued in Chapter 3, contrary to conventional static spatial equilibrium analysis, spatial arbitrage may produce dynamic disequilibrium and even chaotic movements in the spatial flow of stocks. This is evidently possible in the incongruous spatial flow of labor and capital (Dendrinos 1986).

MACRODYNAMICS, STABILITY, CYCLES AND RELATIVE DYNAMICS

In the previous sections of this chapter data were supplied to draw the picture one obtains by looking at the economic growth of nations through the absolute lens and direct linkages. International trade data were primarily used, together with population, employment, and product counts. Glimpses of what can be obtained by looking at economic growth through the relative lens were also provided. The two pictures were shown to be different for certain spatial units at certain time-periods.

The point was made that the dynamics of the US economy, for example, which currently accounts for about a quarter of the world's gross product, must not be studied in isolation. Its past, present, and future course is tied to the world's economy – its meaningful environment. It was suggested that spatial units are globally linked and interdependent. A composite relative parity was proposed to study these global interdependencies. Absolute dynamics are limited in their capability to address these issues. Relative dynamics are much more appropriate than absolute dynamics in this instance.

In this section, closer focus is drawn to the subject of cycles in urban and regional analysis. It is demonstrated that socio-spatial cycles are largely unstable and chaotic. A few cases are discussed which show cycles to be stable or almost periodic (regular).

Absolute growth and business cycles

Economic studies of the US and other large and medium-size contemporary economies are routinely examined by economists in non-spatial absolute terms. Examples include the macroeconomics, the various disequilibrium medium-period cycle models,[72] and the classical studies by Leontief (1953) and Kuznets (1971). These studies address the direct, absolute links among national economies and the world economy.[73] In approaching the subject of international linkages through the absolute-growth direct-interaction lens the general processes underlying causes and effects in stable or unstable developmental growth, and in cycles and other dynamic events are obscured.

A major theme in empirical research and in theoretical speculation about national regional or urban (i.e., non-spatial and spatial) economic growth is whether or not cycles (more precisely, oscillations) are present at an aggregate or disaggregate level in production, income, wealth, population, and other stock accumulation patterns, and whether or not these cycles are periodic and of a long-term nature and not simply ephemeral and random fluctuations driven by unpredictable oscillations in interest rates. Further, it is asked whether the dynamics of stocks and locations are stable or unstable, meaning that they either lead to some steady state (dynamic equilibrium) or to explosive growth or extinction in some or all of them, or pronounced fluctuations at specific points in space–time.

From a locational analysis viewpoint one would like to know how nationwide or worldwide cycles are connected to regional and urban cycles, how sectoral cycles are connected to national and spatial cycles, or how the existence of spatio-sectoral temporal hierarchies are linked to individual sector or location cycles: to shifts in dominance and to collective smooth or abrupt changes in these hierarchies.

The subject of national (non-spatial) cycles has been addressed in the twentieth century by a variety of authors.[74] The focus is on the business cycle, over different time horizons and exclusively through the absolute growth or decline lens. These analyses attribute the cause for cycles to exogenous factors, either technological change (Schumpeter, Kondratieff), or to unsystematic fluctuations in monetary stocks (Lucas). Whereas the earlier work focuses on some form of regular, long period, and wide in amplitude variations in output (Kondratieff's long waves) and the rather more frequent, but still long term oscillations of the so-called Schumpeterian clock, the later work focuses on much shorter cycles involving fluctuations in prices and other monetary variables, including interest rates, producing stochastic variations in quantities of commodities about a mean trend.

Theoretical work in economics on business cycles is much more technically elaborate and detailed than the earlier mostly qualitative work which was presented without much supportive empirical evidence.[75] This work views speculative behavior (in its broader and strict definitions) and signal processing as the key elements in cyclical business patterns. According to the recent theory, agents are receivers and senders (as well as processors) of the signals provided by both expected prices and associated expected quantities demanded and supplied currently or scheduled for some time in the future. All agents are assumed to act on the basis of 'rational expectations.' Under smooth reaction functions and 'perfect information' the outcome is always for the economic system to convert to a stable equilibrium. As a result, according to the recent theory, business cycles are the outcome of either exogenous shocks (fluctuations occurring randomly in time and in magnitude affecting the dynamic path of the system on its long-run equilibrium trajectory), or (systematic) errors in predicting such things as interest rates.

Business cycle theory is partly tested through various linear econometric techniques. It is empirically grounded, as opposed to other non-linear dynamic theories which are very difficult to statistically test in view of methodological and lack of data limitations.[76] It contains, however, a number of features which raise concern. Central among them is the exogenous source of the price fluctuations. Why would outside agents – for example the government – be motivated to shock the system by changing interest rates with motives outside the domain of the model? The implicit dynamics assumed are also disturbing: the shocks perturb the system at the neighborhood of a (static) equilibrium, toward which the system tends to always move. The latter is in accordance with Samuelson's dynamic adjustment theory,[77] currently a very restrictive element in dynamical analysis in view of recent developments.

Ultimately, the emphasis on prices of stocks and interest rates, both very fast adjusters, restricts the descriptive power of business cycle theory. It shifts the emphasis from the dynamics underlying the relatively slow adjusters (quantities of physical stocks), to the fast dynamics underlying the very quick adjusters (prices and monetary variables). In a final analysis, however, it must be noted that the relevant cycles in the long term are recorded on population, capital, output, wealth, and other similar kinds of slow adjusting stocks.

Even more importantly, it denies the analyst the opportunity to analyze the richness in social spatio-temporal cycles. It merely delegates these cycles to exogenous factors or errors; and it does not recognize the internal to the socio-spatial system dynamic processes giving rise to these cycles. The exogenous source of fluctuations also limits the value of its empirical foundation.

The form and inner structure of cycles

Cyclical behavior varies in its form and inner structure according to the level of spatial analysis considered, ranging from the macro to the micro scale, or by how fine or coarse the industrial sector disaggregation is. In general, it will be argued that the larger the spatial unit of analysis, and the more aggregate the sectoral breakdown, the longer the period of the cycles.

Cycles manifest themselves in a variety of stocks, locations, and frequencies (i.e., periodicity). In the case of economic variables, for instance, oscillations in output, wealth, employment (and unemployment), quantity of built capital stock, finance capital, money supply, interest rates and other variables occur continuously. Oscillations occur in political, sociological, cultural, and other variables as well. For example, in the quantity of power, status and cultural dominance, specific social or individual agents possess at given points in space–time. Cycles are also present in the quantity of specific psychological forces individuals or collectives possess in space–time. In the case of individuals, examples of such cyclical variables include greed, love, fear, guilt, exuberance, etc.; in the case of collectives: cycles in mass opinion or hysteria, panic, hate, compassion, quantity of collective risk-taking, mean time-period when speculative action is undertaken, frequency of and oscillations in hate or compassion, and many other variables.

Analytically and substantively the form and inner structure of all these stock oscillations vary. To a large extent, at the product (or individual firm) level there are unique life cycles occurring once in the life span of the product (or firm). These cyclical events differ from the cycles characterizing the various money supply measures, for example. The latter exhibit continuous oscillatory motion and possibly chaotic dynamics. Appearance, maturity, and then disappearance of particular products or firms are successive phases of non-reproductive evolution –

unique events in the life of an individual product, process, or firm.

Very similar to a firm's dynamic life-cycle behavior is the cycle involved in the use of built capital stock of a particular type and its turnover into another type to accommodate different functions. This is referred to as capital stock conversion. Examples of built capital stock subject to conversion include private or public buildings, bridges, roads, parks, or any other type of private or public capital infrastructure (such as utilities) which could undergo conversion. Individual capital stock life cycle dynamics do not contain regular periodic motion. Some attributes of logistic growth are apparently hidden in capital stock dynamics. After the ecological usefulness of the specific built capital stock entity is reached a sharp and abrupt decline occurs. A discontinuity normally follows, leading eventually to the extinction of the stock. Thus, individual units (humans, capital stock entities, firms, products, etc.) exhibit life-cycle dynamics. Use of capital stock, when looked at from a daily, weekly, monthly or yearly basis can be highly periodic.

Monetary stocks or prices in market exchanges seem to demonstrate oscillatory but not periodic, possibly chaotic cycles. They exhibit highly unpredictable irregular cyclical motion. This is a type of aperiodic movement, whereby the current and past levels are not sufficient (or necessary) to predict their future levels. No observer or participant can obtain a predictive model of their behavior, although one may perceive that it is feasible to do so for a limited time horizon, on the basis of which this agent might take a bet. Perceived order in chaos forms trends.

In their qualitative dynamic behavior (frequency, speed of motion, stability, etc.), relatively aggregate spatial cycles of social stocks lasting over extended time-periods differ from individual firm, capital stock unit, or product life cycles and from aperiodic movement of ephemeral variables. An example can be seen in the growth and decline patterns of aggregates of human populations and their average per capita built capital stock size, or age, at various urban locations in time. Their dynamics do not seem to exhibit the life-cycle behavior, or at least in not as simple a form as in the case of individual product, process, or firm life cycles in time-spans of one century or so. Neither do these socio-spatial cycles exhibit the almost random volatility of frequently oscillating variables such as prices.

Causes for such variance in cycle types among stocks must be found in durability, divisibility, and transaction cost differentials. The durability of the built capital stock and its locational fixity, i.e., its very high conversion or transportation costs, reduce its volatility. The source for the regularity observed in human population cycles must be attributed to the relative collective immobility of population stocks, particularly of a large agglomeration, due to a preference for locational commitment by labor and the existence of transportation costs in relocation.

The simultaneous presence of capital stock and the population of different generations (vintages) in space–time must contribute further to the orderly motion of cyclical movement (i.e., dynamic stability) in urban areas. A regular periodic (cyclical) motion in these stocks would imply a continuous economic growth process regularly fueling the build-up of capital stock and replenishing population abundance. Presence of single age structures, or populations dominated by a single age cohort may be indications of dynamic instability at a particular location in space–time. Monotony in vintage may signal the presence of abruptly appearing and short-lived settlements. The rule may be of interest in the study of archaeological remains of human settlements.

It is of interest to speculate on the nature of the various forces at work in socio-spatial cycles and the central variables they are recorded over, as well as their various in-length periods. Forces, variables, periods, longevity of the cycle and the socio-spatial system's size are intricately related. Conclusions can be drawn indicating that the longer the length of a cycle in period, the more fundamental the cycle is, and the more dominant macrolevel forces have become in shaping these socio-spatial dynamics. At the same time, by tracing their origin one moves back to a particular point in time and space where the start-up of these cycles occurred, and when large-scale long-term evolutionary change was triggered.

A city's population changes daily by people moving in or out, and by the excess of births over deaths occurring daily. At times these changes may be significant, although irregular within bounds. These daily (and even hourly) irregular cycles in city size may extend to weekly, monthly, or annual oscillations which are largely not recordable. Annual census estimates of expected city size thus have a random (stochastic) component associated with otherwise deterministic dynamics. Further, the decennial census counts are indeed snapshot estimates of a chaotically behaving count over a variety of time horizons. This oscillatory movement, along with a lack of spatially sharp boundaries for an observer, may be at the core of fuzziness in measuring and recording socio-spatial behavior. At the same time, these considerations might be the initial components of a 'quantum type' theory of spatial and temporal social dynamics.

One can make a quite informative distinction between the *de facto* and *de jure* population size of cities. Their *de facto* size varies continuously, and depends on the hour (and maybe even the minute within an hour of a day) one considers. Such changes in the *de facto* city size render the use of *de jure* measurements questionable in many instances when *current* city-size counts are heeded. On the other hand, the accuracy of the *de facto* size, being closely tied to the time interval considered, is also suspect.

Minute by minute oscillation may be confined to the last few significant digits of the total count. Indeed, the variation may be less than 1 percent. One may erroneously conclude that the finding has no 'practical' meaning. However, it may signal economic and social vitality for a city. A calm, dormant city may be identified with calm dynamics in city size, at very small time intervals and with very small overall oscillations. On the other hand, a vibrant city may be characterized by chaotic dynamics even though the time-scale to record them may be very small. None the less, such cities may be at highly excited states, indicating periods of relatively high growth or decline rates.

There are many more regular hourly, daily, and weekly cycles observed in urban areas. For example, a daily cycle is that recorded in the use of highways by automobile trip makers, or by workers in various buildings during a 24-hour period. These are very regular (high frequency, highly reproductive, i.e., repetitive) cycles equivalent to the circadian cycles of biology. Then there are monthly, quarterly, and annual cycles, prominent among them being the budget cycles by many firms and government agencies. The economic cycles, or more particularly the business cycles forming over these periods, are reflected in the frequency with which a very large quantity of economic data are collected. However, they do not display the regularity in reproduction exhibited by the earlier cited higher-frequency highly periodic cycles, although some seasonal and rhythmic elements are found.

Next, the much less obvious five- and ten-year cycles are reviewed. With the possible exception of political turnover, which is recordable and directly observable, there are numerous political and economic cycles extended over this time-span. They are less obvious because they do not display the regularity of the highest frequency daily cycles, or the seasonal elements of the annual cycle. Certain discreteness, i.e., sudden change, appears in this type of cyclical motion.

Prominent among them are the expansion and recession movements in national economies obeying roughly five- to ten-year highly irregular cycles. Cycles of inflation and deflation are also examples of these medium-term oscillations, and they could be partly congruous (under time delays) with interest rate fluctuations and expansion and recession cycles. Abrupt change may be the opportunity for new socio-economic activity to be introduced and for triggering an extraordinary amount of entrepreneurial behavior. New products, information, and built capital stock are introduced, coexisting at the beginning and then gradually replacing the older stocks.

Longer-term cycles, extending up to one century in period, have been proposed to exist by Schumpeter and Kondratieff for economic systems. These authors have suggested that technological innovation is the key element triggering such cycles. It is not clear in their analysis what are the specific interdependencies among socio-economic variables in depicting these oscillations. Moreover, severe data limitations hindered empirical verification at the time they were proposed. Large-scale population migration movements are recorded over these time-spans, expanded over large areas of space.

Today, it is not even clear that regular oscillations in specific economic variables do exist at such time-spans. Discrete rather than continuous motion in the dynamic behavior in economic variables becomes more apparent. Above all, to the extent that one can detect such phenomena with current data available over half a century,[78] economic behavior undergoes irregular, non-periodic, seemingly chaotic change in these time-spans. Oscillations are clearly detected, but they are not regular.

Cycles of such periods (a quarter to one century long) are also the subject of the empirical work on urban ecological dynamics.[79] Depicted there are economic–demographic (ecological) regular periodic cycles on relative income and population of urban (spatial) economies in the US with respect to national totals.[80] This approach may lay the theoretical groundwork for describing the regularity observed in long term socio-spatial cycles. It also supplies a link, through the mathematics of discrete dynamics in bifurcation theory, to account for turbulent and chaotic motions.[81]

Although some social scientists may argue that transportation (inter-spatial) and transaction (inter-sectoral) mobility costs dampen the potential for turbulence in socio-spatial dynamics within relatively short time-periods (say, a decade), one can argue that declines in both types of costs over such time-frames enhance the spatio-temporal liquidity of social stocks and thus the chance for chaotic motion. It can also be argued that such impedances are even likely to cause chaotic dynamics, particularly when congestion externalities are present or through (shorter in time-scale) periodic forcing. Further, extended spatio-temporal horizons increase the likelihood for observance of turbulence.

Beyond these short-term periodic and longer-term (spanning centuries) aperiodic cycles one might come across cycles in the growth and decline of nations, regions or cities taking up to one millennium or longer to play themselves out. In tracing those cycles and in documenting them one is faced with an even more severe lack of data. No matter how clear the macrovariables may be, vagueness in marginal microvariables, fuzziness in spatial extent, and uncertainty about the form and inner structure of these dynamics further complicate the task of identifying them and recording their effects.

Paleoanthropology and archeology, in attempting to study events of such enormous proportions involving much longer time constants and monumental events in evolution, are forced to come up with very strong assumptions governing these macrodynamics' relatively fast and slow changes. For one, such dynamics must be highly aperiodic. They must be subjected to considerable changes in their form and inner structure, and such changes must be linked to geological time constants. The records are too thin to allow one anything more than a rough approximation of the time-period where appearance and extinction of various stocks occurred. One can only come up with very speculative theories when addressing such dynamics.

Safely, one can say that the forces underlying the form and inner structure of these periodic and aperiodic cycles are different in each case, as these cycles' regular or irregular periods undergo change. The very short term, very high frequency and very low period cycles of hourly, daily, weekly patterns can be recorded along many variables and they are to a very large extent due to habitual behavior, i.e., strong behavioral traits not easily altered in the very short run (up to one year or so). Time budgets are appropriate to document them. Psychological forces are dominant in the bundle of forces at work at this level, where speculative behavior on the part of individuals and collectives is quite important.

In monthly, quarterly, or yearly recorded cycles economic forces seem to dominate, with fluctuations in prices, interest rates, monetary stocks, and seasonal adjustments in the quantity of many commodities demanded and supplied being at the center of these forces. For cycles with periods varying between one and five years, political changes and economic forces in combination may underlie dominance in these dynamics. Here the economic forces are mostly tied to the adoption of marginal technological innovations by firms, and firm or industry restructuring. Dominance of forces might not be stable at this range.

Political, economic, social and related forces may alternate in importance, at different calendar time-periods, at different locations, for different stocks. Social, collective, and individual speculative behavior are still of importance, but these are not as pronounced in importance as the forces in the shorter frequency cycles addressed above.

As one moves to even lower frequency socio-spatial cycles, those of a decade to a century, larger-scale economic and demographic forces become important. It is this type of cycle that was studied in reference to the urban areas of the US and, in part, for Europe.[82] Broader social forces shape these cycles. Cultural norms and changes of a microcultural type are dominant. These oscillations seem to be adequately described for urban dynamics along spatial population and per capita wealth shares of national totals. A nation's economy acts as the relevant environment during these cycles. Industry-wide technological innovations may also play a role in the broad socio-spatial codes found appropriate to describe these cycles. Speculative behavior has now diminished to a trickle in these dynamics.

By moving now to oscillations with even longer time-periods – those exhibited over a millennium or so – one meets with basic or fundamental technological innovations in production and consumption which are very large in scale. These innovations are associated with fundamental shifts in macro socio-cultural norms and very rapid knowledge accumulation rates (i.e., explosive bursts in the amassing of information). In these cycles one can hardly find traces of speculative social behavior.

Some Marxist analysts argue that cycles of long waves sweep the rate of profits and they affect the cycles in capital accumulation.[83] Their analysis highlights the 'single force,' 'strictly disciplinarian' view of dynamics held by economists no matter what their ideological persuasion. They stand in contrast to the comprehensive bundle-like force shaping socio-economic events of such scale and duration, especially since the notion of profit must be construed as multifaceted rather than be held strictly accountable to its economic component. This theoretical failure may have imposed a high price on the peoples governed by regimes adhering to such limited and shortsighted notions.

These technological shocks originate at particular locations and from there, and over the length of the cycle's life, they propagate to the rest of the spatio-temporally interacting regions. Spatial expansion is one important event during these cycles, which are characterized by innovation diffusion during the lifespan of the cycle. Cycles induced by non-human action, but instead through natural events, also occur.

Examples of such environmental shocks may include severe earthquakes and large-scale, long-term climatic changes, or other environmental catastrophes. These events might be reflected as (abrupt and slow) changes in the parameters of the dynamic model or as changes in the form of the model itself.

In spite of the speculative nature of the above statements, the more so the longer the period of the cycle, a few things can be said with some degree of confidence. As the period of the cycle under analysis is augmented, the periodicity (meaning the number of precisely replicated cycles of a given period and amplitude in the oscillation, at a particular location at a particular historical period) must be expected to decline. System size and length of period on the one hand and observed number of cycles on the other, must be inversely related. The slaving principle may link cycles of various periods: longer period cycles and the central forces shaping them (i.e., slower motions) might determine lower period cycles and their forces (i.e., faster motions).

The number of pertinent (central, necessary and sufficient) macro-variables depicting these varying-in-period-length cycles must be inversely related to their period's length and directly related to their periodicity. On the other hand, the actual number of variables related (to whatever degree) to these cycles must be directly proportional to their period and inversely proportional to their periodicity. In other words, the daily traffic pattern on a road could be described by very few variables (in fact a sinoid function is at times enough), although the actual number of (marginal) variables related to the highway use may be very large. Further, the number of variables over which such cycles are observed is large. Contrary to this example, the invention of the wheel, the steam engine, or the microchip were a few central events which ignited very long-term (slow, and thus very few over a particular time-period) cycles. Fundamental technological innovation was central in cycles recorded over a few basic variables.

Thus, there are more variables (stocks) in regular oscillation and more marginal ones, as one moves to higher frequency cycles, than there are in lower frequency cycles. There are many locations experiencing high frequency cycles, as for example the morning and afternoon rush-hour traffic widespread among very many urban areas throughout the globe and throughout the past quarter century, whereas the locations and time-periods initially igniting lower frequency cycles are far fewer in number. The latter simply says that fundamental technological innovation is likely to appear at one rather than many locations simultaneously and only a few times in the course of human history.

As one moves toward larger period cycles, one moves back into history. At the same time one moves toward specific locations and events on the globe where the birth of fundamental technological innovations and bursts of knowledge took place, responsible for these slow cycles. In doing so, one moves back into earlier, and more shallow, stages of human knowledge and to knowledge stocks on specific locations which have vanished or have been significantly depreciated since. Finally, the longer the period of the cycle one wishes to examine, the more basic are the forces at work shaping socio-spatial evolution, and the greater the intellectual effort (energy) by the social system to record and analyze these cycles.

Intranational urban cycles

Next, the conceptual framework will be set for looking at spatial dynamics and cycles extending beyond a quarter century and at a rather macro spatial scale, and the relative dynamics base for tackling the subject at hand will be laid out. Mathematical ecology will be used to provide the analytical and theoretical foundation. It will set the stage for presenting the available empirical evidence in support of the ecological formulation. The relative macrodynamics of regions and of a few large urban areas in the US in reference to the US economy, and then the US macrodynamics in reference to the world economy, will be presented to demonstrate how spatial size and dynamic (cyclical or nodal) qualities are interrelated.

The relative aggregate dynamics of urban settings are well documented with empirical evidence by urban mathematical ecology.[84] Looked at from a relative framework and under a continuous dynamics lens, urban settings (being concentrations of a variety of stocks) exhibit clock-type cycles when viewed within an environment's relative macrodynamics. Such clocks produce a somewhat regular periodic continuous dynamic motion in relative population and per capita income dynamics, both within a national economy and within regular real time intervals extending between half century and one century periods.

Spatial dynamics and cycles are viewed according to this framework, not as temporary or erroneous responses to environmental fluctuations. Rather they are considered to be built-in inevitable cyclical adjustments of these two aggregates to a deterministic environment, taking a century or so to manifest themselves. The inner workings of these macrodynamics are subject to a multiplicity of forces. A macro-spatial aggregate ecological determinism links an individual urban area's dynamics to that of its relevant environment. Individual and collective speculative behavior has been discounted in the inner workings of these cycles.

Spatial units contain a multiplicity of agents with different real time entry or exit schedules in the system. These agents possess different objectives, constraints, time and space horizons in their foresight, and different degrees of aversion to risk and other behavioral traits. All these heterogeneous agents (producers, consumers, governments) hold options on the future state of the broader system they belong to, according to a variety of actions they perform, with different expiration dates. These agents in their actions are senders, processors, and receivers of a variety of signals. To an observer, these agents seem to exhibit certain dynamically regular spatio-temporal patterns in their collective actions. Despite the constituent heterogeneity in agents and actions these patterns result in relatively regular urban macrodynamics with periodic cycles. Aggregate macrodynamic periodic motions are seemingly well-replicated by simple dynamical models.

In these cycles, depicted over these two macrovariables, two continuous dynamic interactions are involved: the interaction between the two central socio-economic macrovariables (elaborated in Chapter 1), and the interaction of an individual spatial unit (an urban area) directly with its relevant environment. The latter is the collective outcome of all other spatial interacting units in space–time. The emerging cycles are such that a zenith in one variable (say relative population) does not coincide with the zenith or nadir in the other (relative per capita wealth). As the cyclical patterns of Figures 2.2 and 2.3 indicate, the maximum in per capita income of urban agglomerations still induces relative population growth, while the per capita wealth starts to decline. At its lowest point in per capita affluence, relative population decline is continuing, while per capita wealth starts to increase.

Although successions of growth and decline have been recorded in time spans of the scale of a century for various urban settings in reference to a broader (national or world) economy they have not been recorded for relative macrodynamics of national subregions in similar time horizons.[85] Apparently, they must take much longer time spans to develop, if regional periodic motion occurs at all. Regional relative continuous macrodynamics, viewed within a national environment, seem to demonstrate logistic growth or decline patterns similar to the product life cycles. Qualitatively similar relative macrodynamics are

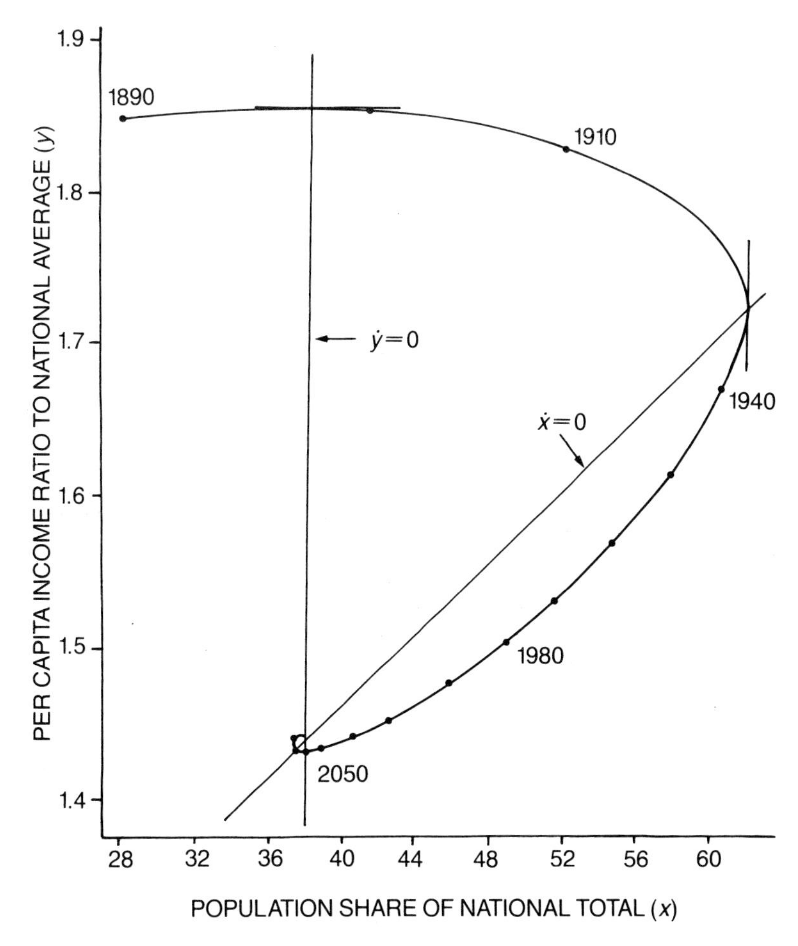

Figure 2.3 The computer-simulated dynamic path of growth for the New York MSA relative to the US, 1890–2050

recorded in this book for the world's largest urban agglomerations when viewed with respect to the world's macrodynamic patterns.

Consequently, the regular periodic motion depicted in the time-scale of one century or so for relative continuous urban dynamics within national economies can be attributed to the life cycle period of individual large and dominant firms in a particular industry in a nation's economy. This cycle can be attributed to the expansion of existing industries or the immigration of new production units being offset by the concomitant contraction or outmigration of firms already located in an urban area. Built capital stock and population (labor) cycles may be synchronous with such product (process or firm) cycles.

On the other hand, the logistic-type dynamics, or cycles with a period much longer than one century, if observed in a nation's relative continuous regional dynamics must be attributed to industry-wide or broader cycles in comparative advantages demonstrated either nationwide or worldwide. Such cycles require much longer time-scales to develop and may also characterize the life cycles of national economies within the world economy. They must slave, as already alluded to, the shorter cycle.

Numerous examples of urban cycles exist at various locations on the globe, at different time-periods. Two cases will be cited here. One is found in the evidence regarding the periodic abandonment and repopulation of the Mayan cities of the Yucatan Peninsula over the last two millennia – particularly the almost periodic colonization, growth, decline, and abandonment of cities like Chichen Itza every two centuries in the AD 987–1441 period.[86]

The other case is found in another peninsula, the Balkans, and it is the population cycle in Athens. Its apparent peak was reached during the Golden Age of Pericles in the fifth century BC, during which the absolute population must have recorded about a third of a million people.[87] It then started to decline, reaching a low of about 12,600 in the mid-sixteenth century.[88] The Athens Metropolitan Region currently stands at approximately 3.4 million.[89] Quite similar is the case of Rome, too.

Both of these cycles, the high-frequency almost periodic Maya cycle and the low-frequency aperiodic cycle in Athens' population, are expressed in absolute terms. Lack of data and gaps in the historical record on the broader environment and the prevailing socio-economic conditions of these settings within these environments, at their respective time-periods, limit the use of the relative lens. The case of Athens in particular underlines the importance of the relative lens, as the current size when compared with the fifth century BC size certainly does not imply superior socio-economic and cultural conditions within Athens' (and Rome's) *de facto* (relevant) environment now and in the past.

Population cycles are events of broader social significance than mere accounting. Cultural anthropologists, archaeologists, and historians associate rises and falls in population or wealth of a social system with different historical periods. Changes in cultural, social, economic, religious and other practices are routinely linked with such population and income (wealth) cycles. Historians still debate the causes for the cycles in the Maya cities,[90] Athens, Rome and other city-states, nations, and civilizations.

It is clear, according to the ecological determinism thesis of this book, that many factors are always involved in these cycles. Interacting economic, environmental, social, political, and cultural forces provide the composite force fueling them. What is (are) the dominant force(s) among them, if any, varies in space–time. Interpretations of the always incomplete and fuzzy historical record only serves current purposes, in line with the arguments supplied in Appendix 1.

Beyond these commonalities, however, there could be a shared force in these cycles at all points in space–time: the phase of the cycle reinforcing itself. The mere decline in one factor – for example, population – in turn accelerates its decline. This is the analytical expression of the 'bandwagon effect.' Such effects are widely recorded in stock exchanges. Cyclical models, where current stock sizes, rates of growth or decline in these stock sizes, and acceleration or deceleration forces interact, could be potentially fruitful extensions of the current literature on cycles of socio-spatial stocks. They could also provide insights into the mass psychology and sociology of populations in cities of the past and present.

Relative dynamics

Cycles and events reported earlier obviously do not play themselves out in human generation time, a much shorter time constant than geological or the biologists' evolutionary time. Most large-scale socio-economic stocks show unstable dynamics in time spans of centuries or millennia. As a result, extinction or explosive growth or urban cycles are not events likely to be detected by looking at individual cases with currently recorded data over the past quarter century. However, elements of such instability may be obtained from snapshots of different cases at a particular time-period, in a similar manner to how cosmologists obtain evidence on long-term and large-scale evolutionary events in their respective fields by single time-period observations.

It turns out that qualitatively expected results on the instability of socio-spatial dynamics are more vividly depicted when indirect interdependencies, relative continuous dynamics growth models are employed. Relative lenses shorten and accelerate the speed of motion in the period of growth or decline, bringing the time horizons of potential dominance or extinction closer. They also amplify the amplitude of cycles.

When the size of population and the magnitude in per capita wealth of a metropolitan agglomeration is normalized with respect to the current world population and prevailing average per capita wealth, then the lens provides insights into the relative location the urban area occupies in the world economy – the more so the larger the urban area. In the relative size of its per capita wealth some (but not all) of the factors in relative parity have been capitalized.

Existence of ecological linkages among metropolitan areas, within and among nations, implies the presence of an hierarchy of interactions characterized by a multiplier effect in their diffusion levels and rates, ultimately manifested as income or population growth or decline impulses. At the top of these multiplier effects (i.e., where most of the impact is felt, and possibly exponentially declining as one moves down the hierarchy) are the primal effects due to the direct interaction among the very large urban centers. Further, these centers experience the secondary waves of growth or decline once the original wave hits their national agricultural (or urban) hinterland which acts as a periphery wall from which growth or decline waves move back towards the center at decreasing amplitudes. They identify the round-about (cumulative causation) direct interactions among these centers.

As a result of these inter-national interactions, at an absolute and direct level, the world's major urban areas grow or decline. Thus, part of the population size of Tokyo depends directly on the size of New York and Detroit. Part of the size of Los Angeles is associated with part of San Francisco, Mexico City, and Hong Kong among other very large or large metropolitan centers, located either in or outside the US. More importantly, the growth of Los Angeles is linked to, i.e., determined by, the collective dynamic behavior of a number of other urban US or non-US agglomerations.

The end result of this soup of interaction impacting the urban sectors of nations is that at any time-period there are nations which grow and others which decline at various rates. In the short run, some spatial (regional, national, or urban) population, wealth, and other stocks grow or decline faster or slower than others in either or both absolute and relative terms. In the long haul, certain nations become extinct and new nations emerge. One observes similar patterns for cities. Over time, within and among nations, cities grow and decline. Their population and wealth grow or decline faster than others. In the very long term, certain cities become extinct and new cities emerge. One can take this evolutionary pattern a step further into the dynamics of particular industries, to various units of government, or to specific social and cultural subgroups within a nation.

Absolute growth in size and wealth of national populations does not take place in a vacuum. Any national economy competes for growth rates with other nations' populations and wealth in ways far more complex and involved than implied only by the direct links. Widespread conflicts arise through individual and collective economic growth or decline. For example, when a nation's population or wealth grows at a pace different from that of another nation's, and the two nations are not isolated from one another, then the observed growth rates must be viewed as the scheme having successfully competed with any other alternative scheme for acceptance as a result of a combined (collective) decision by both nations. Similar arguments hold for changes in a nation's political system of governance, or even borders.

The collective decision may not have been, and in fact never is, consciously or directly made by either nation. Reaction or lack thereof by either or both nations demonstrates acceptance or rejection of the scheme. This reaction, or lack thereof, may be the outcome of a very involved and complex interplay of many variables. In a game theoretic framework many interactions are involved, by many players, on far too many issues too fuzzy or unclear to model explicitly. International accords, military interactions including war, economic exchange, and migration movements are partial and simple examples of individual and collective spatio-temporal conflict resolution or ecological interaction of such game theoretic actions. There are other linkages involved far more unclear and indirect. The outcome, however, is simple and clear so that it can be directly modeled.

All such simple and composite links are picked up collectively when relative dynamics of stocks and locations are derived within prespecified ecological environments. Thus, parallel to the direct interactions among the world's largest metropolitan agglomerations one finds a much stronger and much less studied interconnection expressed in relative terms. When relative size in population and wealth are examined, within a worldwide and continuous or discrete dynamic framework, these linkages account for elasticities of growth or decline. These elasticities identify spatial units possessing locational comparative advantages, worldwide, for growth or decline relative to all other competing unit dynamics.

One might consider national, regional or urban socio-economic systems linked together as a set of non-linear oscillators in per capita income; linkages are the multiple spatial interactions among these systems, one of which is trade in commodities. Indirectly, such linkages can be picked up by assuming that the per capita income and its rate of change of one nation affects that of another. Unstable dynamics can ensue as a result of such coupling. Coupled oscillators exhibit the phenomenon of resonance, whereby slow or high per capita income growth rates alternate (move, commute) among spatial units, or where growth and decline rates migrate (through the coupling or interaction mechanism) from one spatial unit to another. Such a coupling may shed some light into the interdependent dynamics of spatial economies. This framework, however, is far more complex to model than the simpler mathematical ecology framework expressed in the relative dynamics adopted here, and potentially far more insightful.

An interesting example of relative growth linkages is the strong interest shown by western nations in seeing in the post Second World War period that the absolute population growth rate of LDCs decreases, particularly that of India, China, and other Far Eastern, African and Latin American nations with relatively high population growth rates. The presence of economic, demographic, and other social theories addressing population issues might be linked to a variety of interests present in the worldwide population distribution game. Despite the power of the indirect interaction, but because it is indirect, the force of relative growth has eluded the analysts' attention. The remainder of the study is an attempt to bridge this gap.

THE MATHEMATICAL FORMULATION OF RELATIVE MACRODYNAMICS

From Chapter 1 it has been continuously asserted that macrodynamics for spatial units (be those urban areas or nations) can be captured by the dynamics of two key indicators of development: population and per capita gross product. In the earlier sections of Chapter 2 it was proposed that the relative dynamics and stability properties of these two variables, when expressed as ratios to the world's total population or average per capita income, are informative. Next, the mathematical ecology framework to consider these dynamics is set up.

At the start, a modification of the standard multiple species non-locational Volterra–Lotka continuous dynamic framework of mathematical ecology will be presented followed by Chapter 3 where the model of the spatial ecological macrodynamics of the subject at hand will be presented. This is done mainly for two reasons. First, to present certain examples from the US urban and national macrodynamics; and second, to draw from its ecological components in laying the ground for the worldwide urban macrodynamics.

Assume a system of simultaneous differential equations, continuous dynamics, defined over the relative population stock size, x_i, for spatial setting i (among I settings constituting the closed environment E, in this case the world economy); and the per capita product ratio to the world's currently prevailing average, y_i, at setting i:

$$\frac{dx_i}{dt} = f_i(x_i, y_i, \boldsymbol{P}_i; i = 1, 2, \ldots, I), \qquad i = 1, 2, \ldots, I$$

$$\frac{dy_i}{dt} = g_i(x_i, y_i, \boldsymbol{R}_i; i = 1, 2, \ldots, I), \qquad i = 1, 2, \ldots, I$$

where t stands for time $\boldsymbol{P}_i$ and $\boldsymbol{R}_i$ are vectors of parameters associated with the particular environment within which location i is viewed. These parameters and the form of the functions f_i and g_i depend on the currently (and in the time horizon T the model is applicable) prevailing interdependencies among the various settings I within environment E. Interdependencies include a bundle of forces, manifested in the multifaceted interactions occurring among the I settings. Their end effects are assumed to be recorded along the two macrovariables x and y.

These kinetics correspond to continuous and simultaneous dynamic adjustments in demand for (the first condition) and supply of (the second condition) relative population share to appropriate population levels and per capita product differentials among the competing locations within the closed environment E. Typically one could assume J different stocks (instead of the two identified here – namely, relative population and product). If empirical testing is possible then one can choose those stocks found most appropriate to state these dynamics. Lack of time series data however, hinders this effort. A substitute to this testing has been the analysis already provided.

It can be argued that the above simultaneous and continuous macrodynamics could be further reduced to a single stock dynamics for particular urban settings:

$$\frac{dx_i}{dt} = F_i(x_i, \boldsymbol{R}_i; i = 1, 2, \ldots, I), \qquad i = 1, 2, \ldots, I$$

where the effect of stock y has been absorbed by the new set of parameters $\boldsymbol{R}_i$ in the new functional forms F_i. Whether such reduction can be achieved is still an open question. Its validation critically depends on the necessary data becoming available. Finally, the ultimate reduction would involve the system of differential equations:

$$\frac{dx_i}{dt} = G_i(x_i, S), \qquad i = 1, 2, \ldots, I$$

where the effect of all other variables and sizes has been reduced to a set of parameters S, identical to all locations i in the environment E, and new forms of functions G_i. Evidence for currently adopting some of the above formulations will be supplied later.

These macrodynamics can be construed as disequilibrium dynamics involving paths moving toward stable, or diverging from unstable equilibria. A steady state is identified when:

$$\dot{x}_i = 0 \rightarrow f_i = 0, \qquad i = 1, 2, \ldots, I$$

$$\dot{y}_i = 0 \rightarrow g_i = 0, \qquad i = 1, 2, \ldots, I$$

where the dot stands for time derivative. Such systems, where particular configurations of f, g, F and G include second degree functions in the state variables (x, y), have been extensively studied in mathematical ecology.[91] Since this is not the appropriate forum to present this work in full only some very brief basic remarks will be made on the dynamic features of the above models which are of interest in this book. The interested reader is directed to the above and other already listed references to further reading.

The continuous and simultaneous Volterra–Lotka dynamic specifications of the above two systems have proved to be of particular interest (and use). If a two-stock, multiple-location relative continuous dynamics model is stated where each location interacts directly with the environment, then:

$$f_i = x_i(a_{0i} + a_{1i}x_i + a_{2i}y_i), \qquad i = 1, 2, \ldots, I$$
$$-\infty \leq a_{0i}, a_{1i}, a_{2i} \leq +\infty$$

$$g_i = y_i(b_{0i} + b_{1i}x_i + b_{2i}y_i), \qquad i = 1, 2, \ldots, I$$
$$-\infty \leq b_{0i}, b_{1i}, b_{2i} \leq +\infty$$

It has been shown that this relative continuous dynamics system is inconsistent and degenerate.[92] That is, it does not meet the relativity condition at all time-periods, and the income variable operates independently from population. However, and this is an important caveat, the system is not necessarily inconsistent when a few settings' dynamics are approximated by these kinetic equations. Thus, although the whole system may be inconsistent, partitioning it can avoid the inconsistency and degeneracy problem. On the other hand, if a one-stock multiple-location model continuous dynamics is formed where interactions among stocks are included as in the standard Volterra–Lotka model, then:

$$F_i = x_i(a_{0,i} + \sum_j a_{ij}x_j), \quad i = 1, 2, \ldots, I \quad -\infty \leq a_{ij} \leq +\infty$$

Coefficients a_{ij} constitute the community interaction matrix of the association. All the important properties of these ecological associations depend upon the magnitude of these coefficients a_{ij}. Furthermore, there are six basic ecological interactions which contain all the pair-wise combinations of all locations. They are: predatory ($a_{ij}, a_{ji} < 0$), competitive (a_{ij}, $a_{ji} < 0$), cooperative (a_{ij}, $a_{ji} > 0$), commensal ($a_{ij} = 0$, $a_{ji} > 0$), amensal ($a_{ij} = 0$, $a_{ji} < 0$), or isolative ($a_{ij} = a_{ji} = 0$) ecologies.

In the above continuous dynamics Volterra–Lotka one-stock, multiple-location formulation the stability versus complexity argument can be directly presented. If I is large and a_{ji} are small, then the dominant part of each stock's eigenvalue (derived from the characteristic polynomiae) is the coefficient $a_{0,i}$, the self-growth or decline rate depending on its sign. The future of the stock depends on the relative magnitude of $a_{0,i}$ compared with other stocks' coefficients. The stock with the largest positive a_0 coefficient will grow to dominate the community of interacting locations or stocks, whereas the stock or location with the smallest a_0 coefficient will become extinct.

A more specific model, widely used in the ecological literature is the predator(x) – prey(y) model:

$$\dot{x} = x(a_1 + a_{11}x + a_{12}y)$$
$$\dot{y} = y(a_2 + a_{21}x + a_{22}y)$$

where $a_1 < 0$, $a_{11} = 0$, $a_{12} > 0$, $a_2 > 0$, $a_{21} < 0$ and $a_{22} = 0$.

Or simply:

$$\dot{x} = x(-a + by), \; a, b > 0$$
$$\dot{y} = y(c - dx), \; c, d > 0$$

which shows that as the prey grows (near the equilibrium) so does the predator ($b > 0$); whereas, as the predator abundance grows, the prey population declines. It turns out that this is a basic model when stability in dynamic interactions is sought. It is repeated that substantive analogies with ecology for spatial systems are not necessary, only methodological (analytical) equivalences exist.

Predator–prey absolute size (expressed in simultaneous continuous dynamics) ecologies are stable, as they result in spiralling sink-type dynamic equilibria where both species coexist (x^*, $y^* > 0$). The dynamics of the two population paths are characterized by four phases in development. There is a built-in cycle in the association, so that on the convergence toward the steady state a period of growth in both stocks is succeeded by a period of growth in one and decline in the other; in turn this phase is followed by a period of decline in both stocks; finally the cycle ends with decline in one and growth in the other stock respectively.

The above system of continuous dynamics Volterra–Lotka equations can be extended to simulate interactions among many species, but at the cost of losing its stability properties. It can also be used to link population and per capita product (or income) cycles. Then the cyclical movement it contains identifies periods where the peaks in population and product (income, or wealth) do not coincide in time. Maximum or minimum per capita product occur at different time-periods than maximum or minimum in population. This is in accordance with findings by archeologists and anthropologists studying the historical record of human settlements, and thus of broader interest. It is also a wider interaction of the qualitative properties of the basic guide to historians.

As was the case with the relative macrodynamics indicated in Chapter 1 (pp. 18–23), the *continuous* Volterra–Lotka dynamics depict non-chaotic *moving mean* (expected) value equations for the urban paths. Urban trajectories oscillate, in a non-periodic fashion, but within bounds well-depicted by the Volterra–Lotka type urban dynamics suggested earlier. Discrete dynamics are necessary to depict chaos in two-dimensional analysis.

In a previous study, spatial (in that instance, urban) notions of relative dynamics within an environment were elaborated and found appropriate for analysis.[93] Specifically, the behavior of all US metropolitan areas was studied from a macrodynamic perspective. The environment used, defined as the broader geographical entity over which the size of a spatial unit is normalized, was the US total economy. The simulated dynamic paths of a number of US urban areas in the phase portrait of relative population–per capita income space (the central variables) were derived and shown in Figure 2.2. The individual dynamic paths of the metropolitan areas were shown to obey Volterra–Lotka continuous simultaneous dynamics under limits to growth. These dynamics are given by the system of ordinary differential equations:

$$\dot{x}_i = \alpha_i(y_i - \bar{y})x_i - \beta_i x_i^2$$

$$\dot{y}_i = \gamma_i(\bar{x} - x_i)y_i$$

where the dot stands for time derivation; x and y are, respectively, relative population and per capita income in city i; $\bar{x}_i$ is a carrying capacity for city i within the national environment; $\bar{y}$ is the national prevailing average per capita income; α_i, γ_i are respective speeds of adjustment parameters; and β_i is a friction (negative net external agglomeration effects) coefficient.

The carrying capacity $\bar{x}_i$ and its ecological meaning is of central interest, in view of the arguments made earlier in this chapter. A carrying capacity represents locational and other factors internal to the city, as well as external factors emanating from its relevant environment.[94] Variable $\bar{x}$ could be specified as a function of other variables, ranging from topographical, climatological, to economic, social, political and cultural; in specific

$$\bar{x}_i(\tau) = qi(\lambda, \tau)$$

where τ is a time horizon the carrying capacity remains constant at a specific value; and λ is a set of variables affecting $\bar{x}$ internally or externally specified in reference to a specific urban area i.

The above dynamics produce a damped oscillatory motion in the phase portrait constructed in the (x,y) space. This is the cause for the internally induced cycles to occur in urban macrodynamics. The $\dot{x} = 0$, $\dot{y} = 0$ linear equations define the system's isoclines. Their intersection is the point where the motion ceases. The individual urban signatures are sinking spirals demonstrating a stable dynamic cyclical behavior for each metropolitan area within the US environment (see Figure 2.2). Note, in particular, the behavior of the New York MSA in reference to the US economy, isolated in Figures 2.1 and 2.3.

The relative urban dynamics, thus depicted, seem to be consistent with a *code* of development resilient over the study period. In this code collective and individual speculative behavior is imprinted. One can identify in such a code a mechanism of selection. At each time-period particular technological innovations are adopted; such changes in the economies of production and consumption are selected; and those political decisions are made nationwide and metropolitanwide (among many possible alternatives) which produce outcomes on relative population and wealth dynamics of urban agglomerations consistent with this code. The code identifies the existence of internal regular periodic clocks in the dynamics of urban centers, when viewed within the framework of a national environment.

Next, a look is taken into the US aggregate dynamics within a worldwide environment, to make the point that such continuous dynamics might be present in larger settings as well. Throughout the 1958–84 period the US gross national product increased in nominal and constant terms. This is the picture obtained from the absolute growth lens. However, a different picture is obtained using the relative lens.[95] Its dynamics are found to be qualitatively very similar to features exhibited by US urban dynamics now in reference to the world economy, and to obey similar type forces. They are shown in Figure 2.4. In combination, Figures 2.3 and 2.4 seem to suggest the following: a nation within a community of nations, traces a path similar to the one its major urban center follows within its community of metropolitan areas.

This seems to supply strong supporting evidence for the arguments made in the previous sections, linking the fate of nations to that of their urban areas and their industries. It may be due to a decline in the world markets of the nation's dominant industry, which in the case of New York and the US was manufacturing. In turn it is manifested by the decline of the nation's major urban area (presumably tied to the nation's major industry) within the nation's urban centers.[96]

The findings from Figure 2.4 are of special interest. Since the publication of the original paper covering data from the 1958–80 period, more time series have become available. They seem to confirm the further deterioration of the US position in the phase portrait as projected in Figure 2.4(b). In 1987, the US relative population share of the world's total stood at 4.85 percent, whereas the ratio of the per capita product to the world's MDC (not market, however) economies had fallen to 1.53.[97] The sharp depreciation in the US dollar during the 1985–87 period relative to all other global (formal) currencies would have affected the course of the per capita product ratio.[98]

The R. May theorem on complexity versus stability, appropriately extended, can be useful and insightful in reference to the US macro urban evolutionary patterns. Transcending animal and plant species, its validity spills over to human spatial social systems, urban or national. It provided the foundation for urban mathematical ecology and its qualitative intra- and inter-urban continuous dynamics. Further, it was discussed earlier in this chapter in reference to international trade, national economic growth, and unstable inter-national dynamics. One could attribute the exsitence of very few dominant nations, with a sphere of influence over smaller and at times relatively less wealthy nations, as a stabilizing factor in international affairs. One could also see its dual version: the fact that only a few dominant, very large, and

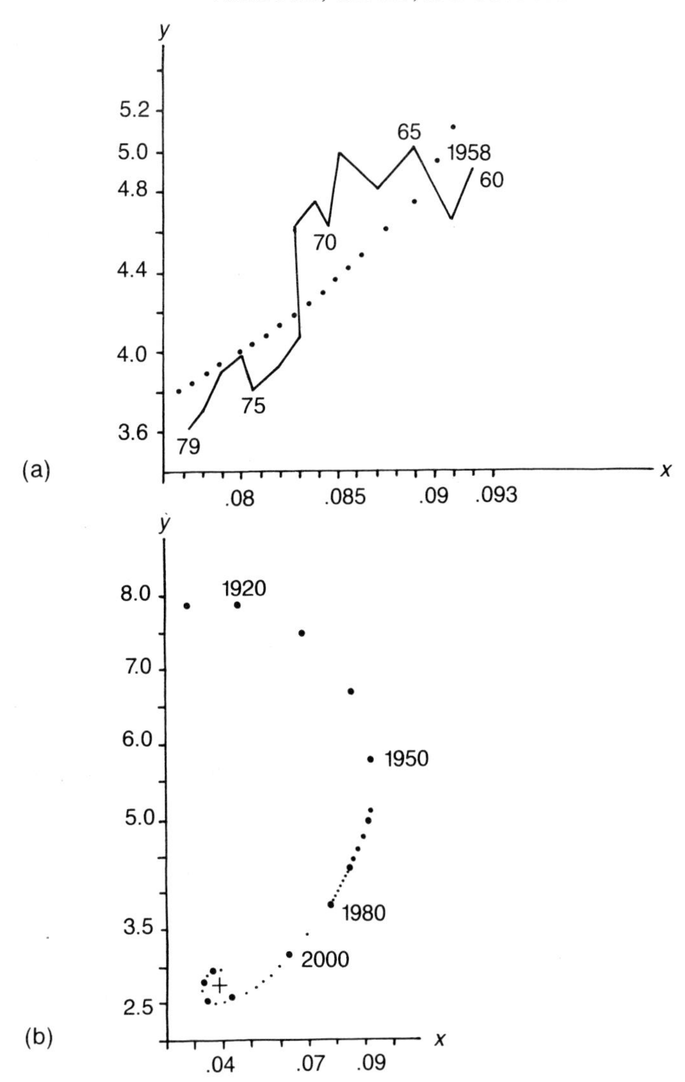

Figure 2.4 The US economy's macrodynamics within the framework of the world's market economies. (a) Actual (continuous line) and simulated (dotted line) dynamics, 1958–79. (b) The simulated macrodynamics, 1900 to steady state. The horizontal x-axis depicts the US share of the world market economies' current population; the vertical y-axis depicts the ratio of the current US per capita income to the world market economies' currently prevailing average

Source: D.S. Dendrinos, 'The decline of the U.S. economy: a perspective from mathematical ecology,' *Environment and Planning A*, Vol. 16, 1984, pp. 651–62.

relatively wealthy nations currently exist may also be viewed as the end result and manifestation of an inherent instability in a previously highly interconnected community of nations. The periods of high interconnectance must not last very long.

The existence of relatively few urban agglomerations with a population abundance currently exceeding five million inhabitants, in view of the total number of urban areas on the globe, can be attributed to the same qualitative elements of the May argument. As was the case with national economies, one could accept the instability part of the argument but also recognize its dual version. Very few and very large urban agglomerations must always be expected to emerge in space–time, be that the result of currently stable or past unstable global dynamics.

Ecological niche is central in population dynamics and in the field of ecology. It identifies a range of values in a multi-dimensional space of attributes (including other species' abundances) in which a particular species develops.[99] This broad definition of niche includes locational socio-economic attributes which, according to location theory, may involve a large array of variables, both exogenous and endogenous to the location and stock in question. Within an absolute growth or decline framework, these attributes may be classified as related to the input factors of the production or consumption process and their corresponding dynamics.

At the input level they might include proximity to natural resources, topography, technological diffusion level, and accessibility to information, capital and labor markets. At the output level they account for access to markets of finished products and relevant prices. They might also include other social, political and cultural attributes. All these factors are captured by the term 'comparative locational advantages' (a concept addressed in more detail in Chapter 3, section 2) in a relative framework. The term identifies a ranking of suitability. It is normally measured in terms of marginal cost and profit or utility extracted from production, consumption and trade at various locations.

In a relative growth framework, niche and niche dynamics are embedded in the bundle type composite (ecological) community matrix and its resulting fast dynamics. Fluctuations or changes in the relative magnitude (values) of locational attributes are translated into slow changes in the composite coefficients of the community matrix. Dynamics are now defined over elasticities of growth in relative population size and the wealth of an urban center, relative to other urban areas, so that the urban center in question can be thought of as possessing a relative size niche within the national environment. Thus, in a relative growth framework, population abundance of a spatial unit is determined by dynamics of its particular locational (niche) attributes.

Relative accessibility or impedance to various spatially distributed populations, the backbone of location theory, is only a part of the attributes affecting the values of the phenomenological (ecological) community matrix coefficients. How severe is the effect of distance upon the magnitude of the coefficients is still an open research question. An earlier reported experiment in identifying the role of distance upon international trade turned out to show a very limited use of spatial impedance for spatial interaction dynamics. Another experiment in testing the role of distance upon the magnitude of the community matrix coefficients for urban areas turned out similarly disappointing results.[100]

CONCLUSIONS

In this chapter, the subject of spatial and social dualisms and disparities was addressed. One of the main analytical causes for dynamic instability was examined, namely that associated with interactions among spatial units. It was demonstrated that dualisms flow inescapably out of complex dynamical interactions. The finding reveals that one need not employ ideologically based theories of socio-spatial dualisms, as simple analytical constructs of social systems would inevitably produce such disparities. Evidence seems to suggest that such dualisms are widespread and multifaceted.

As is the case with all social science theories, two conflicting interpretations of this hypothesis regarding a source for instability were obtained and then an effort was made to reconcile them. A currently high degree of selectivity and dualism in spatio-sectoral structures was looked at as the end result of a prior unstable spatio-temporal interaction pattern. At the same time the two currently observed phenomena could be interpreted as the requirements for a stable spatio-temporal hierarchy. Reconciliation of the two viewpoints was attained by not rejecting the dynamic instability hypothesis, no matter what the view adopted regarding the current spatio-sectoral structure.

The presence of very few dominant sectors in a national economy, the presence of very few dominant urban settings worldwide at present, and the waning of past dominant centers of power and wealth, the presence of dualism in the labor force of an urban economy and the abundance of slums, the presence of formal (global) and informal (local) national currencies – all these are phenomena attributed to inherently unstable, highly selective, socio-spatial, multiple stocks interactions. They are examples of a general dualism characterizing the dynamics of selectively interconnected stocks in a complex ecological community.

Cities, independently interacting in space–time, are more likely than nations to cause dynamic instability as they are much more numerous than nations. National boundaries can be viewed thus as attempts to decrease the number of independently acting spatial units, and consequently as an effort to decrease the likelihood for instability. Cartels among nations formed around particular sectors (say, petroleum) can also be viewed as attempts to decrease the likelihood for dynamic instability in these sectors' markets by limiting the number of independently interacting agents. The analysis might explain, to an extent, why a few supra-national units (like the European Community) have a tendency to be formed with a much smaller number of currencies (with the common European currency unit being a case in point), and why there are so few superpowers worldwide at any point in time. Splitting up of nations and the breakdown of alliances are manifestations of unstable dynamics. Similarly, the merger of nations and formation of alliances also are testimony to instability.

To an observer, a worldwide relative parity in net attraction among locations seems to be a principle linking the dynamics of nations, regions, and cities. Through a spatial arbitrage process of flows, social stocks grow or decline in space–time in response to an effort toward relative parity. Spatial arbitrage occurs in fluid factors, like population capital commodities and information.

Macrodynamics of the world's largest urban agglomerations demonstrate the outcome of this inter- and intra-nationally sought after relative parity. Socio-temporal spatial arbitrage fills, according to an observer, perceived relative parity in net attraction imbalances (vacuums forcing change). It is one element of the ecological determinism governing urban macrodynamics. Dynamic instabilities are the reason why under a relative parity principle sharp spatio-temporal dualism emerges, as it will be further discussed in Chapter 3.

The relative parity principle points to a complex bundle of factors producing spatial interaction. The effect of physical distance is only a minor component. There are economic, social, political, cultural, environmental and other components in this relative parity, each balanced out against the others through spatial arbitrage. Many examples of events and flow of stocks in space–time have been observed confirming the notion of spatio-temporal arbitrage. They involve among others the spreading over space of technological innovations and major shifts in the industrial base of urban agglomerations, rural to urban migration, urban to suburban movements, industrialization, the increase in the size of the services sector, and other major spatial and sectoral shifts.

Macrodynamics contain the possibility for violent and calm dynamical events and cycles. Presence of cycles was demonstrated for certain spatial stocks under specific environments. The analytical foundation for a theory of spatial cycles was presented, drawing from the theory of mathematical ecology which is next used as the background for setting the analytical model of urban macrodynamics among the world's largest urban agglomerations in reference to the world's economy.

3

THE DYNAMIC CODE OF A GLOBAL URBAN HIERARCHY

INTRODUCTION

A relatively few and mature urban areas – comparatively old, large, and wealthy – coexist globally with the much larger in number, young, less wealthy metropolitan agglomerations: some at the first stage of their life cycle motion toward a more abundant population and higher per capita income. Cities at in-between stages are also found alongside these two polar cases. In combination, all of them constitute a global urban hierarchy founded on an age, wealth, and population distribution structure. They constitute a *demography of urban settings*. In this chapter, the analysis penetrates deeper into the size–wealth demographic urban hierarchy, as the exposition becomes considerably more technical.

Individually, large urban regions move within this age–wealth–population urban demography restrained and propelled by the collective behavior of all other urban agglomerations, as well as by their internal conditions. Simultaneously, individual urban macrodynamic paths trace and are drawn by the structure of collective behavior. Global interdependencies link the dynamics of all urban and rural settings of the world. This chapter elaborates on the unstable dynamics of global urban paths. These dynamics are attributed by an observer to the second major source for spatio-temporal instability – namely, that of interdependencies picked up by a *relative parity in attraction principle*. The other source of instability, due to extensive inter-spatial interaction, was analyzed in Chapter 2 through the representative case of inter-national trade. Relative parity can be construed as a 'potential' giving rise to individual urban macrodynamic paths. Dynamic instability is viewed as the reason why sharp spatial dualisms exist while to an observer a drive toward relative parity is underway.

Being so large at this spatio-temporal scale the system does not allow much room for each individual urban area's macrodynamics to significantly affect collective behavior, which changes marginally and very slowly in response to individual urban area macrodynamics. Individual urban settings macrodynamics, on the other hand, react very quickly to collective behavior within this urban hierarchy.

Analytical focus is now placed on the urban dynamic paths of the largest urban areas on the globe in reference to the world's total population and average per capita product (income). The notion of composite relative parity in net attraction among locations, its constituent elements and its function, are analytically stated. What occurs to a background level of attraction in reference to which the attractivity of individual cities adjusts by changes in spatially liquid stocks is outlined, as this dynamic process provides the linkage among metropolitan (and rural) settings of the globe.

Regional and urban hyper-concentrations and hypo-concentrations are defined and analyzed as to their internal dynamics. To examine such phenomena, the subject of spatial agglomeration economies is closely reviewed. Empirical evidence supplied clearly demonstrates that the relative macrodynamic lens provides rich insights into this relative parity principle and the associated dynamic behavior of individual large urban agglomerations in the contemporary world. Among other things, using the empirical evidence from the past twenty-five years one can deduce the existence, from the inner workings of this effort toward relative parity, of a relative developmental *threshold* level restraining per capita product in the evolution of urban agglomerations (and nations) belonging to the LDC category.

Existence of an aggregate *code* of development is inferred, which evidently contains the dynamic paths of the world's largest agglomerations. This code provides clues as to the dynamics of all, and not just the largest, urban areas. With sharper focus placed on the empirical evidence available, a thesis is advanced on pp. 215–18 whereby a pitchfork-like bifurcation tree is suggested as containing the deterministic equivalent to stochastic hierarchical macrodynamics of all urban areas on the globe. Scenarios on probable outcomes in the long term at this macroscale are discussed in Chapter 4.

Considered in this chapter are nineteen metropolitan areas with a population greater than 5 million in 1975: Bombay, Buenos Aires, Cairo, Calcutta, Chicago, Jakarta–Utara, London, Los Angeles, Mexico City, Moscow, New York, Paris, Beijing, Rio de Janeiro, Shanghai, São Paulo–Santos, Seoul, Teheran, and Tokyo.[1] Three other metropolitan areas had a population close to the adopted threshold according to the UN data in 1975: Leningrad (which in October 1991 reverted back to one of its former names: St Petersburg), Lima[2] (each with approximately 4.5 million inhabitants), and Tianjin (with a population of 4.3 million in 1970). Reference will be made to other very large urban agglomerations which have surpassed the 5 million mark since then. See also note 12, Chapter 1 and Table N.1, where a number of additional urban agglomerations and their 1985 estimated population is given, together with a classification based on their size.

There seems to be a significant variation among different data sources on the population size of the world's largest urban agglomerations, see Appendix 3, pp. 310–11 and Table N.1, pp. 339–40. This variation is partly due to varied political jurisdictions, time-periods of the recordings, and the recording sources. Often data on 'city' size are given, and rarely on a 'metropolitan area' scale. Still, even under the second category, sources seem to differ in their *de jure* definitions, as different urban areas loosely associated with a large metropolitan region are sometimes lumped together. This is done, for instance, in the case of the US MSAs, particularly New York and Los Angeles.

In view of the discussion regarding data in the social sciences (see the Introduction, pp. 8–15) one must not be overly impressed by this fuzziness. It is, as suggested, a necessary component in the internal growth processes of each of these urban agglomerations. To an observer viewing these metropolitan areas collectively and under a relative lens, the effect of this fuzziness is somewhat diminished, although not totally eliminated, in the ecological determinism postulated.

The full extent of the problem for the purpose of the current macrodynamic analysis cannot be adequately assessed at this time. For indicative purposes, and in order to illustrate the widely varied population estimates, data from the Mexican census on urban agglomeration size in the early 1980s are supplied in Appendix 3, pp. 310–11. Further, data on the twenty largest Chinese cities are supplied from the People's Republic of China 1984 census as well (see Appendix 3, pp. 312–16).

It should be noted here, that preliminary quantitative evidence using the Mexican and Chinese census data seems to indicate that *although the position of the nineteen urban areas in the phase portrait of Figure 1.1 are slightly altered, the qualitative properties of their dynamics are not.* Evidence also suggests that adding other metropolitan areas that may have reached the 5 million population threshold after 1975 does not alter the qualitative properties of the model to be presented.

HYPER-CONCENTRATIONS, HYPO-CONCENTRATIONS, AND AGGLOMERATION GRADIENTS

The difficulty in precisely identifying the relevant spatial composition of a metropolitan area is not merely one of optimum, prevailing, or employable political jurisdiction for data collection purposes. Difficulties abound in specifying the various forces at work affecting the spatio-temporal *de facto* form and structure of metropolitan areas. *De facto* extent and form or urban settings is closely related to a field of forces attracting or repulsing urban activity in space–time.

In spatial economics the bundle of forces pulling economic activity toward urban core regions is sometimes referred to as *agglomeration economies.*[3] The notion of agglomeration economies or forces and its implied effects will be used here, although in an enlarged spatio-temporal frame of reference as an 'agglomeration forces gradient,' within a dynamic and static framework. It will be viewed as a temporal force-field distributed in space. It is noted that the analysis which follows is general, and applies equally to present-day urban agglomerations as well as to those of the distant or close past at all places on the globe.

Agglomeration forces can be classified under positive (attractive) and negative (repelling factors). They are not directly observable but only detected by their end effects in space–time. These end effects are referred to as 'agglomeration effects' and are categorized as local (in absolute terms) or comparative (in relative terms). When a number of locations are examined (instead of looking at a single location in isolation, a condition that very rarely is of any interest if it exists at all) a *comparative* context must be adopted. Agglomeration economies can only be tied to *locational comparative advantages* and do not exist outside such advantages.

Agglomeration effects are partly the result of population and capital stock concentrations: the 'endogenous' forces; and they are also the result of the relative position and comparative advantages a location enjoys within a national and international landscape: the 'exogenous' forces.

Locational advantages must be attributed to a ubiquitous *locational heterogeneity.* Locational heterogeneity is partly due to the difference in availability of and access to natural resources, topography and climate, and of access to national and international markets (i.e., differential transportation costs). Space as an impedance measure is *a priori* a factor forcing economic activity to spatially concentrate rather than disperse – in other words, resources must be expended in the form of transport costs in order to overcome it. Declines in transport costs, through investment in the transportation infrastructure (at all modes), is a component contributing to a breakdown in agglomeration economies and spatial dispersion. Prior existence of such advantages linked to heterogeneity give rise to the initial attraction of stocks at *particular* locations and not at any other location. Once inter-spatial population density differentials set in, then they are reinforced by scale factors in agglomeration effects triggered by such density differentials. In turn, they affect locational comparative advantages. The argument holds within an inter-urban as well as an intra-urban (or regional, national) context.

In the popular press an urban agglomeration is referred to either as a 'megalopolis' (a term attributed to Gottmann) or as a 'ecumenopolis' (to use a Doxiadis term). Neither term is precise enough, nor useful for the purpose here. In any case, they are manifestations of the fuzziness involved in the way urban agglomerations have been perceived.

To assume the existence of agglomeration economies in a homogeneous landscape (something that no evidence that ever existed on the globe has suggested) *prior* to the appearance of inter-spatial comparative advantage differentials begs the questions: where exactly in space do significant spatial population concentrations start, and where did the agglomeration economies (forces) originate from? An argument is sometimes posed according to which perturbations (or random fluctuations) cause cities to commence at random points in space–time. However, this argument is without sound etiological foundation. It fails to explain why cities started *at their particular locations* – that is, where significant topographical features and comparative advantages are usually found.

Fluctuations do play a significant role in non-linear dynamics. But they do not play a significant role in the start-up of cities. It is noted that this argument does not imply that agglomeration forces are not necessary for cities to exist; indeed they are the necessary conditions, with the sufficiency conditions being that spatial heterogeneities make space differentially advantageous. Cities did not start out of a uniformly distributed-in-space population. Human population always clumped together in families, tribes, or colonies; from caves to lake settlements the archeological record does not supply any evidence that population was at some point in time uniformly distributed in space and that through some minor fluctuation cities, or other human settlements, started in a homogeneous landscape. Indeed, there is no geological evidence that the globe's landscape was ever homogeneous in the time-framework of human evolution. Thus, theoretically appealing models of agglomeration like those by Weidlich and Haag (1987) or Papageorgiou and Smith (1983) postulating such agglomeration dynamics are off the mark.

In agglomeration models which postulate that the uniform spatial population distribution be subjected to a fluctuation setting forth forces of agglomeration and thus creating human settlements, the process is irreversible. But such is not the case according to empirical evidence whereby the event of agglomeration is not irreversible. Processes of agglomeration are succeeded by processes of deglomeration (deconcentration, dispersal). It is rather safe to assert that such reversal is not expected to lead to socio-spatial configurations even remotely resembling a uniform spatial population distribution in the future. A cyclical motion in spatial population distributions is more likely.

Such a cyclical motion is to an extent equivalent to reversing the arrow of time in modeling spatial population distribution dynamics. Models postulating the birth of human settlements as the result of a stochastic process involving fluctuations of a uniform prior spatial population distribution could be 'locked out,' never reproducing (let alone cyclically replicating) such even distribution. Thus, the reasonableness of such initial perturbation must be questioned on the grounds that if it could never be encountered in the future, nothing must have produced it in the past. Next, a brief discussion of these agglomeration forces and effects is presented to set the stage for a spatial relative agglomeration model to be constructed allowing one to view the inner growth or decline mechanism of core regions.

Economies (or diseconomies) of agglomeration can be indirectly measured by two different, but combined effects: net attraction (or repulsion) of socio-economic activity (stocks) from (to) peripheral areas to (from) the core of national and regional settings, and the ability (inability) of the core region to retain a positive amount of the endogenous socio-economic growth.

Positive economies (forces) of agglomeration are due to the presence of scale economies from current ongoing local, economic, social, political, religious, and other production, consumption, administrative, and market processes. Negative economies of agglomeration are attributed to negative environmental externalities resulting from these processes, like various forms of congestion, pollution, etc. Current net positive (or negative) economies of agglomeration is the relevant factor for future urban development.

The hypothesis is that net positive economies of agglomeration cause spatial concentration, whereas net negative economies of agglomeration cause spatial dispersal of economic and social activity. Thus, net attraction or retention of socio-economic activity depends on the magnitude of net positive economies (forces) of agglomeration. The above discussion defines the local (absolute) agglomeration forces. As locations compete for socio-economic growth, however, relative (i.e., comparative) agglomeration effects become important. In this case, differentials in agglomeration forces among competing locations become the driving (or repulsing) factor.

Further, the analysis can be structured in either long-term static or dynamic terms. In a static long-run equilibrium framework, the assumption is made that in time the net (positive or negative) agglomeration effect differentials diminish and eventually are eliminated as spatially fluid factors (continuously and smoothly flowing among regions in a nodal fashion) adjust in response to these differentials. In a dynamic framework, one must look at these net agglomeration forces as current expectations about a future state of affairs.

In an effort to eliminate them in response to these differentials spatio-temporal arbitrage occurs by adjusting (discontinuously or continuously in a nodal or cyclical fashion, and smoothly or abruptly) spatially liquid stocks (labor, capital, information, commodities). The difference between the two frameworks is that the static, unique, equilibrium-based analysis assumes a smooth process always driving toward a unique long-term equilibrium state, dynamical analysis, on the other hand, assumes a process driving the system in a variety of ways mostly toward and at times away from one or multiple dynamic equilibrium states.

Due to complex interactions in the speculative dynamics involved, the process may not converge toward an equilibrium (fixed point, steady state, or periodic motion) in the long run. Instead, the long-run dynamic equilibrium state(s) may be unstable, as speculative dynamics may be such that they enhance the opportunity for the system to oscillate in a non-periodic or chaotic manner. In a speculative frame of reference, expectations among the different socio-economic spatio-temporal agents may not be always (rarely are) congruent. Non-linear interactive speculative behavior may underlie violently oscillating states of adjustment in this succession of economies and diseconomies of agglomeration. Oscillatory behavior may be more profitable than a smooth drive to a steady state.

Agglomeration forces are formidable when the net positive (driving for growth) differentials are relatively high, these being expressed as either multifaceted (not merely economic but rather ecological) currently expected profit (for producers) or expected utility (for consumers) between the core (as a point in space) and the periphery. In this case of inter-urban comparative net positive agglomeration differentials, the greater the current difference the greater the current speed of concentration into the core. As in the static case, the movement of stocks into or away from the core ceases, and stock sizes stabilize when the differentials are totally eliminated.

These kinetics identify a basic lack of synchronicity between the maximum in expected profit or utility differentials and the maximum level of stock concentration in the core. When the driving differentials are at their peak, instead of stock sizes being at their maximum the *speed* of concentration is at its peak. As the differentials start to diminish, the speed of concentration is still positive. Such a rule must have been applicable in past cases where the golden ages of very large cities must not have been the time-periods when their population levels were at their maximum. Instead, population levels reached their peak when the differences in wealth between these cities and their spatial competitors were eliminated. By then, it could have been too late, as a period of decline may have commenced.

As a consequence of high inter-urban net comparative positive agglomeration differentials, absolute or relative hyper-concentrations occur. Under these super urban gravitational conditions, the speed by which socio-economic activity from the hinterland enters the core increases as the core size increases under a bandwagon effect. The activity potentially attracted by the core with greater strength is located 'ecologically closer' to the core.

Ecological distance is a composite notion measured by many distance indices. Topographical distance is only one element in the labyrinthine bundle; social, ethnic, cultural, and economic distances are other types of socio-spatial impedance entering into the notion of ecological proximity or distance. The bundle nature of distance is equivalent to recognizing ecological forces and locational niches at work, determining attraction or repulsion effects in the bundle of agglomeration forces.

The presence of net comparative positive agglomeration forces at the inter-urban context results in growth of the socio-economic stocks at a particular point in space–time enjoying these comparative advantages. At the same time, the presence of socio-economic stocks concentrated at a specific point in space–time gives rise to comparative agglomeration forces (positive and negative). Cause and effect between the two (stocks and their growth or decline on the one hand, and agglomeration forces on the other) is merged in the presence of these stocks in the core region.

This is the positive feedback or bandwagon interdependence effect in the process of agglomeration economies and growth.

Another element in the linkages between the two, and associated with change in the size of stocks at the core and its periphery, is the spatio-temporal interactions between the core and its hinterland. These interactions are mostly tied to migratory flows of population (labor), capital, and other factors including environmental externalities (pollution for instance). They are the direct effects of the underlying interdependencies, and the cause for considerable instability as discussed in Chapter 2.

What is the triggering mechanism for the start-up of the spatio-temporal interdependencies (i.e., the inter-urban net positive comparative agglomeration forces)? They must originate both endogenously and exogenously in these regions. The former must be more often the case than the latter. Internal topographical, and initial economic, social, cultural, environmental and other relevant factors at a specific point in space–time can be among the central causes. Comparative agglomeration forces enter the picture as a result of broader spatio-temporal factors often exogenous to a particular location. There must not be a general rule of start-up for agglomeration processes at all points in space–time, as different factors in combination may play a role during different circumstances.

The end effect, however, seems to be the same. It is connected to spatial dualism and at its extreme takes the form of primate urban settings in national or regional economies. At their very center, these hyper-concentrations contain very high shares of the national or regional employment, population, finance capital, built capital stock, output, wealth, and political, social, and religious power, etc. They are few and they occupy in most cases the intra-nationally prime location, the attraction of which they may have further enhanced by their very presence. They have in the past successfully competed with other urban centers in their region or nation for primacy, and they are currently in the process of competing for primacy with other rivals. Next in line and lower in size are urban areas found relatively far from the core. Although they may compete with the core to an extent, they are not, by and large, serious contenders. Primacy is durable, meaning that it is not spatially mobile but rather sticky. Temporally, frequency of change in primacy is very low. Primacy is the result of intra-national unstable dynamics favoring the growth in size of the primate city and the relative decline of the rest of the nation's settings over prolonged time-periods. On the other hand (see note 3 of Preface), primacy may be a requirement for spatio-temporally stable nations to exist, i.e., for nations to function without significant forces applying pressure toward the splitting up of the country into smaller entities.

Agglomeration effects and hyper-concentration of stocks at specific points in space–time must be stronger and more pronounced when per capita income (or wealth) is lower; conversely, the hypothesis is that agglomeration effects weaken when incomes are higher. This assumption implies that as per capita income grows agglomeration effects must decline and hyper-concentrations become less pronounced, while the opposite must occur when per capita income declines. One certainly observes this dynamic connectance between strength of agglomeration and wealth at the individual average family size within an international (as well as national) context: family size decreases as income per capita increases. This force could be construed as a basic force in the declining family size of highly developed western nations. It could also account for the very large (nuclear as well as extended) family size of the less developed nations of the Third World.

Hyper-concentrations of wealth and political power are not recorded exclusively over space. They are also recorded over different socio-economic groups. Hyper-concentration of wealth and power characterize the local, regional, or national socio-cultural elites. For more on this and the phenomenon of dualism, see Appendix 1, pp. 260–8. In summary, hyper-concentrations are pronounced spatio-sectoral monopolies or monopsonies.

It seems that spatial hyper-concentrations and very uneven income, commodities, capital, information, and political power (among many socio-economic stock) distributions coexist. Spatio and sectoral monopolies or monopsonies are closely connected, in both market and in centrally planned economies. Presence of any one is necessary for the rest to occur, and their collective presence is sufficient to guarantee the presence of any one in particular. From a testable hypothesis standpoint, it must be difficult to attribute cause or effect to any one. They seem to be all strongly interrelated and simultaneous.

Hyper-concentrations must also be viewed, along previous arguments, as the result of unstable ecological dynamics characterizing the interaction of various locations and socio-economic stocks. Whereas most stocks in the socio-spatial configuration become extinct under these unstable dynamics, a few grow exponentially fast to dominate all the rest. Such a process is depicted by the concentration of wealth to a few, and attraction of employment (and population) as well as capital stock to a small number of urban centers, and by the repulsion of these stocks from the rest of the national or regional economy. Activities and locations with the strongest and most selective links survive and grow. These are the analytical aspects of spatial monopolies (monopsonies) and oligopolies (oligopsonies) formation.

So far the discussion of agglomeration forces has focused on the inter-urban scene. Next, a static long-term view of the intra-urban field force will be taken. The static approach has been chosen to synthesize previous work with the above analysis.

In a similar fashion the center of the core can be considered as a primary gravitational source with its suburban space (its periphery) as a secondary gravitational source. Comparative positive and negative agglomeration force gradients appear as declining functions of distance from the center of the core region (see Figure 3.1). A point on these gradients represents a composite willingness (or unwillingness) to locate at this distance from the center of the core as a result of comparative positive (or negative) agglomeration forces and current stock size distributions prevailing at that point. Composite willingness (or unwillingness) could be translated into a composite index of expected benefit (or cost) of location at that point in space–time. Land price or rent is only one component of this composite index.[4] For simplicitly, physical distance is assumed to be relevant here only, although the discussion can be easily extended to accommodate the more elaborate notion of ecological proximity or niche as defined earlier.

Gradients evolve in space–time as the core itself grows. At the start-up of the process, the speed of upward motion in the positive comparative agglomeration forces gradient is much faster than that of the negative. As these spatial hyper-concentrations expand, negative comparative externalities start to pick up as these core urban areas expand, both in terms of area and population as well as in per capita income or product. Comparative agglomeration economies of higher economic production densities are overcome by various diseconomies and negative comparative externalities of high density, pollution, congestion and other related effects. As this occurs, neighboring centers develop at the periphery of primate agglomerations where economic growth spills over.[5]

These main and subsidiary centers are still ecologically inter-connected. One center's employment, residential, administrative, religious, social or commercial market area is found partly in the other centers' residential, employment or other cultural concentrations. A center's part of the space may be speculatively held awaiting another center's growth or decline. In present-day urban centers, the strength of these links is manifested in, among other factors, their direct absolute levels of passenger and freight movement, as well as by more diverse factors such as regulations, information, and capital flows.[6]

In all these conglomerations a dominant center is always present. This is the reason why among the urban regions used in this study the Milan–Turin and Rhein–Ruhr metropolitan regions were not included, although they are at times considered urban agglomerations of more than 5 million inhabitants by World Bank publications. These two urban concentrations are assemblies of large individual urban areas not possessing a common high-density core. They do not collectively form a spatial hyper-concentration of urban activity in their national space. After Germany's unification, however, one may consider the rise of Berlin as a primate urban agglomeration within the unified Germany.

A phenomenon intrinsically associated with the formation of hyper-concentrations is that of *hypo-concentrations*. It is due to opposite forces than those identified in the formation of hyper-concentrations. Equivalently to the case of hyper-concentrations, there are positive and negative effects in hypo-concentration factors. A striking example of the latter is the desertion of rural (productive) lands by non-surplus rural labor migrating into urban areas as surplus urban labor and thus becoming counterproductive (and not merely unproductive) from a national standpoint.

Hyper- or hypo-concentration and the associated forces could provide a useful framework to view concentration conditions in (relative or absolute) spatio-sectoral growth (decline), as the subject of future research. In a cycle, hyper- and hypo-concentrations may succeed one another in a state of perpetual disequilibrium dynamics. This succession will be further discussed on pp. 195–201, when the relative parity principle will be analytically stated.

Much has been said on the 'explosion' phenomenon of the core regions of nations, usually associated with their largest urban agglomeration. Having already been observed in developed economies, it is currently underway in developing nations.[7] This 'explosion' is linked to a phase in national development for both LDCs and MDCs.[8] Connections between urban and regional growth, primacy, and national development have been studied particularly in the context of the Third World.[9] In the next few paragraphs this 'explosion' phenomenon is analyzed in more detail, as it relates to the previous discussion on agglomeration economies.

At the outset it must be noted that 'explosion' is a misnomer in this case. It is only accurate to the extent that the space close to the core is occupied by a stock mass which is increasing in size (population, employment, built capital stock), and thus it experiences an increasing density during the event. But this mass is not emitted entirely from the core, only part of it is. The spatial expansion is not a pure internal event of the core. Stocks from the hinterland (population, employment, capital, resources, etc.) are still attracted to it. What actually occurs in a static framework is this: a stock's distribution over space under the influence of the comparative agglomeration gradient has undergone a transition, so that its density gradient originally becomes steeper toward the core.

Later, and after having reached a maximum, it starts becoming less steep in space–time. It is then that the spatial expansion of the core reaches its maximum speed. Forces of negative effects associated with further spatial concentration outwardly push the willingness to locate curve close to the center of the core region. At the same time, positive relative effects of agglomeration pull closer to the core activities from the core's hinterland.

These two forces, their spatial field and their interplay, are shown in Figure 3.1 in the shape of two comparative agglomeration force gradients. Growth or decline in local density at any distance $\bar{d}$ from the center of the core is determined by the net difference in relative agglomeration forces gradient s. Up till distance $\bar{d}$ there is a net outflow of socio-economic activity and thus a decline in local density.

Beyond this point, conditions start to reverse: there is a net inflow of activity and an increase in the local density. The inflow reaches a maximum at point d^* where s attains a maximum, s^*, as at this optimum point in space the outwardly moving, as well as the immigrating, economic activities' willingness to locate is at its peak. Point d^* moves away from the center in time as the two gradients shift. Distance d^* is the point where the tangents to the two functions (in the diagram showing the willingness to locate, depicting the negative and positive effects of agglomeration) are parallel. There, their difference (i.e., the net effect of the relative agglomeration gradient, s) is at its greatest, s^*, as the willingness to locate (or attraction) is at a maximum.

There is, of course, an opportunity cost to a socio-economic activity in its willingness to locate close to the center of the core. It is shown by the horizontal line $\bar{s}$ in Figure 3.1. It represents the willingness to move to the next most attractive urban center in the nation or region. If $\bar{s} > s^*$, naturally the economic activity in question will not be motivated to move into the core (and at a distance d^* or any other distance within its market area). A discontinuity is built into this mechanism: as either the opportunity cost rises or the peak s^* declines relative to $\bar{s}$, then there

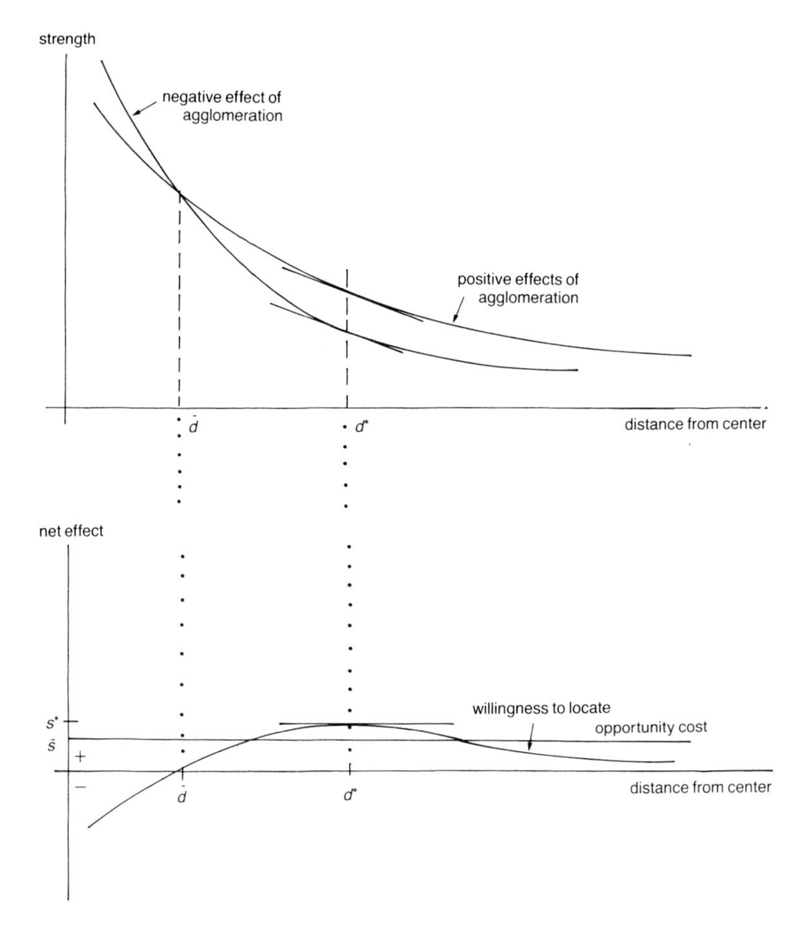

Figure 3.1 The negative and positive effects of agglomeration versus distance. At distance $\bar{d}$ away from the core of a metropolitan region the positive effects of agglomeration start outweighing the negative, the net reaching a maximum at d^*

comes a time-period where $\bar{s} > \bar{s}^*$. At this point, the last settlement attracted to the core is reached, beyond which the core's attractive power is no longer effective. The core could then actually start to jettison economic activity away from its region.

These mechanisms could depict the birth (and disappearance) of both spontaneous settlements (mainly driven by the economics of location) or planned settlements (mostly driven by the politics of location). An interesting example of the first type is the spontaneous city of Netzahualcoyotl near Mexico City in the Mexico City Metropolitan Region.[10] Most of the residents in this satellite town are recent rural migrants. An example of the second type is the location chosen by the Egyptian government to locate a ring of new towns near the Cairo metropolitan region.[11]

One could use the above abstract and general processes to depict changes in the size of very large urban agglomerations not only in reference to their national hinterlands, but also in reference to their worldwide environment. A worldwide spatial relative agglomeration forces gradient is now at work so that world urban hyper-concentrations are formed. In this case, the opportunity cost line $\bar{s}$ is not drawn in reference to a nation, but in reference to worldwide prevailing conditions. This is a constituent element of a worldwide relative parity, to be more fully addressed in the next section.

At the inter-urban scene, the precise time when a specific urban center reaches the stage where the positive forces of relative agglomeration are surpassed by the diseconomies of further concentration, thus signaling the start-up of overall absolute or relative population declines in the urban area, depends on local conditions. It is also a function of the time-period in the evolution of the other urban areas with which the city in question ecologically interacts (competes, cooperates with, or preys on) in space–time.

From an intra-urban perspective, if the willingness to locate curve flattens with distance it signals the beginning of an equalization process among intra-urban densities. Whether a lower overall intra-urban density (the suburbanization phenomenon) precedes or succeeds the start-up of decline in the absolute or relative urban population or per capita income again depends on local conditions (regional and national) and the time-period in question. For US MSAs, the available evidence seems to suggest that a more even distribution of densities preceded overall urban decline during the last quarter century in metropolitan areas of the north and north-eastern regions.[12] It also succeeded overall rises in per capita income levels.

Whether changes in intra-urban density precede or succeed changes in absolute or relative population, or whether these changes are simultaneous, is still an open question. Are these changes smooth or discontinuous? The answer obviously affects the manner in which these changes are modeled, whether in a discrete or continuous (with or without delay) mode.

During the third quarter of the twentieth century, globally, one finds that large urban areas experienced the phenomenon of urbanization and a decline in their average prevailing densities. Concomitantly, one also comes across large urban agglomerations which have undergone the phenomenon of urbanization and an increase in their average densities. Indeed, the individual urban macrodynamics and their intra-urban density changes were only dependent upon their starting point on a phase portrait about twenty-five years ago. Next, the focus will shift onto the global inter-urban dynamics.

AGGREGATE DEVELOPMENT CODE, RELATIVE PARITY, AND THE EMPIRICAL EVIDENCE

The evidence (1958–80)

The analysis now turns to a key task at hand, the analytical specifications of the evidence presented in Chapter 1, pp. 18–23. This evidence provided both the impetus for this study and its perspective. Its theoretical implications provided much of the justification for setting up the ecological macrodynamic framework of reference. First, a brief note regarding the data sets used is made.

A complete description of the data used and their sources is supplied in Appendix 3, pp. 303–9, with a number of qualifying statements. A few comments are in order here. Data for the US metropolitan areas are taken from the author's previous studies which employ US Bureau of the Census statistics. Although the data differ from that supplied by the UN on population for New York, Chicago, and Los Angeles, more confidence is placed on the US data regarding their spatial composition given their Metropolitan Statistical Area (MSA) definition. As a result, in this work these MSAs have a smaller size in the study period than that provided by the UN series.

This impacts the ranking of the nineteen urban areas as far as population is concerned, although this is not considered a particularly serious problem since the focus is on the paths of aggregate dynamics. Use of the US statistics also eliminates three other urban areas in the US with a population, according to the UN, of over 5 million in 1980: Detroit, Philadelphia and San Francisco.

Whereas in all other cases the national average of per capita gross domestic (or, in certain instances, national) product is used, for the three US metropolitan areas their per capita income from US Census series is employed. Since the bulk of data on population size and per capita Gross Domestic Product (PCGDP) for the sixteen urban agglomerations outside the US comes from UN sources, the PCGDP for *market* economies is used to derive the per capita income ratio. Evidence suggests that the proximity of the PCGDP for market economies to the PCGNP for nations worldwide, as reported by the Population Reference Bureau's *Annual Data Sheets*, supplies results not markedly different than if PCGNP for the world were exclusively used.

In Figure 3.2 the dynamic paths of the world's nineteen largest urban agglomerations are recorded for the period 1958–80, for the years that data are available. The urban level aggregate ecological macrodynamics are recorded over the two central variables: relative urban population ratio to the world's current total, x, shown in 10^4; and per capita product to the world's (market economies) current average y, shown tenfold in the vertical axis, in Figure 3.2(a), and its logarithmic (of the tenfold) transformation in Figure 3.2(b). Lack of data for urban per capita income for non-US urban agglomerations forces the use of corresponding national averages. Evidence from the US urban areas seems to justify their use; for US MSAs, variation from the US nationally prevailing average was not pronounced in view of the overall variation in national per capita incomes. For some additional information on Shanghai and Beijing see Appendix 3, p. 304.

A number of qualitative features become immediately apparent. The main finding is that indeed a 'relative developmental threshold' is present in the form of $\bar{y} = 7.4$ ($\ln \bar{y} = 2$) below which all LDC's urban agglomerations are separated from those of MDCs. It represents a 74 per cent level of the world's current average per capita product. It is not the Malthusian trap which locks in population and income growth rates. The line simply identifies a threshold in relative income level which LCDs have not been able to exceed in the last twenty-five years. In fact, as will be seen later, this threshold is a window for growth separating different dynamic paths. More specifically, the following observations can be made. On the relative population size (x):

1. *All* urban areas with a ratio y below 7.4 (or, in logarithmic scale approximately 2.0) show their relative population size to increase ($\dot{x} > 0$) in the study period;
2. *All* urban areas with a ratio y higher than 7.4 (or $\ln y = 2.0$) decline in their relative share of the world's population ($\dot{x} < 0$);
3. The speed of growth/decline in relative population size for LDC/MDC urban centers varies: it is higher for urban areas belonging to the LDC category than the speed of decline in the share of population for urban agglomerations belonging to MDCs. Further,

the speed seems to vary significantly at times among LDC urban areas;

4 The vertical lines in Figures 3.2(a) and (b) indicate the location of a hypothetical urban area with a 5 million population at steady state (in 1950, 1960, 1970 and 1980). In this period the 5 million spectral line seems to be moving to the left at a geometrically declining magnitude of x. Its leftward motion indicates the absolute growth rate of the world's total population.

As indicated in 4 above, the movement of the 5 million population line to the left at declining magnitudes of x clearly demonstrates the presence of the first fundamental force, the Malthusian biological (B) force at work. It indicates that in the 1950–87 time-period, (see data on the world's population in Appendix 3, pp. 321–4) there is exponential growth in the world's population, expanding at a rate of approximately 2.08 percent over the 37-year period. This is a higher rate than the 1.7 percent estimated by the Population Reference Bureau for the 1986/7 year, and slightly higher than its 1970 rate of 2.0.[13]

The first two observations (1, 2, above) empirically confirm the second fundamental force, the economic–demographic (E–D) force as discussed in Chapter 1, pp. 29–33. They also provide for a clear-cut threshold one can use to partition LDC and MDC urban dynamics: a per capita income higher than the 74 per cent of the average prevailing in market economies seems to distinctly separate different developmental dynamics. The fundamental difference between MDC and LDC metropolitan centers seems to be the behavior of their relative population size, declining in the former and increasing in the latter during the study period.

On the per capita ratio (y) one can observe the following:

1 In its algebraic form it indicates that eleven urban agglomerations (and the vast majority of the LDCs' metropolitan areas) are contained within a very narrow band ($0 \leq y \leq 7.4$); whereas, in a much larger space ($7.4 \leq y \leq 55.0$) eight urban areas and the vast majority of the MDCs' metropolitan areas lie;

2 No metropolitan area, LDC or MDC type (with the exception of Teheran, which will be discussed later), has crossed the $\bar{y} = 7.4$ threshold in the 25-year period;

3 In both bands, there are urban centers with rising and/or declining y;

4 In the MDC urban areas' band, the *higher* the y the more likely it is

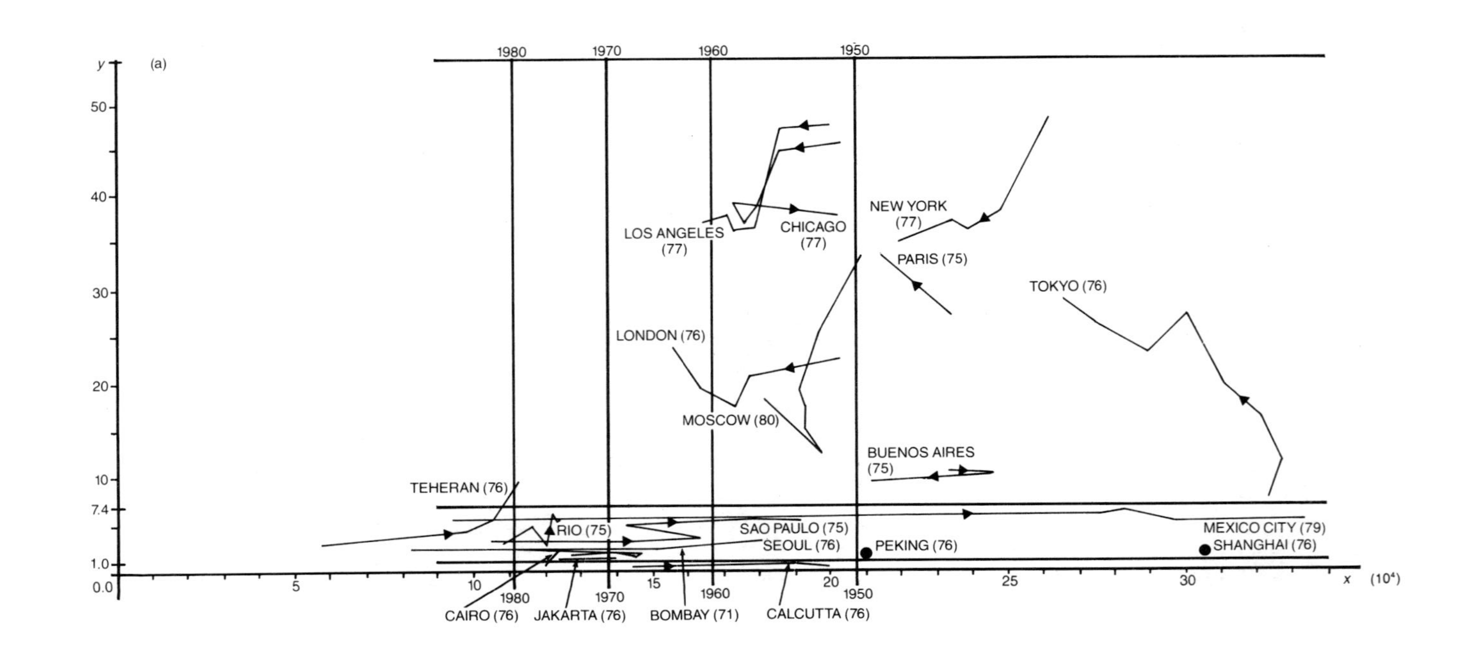
(a)
y
50
40
30
20
10
7.4
1.0
0.0
5
10
15
20
25
30
x
(10⁴)
1980
1970
1960
1950
LOS ANGELES (77)
CHICAGO (77)
NEW YORK (77)
PARIS (75)
TOKYO (76)
LONDON (76)
MOSCOW (80)
BUENOS AIRES (75)
TEHERAN (76)
RIO (75)
SAO PAULO (75)
SEOUL (76)
PEKING (76)
MEXICO CITY (79)
SHANGHAI (76)
CAIRO (76)
JAKARTA (76)
BOMBAY (71)
CALCUTTA (76)

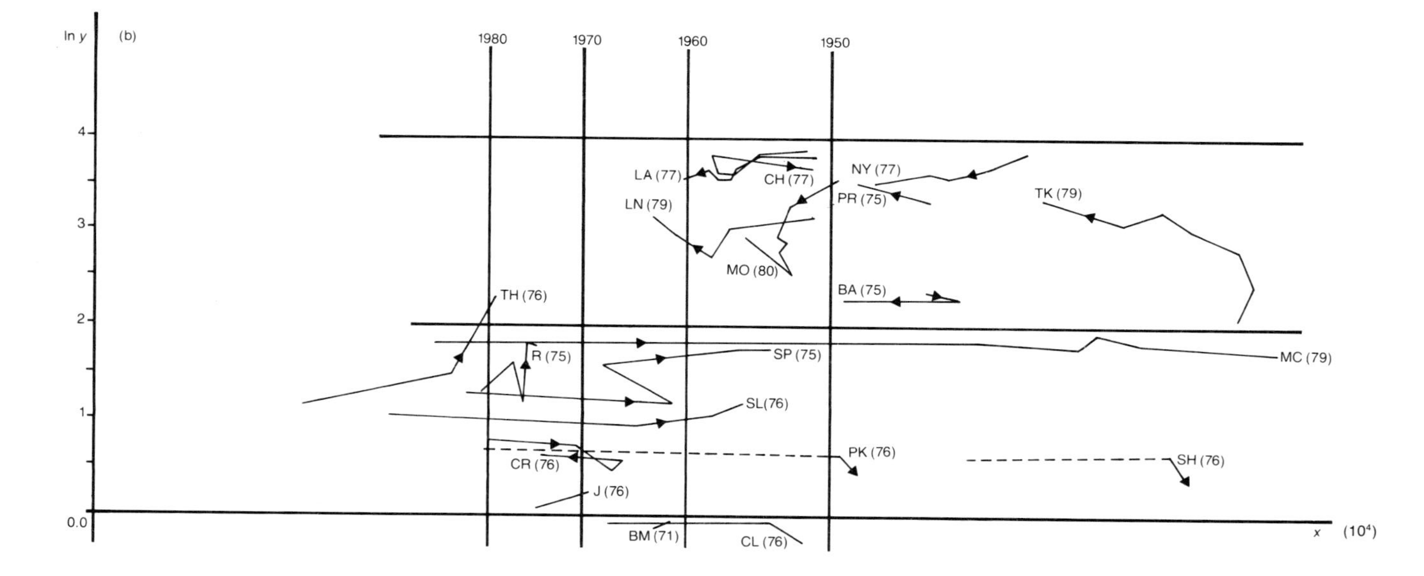

Figure 3.2 Urban macrodynamics and world population growth: (a) contains the relative population size x, plotted against the current ratio of national per capita product y to the world's average; (b) shows the natural logarithm of y. The 5 million mark in population size is shown for the four decennial years by the vertical lines

that y declines over time; in the LDC urban centers' band, the *lower* the y the more likely it is that it declines over time;

5 The speed of motion in y (when y is expressed in actual terms) for MDC urban centers vastly exceeds that of y in LDC urban areas; in the logarithmic scale, the two speeds seem to be about equal, as the space spanned by MDC and LDC metropolitan agglomerations are approximately equal.

In combination, observations on the per capita product and population paths strongly indicate that the urban relative macro-dynamics studied are subject to instability. There is a particular point in the diagram of Figure 3.2 where a combination of urban areas converge toward competitive exclusion, i.e., where x is approximately zero. There is also a part in the phase portrait where the population share becomes extremely large. However, both points are temporally and spatially (in the x and y axis, respectively) remote.

Although the data limitations on per capita product for these agglomerations make some of these observations somewhat vulnerable – more so than those associated with relative population dynamics – the fact that relative dynamics are analyzed rather than absolute, softens the strength of potential data related shortcomings. Clearly, the perennial caveat applies: continuous empirical studies are necessary to further strengthen these findings as more data become available.

The code

Combining all the qualitative features of the phase portrait, an *aggregate developmental code* (ADC) emerges. The code gives rise to the individual dynamics of the nineteen agglomerations considered. It also suggests the possible location of the macrodynamics of other urban areas not recorded in Figure 3.2. In contrast to simple Malthusian dynamics, depicted by the Malthusian trap of Figure 1.2, the ADC allows for various motions to unfold in the dynamics of metropolitan areas without tying growth rates to current levels in relative population and per capita income. It points to the different qualitative dynamics associated with LDC and MDC type urban agglomerations.

This hypothesis contains two interdependencies: one directly links the two macrovariables (per capita product and population); the other attaches each metropolitan area's macrodynamics to its (collective) environment through the use of the relative lens. Use of the code renders the modeling of detailed interactions among the world's largest urban agglomerations unnecessary. Data available allow one to directly but qualitatively test for the existence of the code through each urban area's macrodynamics within the global environment.

According to the urban macrodynamics ADC code, each metropolitan area's *mean* dynamic path in Figure 3.2(b) (and through appropriate transformations, Figure 3.2(a)) can be depicted by the following set of simultaneous ordinary differential equations:

$$\dot{x}_i = \alpha_i(\beta - w_i)\, x_i$$
$$\dot{w}_i = \gamma_i(aw_i^2 + bw_i - x_i)$$

where: α_i and γ_i are the speeds of adjustment parameters for relative population and per capita income respectively; and $w_i = \ln y_i$, $\beta = 2$, a $= -10$, b $= 40$, $x_i = 10^4(X_i/X)$, where X_i is the individual urban area's absolute population at time t; X is the world's current total population. The maximum of x along the second differential equation's isocline is at 0.004, i.e., equal to the $[b/a]$ ratio in units of relative population share. The interpretation of this ratio is informative, as will be seen a little later. It is noted that a, b, and β are not subscripted by i, being independent of the individual urban paths. *Having set the values for these parameters*, one can estimate the speeds of adjustment for each setting using differential equation simulations. The dynamics of each urban setting depends upon the eigenvalues of the characteristic polynomial of the dynamic system.

All these parameter values are obtained through calibration by numerical simulations, using both a Runge–Kutta and an Adams type integrator. The dynamics of the two differential equations stated above, and the qualitative features in the (x,y) phase portrait, are shown in Figure 3.3 for the nineteen urban agglomerations. Modeling the average (expected) paths of each city's macrodynamics with differential equations takes away the possibility of replicating their capped (within bounds, that is) chaotic motion. To capture this type of a motion one must utilize difference equations. The form of these two kinetic equations exactly identifies the effect of the two fundamental forces at work.

Looking at the form of the differential equations, one finds a number of quite informative features. Exponentially fast relative population growth is depicted by the first differential equation. It is of first degree in x_i. Contrary to Malthusian growth, in this case the relative population growth rate is not constant in the long run as the term inside the parenthesis undergoes fast dynamical change. The first equation's isocline is linear and horizontal. The line identifies the per capita income

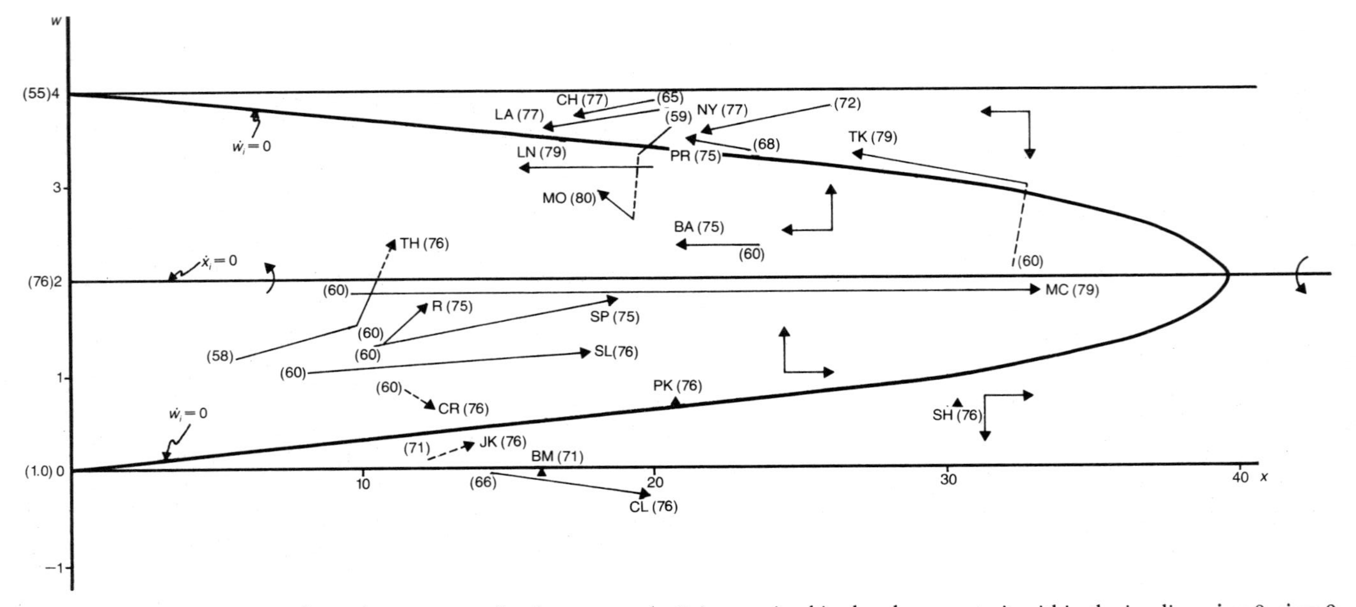

Figure 3.3 The urban macrodynamics aggregate development code. It is contained in the phase portrait within the isoclines $\dot{x} = 0$, $\dot{w} = 0$ and defined by them

required to halt population growth at a specific relative population share level. This implies that for the case at hand, and within the time horizon considered (the third quarter of the twentieth century), that level was independent of relative population size. Motion ceases in the first differential equation, that is population stops growing, when the term inside the parenthesis becomes zero or the population size itself is zero. This is the first dynamic equilibrium condition for the system at hand.

Equation $\dot{w}_i$ is of second degree in w_i, without implying exponentially fast growth in the per capita income ratio. Lack of exponential per capita income growth indicates a slow motion in relative wealth size. Motion ceases in the second differential equation, that is per capita income stops changing, when the term inside the parenthesis becomes zero. There are scale effects in per capita income growth. They are picked up by the second degree equation inside the parenthesis. Every point on this isocline identifies the relative population size which would stop per capita income growth at a particular per capita income level.

The isocline of the second differential equation implies that at the start-up of the development process, per capita income takes progressively higher relative population sizes to retard it as it grows. Beyond a point, congestion effects settle in and developmental bottlenecks commence: increasingly higher per capita income levels need increasingly lower relative population sizes to stop their growth. In reference to this condition, the interpretation of the $[b/a]$ ratio becomes informative: it represents the maximum feasible level of population able to affect changes in per capita income, the maximum quantity of relative population size that matters not making any difference at all on per capita income. This level is estimated at present to be about 0.004 and close to Mexico City's relative population in reference to the world's population level. It is also noted that at present the isocline of the first differential equation coincides with the line going through the maximum point of the second isocline. This might be simply a historical coincidence or due to fuzziness in the estimation procedure.

Negative external effects of agglomeration due to relative population size are depicted by the negative coefficient of x_i in the $\dot{w}_i$ equation. Negative external effects of per capita income scale upon per capita income growth are picked up by the second degree term in w in the second differential equation. Weak positive effects of scale are depicted by the positive first degree in w term in the same equation. It is noted that these two differential equation forms are only found to hold for the particular time horizon examined. How they may be changing in time is a subject addressed in Chapter 4.

In this ecological macrodynamic framework, the first differential equation contains the modified B force, whereas the second differential equation identifies the modified E–D force at this level of analysis and at this particular time horizon. The individual city's overall *system dynamic equilibrium* is reached when changes in both state variables (x,w) are zero, that is the right-hand side of both differential equations become zero simultaneously. This condition does not necessarily imply that the dynamic equilibrium state(s) is (are) stable. Indeed, they are shown not to be stable further below.

The first differential equation can be viewed as a relative *demand for* worldwide relative population by urban setting i. Parameter β is the *carrying capacity in terms of per capita product* for urban area i. If this carrying capacity is not exceeded, then relative population rises in city i, whereas it declines if this level is surpassed. Evidence indicates that it has been surpassed by large urban agglomerations in MDCs, and that it has still not been exceeded by the giant urban areas of LDCs.

Isocline $\dot{x}_i = 0 \rightarrow w_i^* = \beta$, i.e., the horizontal line of Figures 3.2 and 3.3, indicates that motion in terms of population growth ceases when the carrying capacity in terms of per capita income limit has been reached at a particular relative population size level. The horizontal isocline implies perfectly elastic demand for relative population by urban area *i at relative population size equilibrium*. Were the isocline to be downwardly sloping, then it would have had the form of a conventional *demand* curve for population: as income rises, less relative population size is demanded. An upwardly sloping isocline, on the other hand, can be viewed as an unconventional demand for relative population curve (a Geffen type demand curve) as higher per capita income levels would commend higher population concentrations.

A relative *supply of* population is depicted by the second differential equation. Negative effects of concentration upon per capita income are reflected in the non-linear isocline $[\dot{w}_i = 0 \rightarrow x_i^* = (aw_i^* + b)\, w_i^*]$, the second degree line of Figure 3.3. The form of a *carrying capacity in terms of relative population size* for urban area i is more complicated than that of a carrying capacity in terms of per capita product. Its form indicates that *at per capita income equilibrium* relative population size is directly related to the current per capita product (income) ratio $(bw_i^*, b > 0)$, and negatively related to the scale effects created by the current per capita product $(aw_i^{*2}, a < 0)$.

To explore the dynamic behavior of the system and its *system dynamic equilibrium* conditions under these two carrying capacities one needs look at its phase diagram (see Figure 3.3). The phase portrait of the system, i.e., the ADC, is defined by the two isoclines (the conditions where $\dot{x} = 0$, $\dot{y} = 0$):

$$\dot{x} = 0 \rightarrow \beta = \ln y_i^* = 2 = \bar{w}$$

$$\frac{d \ln y}{dt} = 0 \rightarrow 10\,(\ln y_i^*)(4 - \ln y_i^*) = x_i^*.$$

In the second degree isocline, $x_i^* = f(\ln y_i^*) = f(w_i^*)$, the part above the $\bar{w} = 2$ line corresponds to a stable quasi-growth ray (see Figure 3.3). Urban areas with a w greater than $\bar{w}$ contain paths with motion converging toward this ray, since near that ray relative population decreases, whereas relative income increases: $\dot{x} < 0$, $\dot{y} > 0$. Thus, this part of the bent ray is an attractor, whereas the part of the second degree isocline below the $\bar{w} = 2$ line corresponds to an unstable (diverging from) growth ray. This part of the bent ray is a repeller: it repels upwardly these urban areas above it, whereas it repels downwardly the paths of those urban regions below it.

Urban areas inside the arch incur both relative income (product) and population growth ($\dot{x} > 0$, $\dot{y} > 0$); those outside the arch incur relative wealth (product) decline and relative population growth ($\dot{x} > 0$, $\dot{y} < 0$). On the arch the motion is along any isocline ($\dot{x}$ and $\dot{y}$) direction applicable at the specific area of the phase portrait. Consequently, near this unstable part of the ray the end state is very sensitive to the location of the starting position. Slight perturbations in population or per capita income for cities close to this section of the second degree isocline may have considerable implications upon their future developmental paths. Whereas, at locations far away from this isocline fluctuations in starting values have little effect.

Thus, the phase portrait has two sections ($\ln y \geq 2$, $\ln y < 2$) which depict distinctly different dynamics and contain MDC and LDC urban areas respectively. There is a window inside the arch which allows for these urban agglomerations currently located above the unstable part of the growth ray to increase their per capita wealth while in a first phase their population share increases. The window is found in the part of the linear isocline inside the arch. This state might be associated with recently industrialized nations as the case of Seoul hints at in Figure 3.3. In a subsequent phase, their population share declines.

Another type of window is located in the part of the linear isocline outside the arch. The window initially causes a decline in both relative population and per capita income to relatively well-off urban agglomerations. Once crossed, it jettisons them to a subsequent phase of rapidly declining per capita income and growing population. This is potentially a most slippery state, a trap which a number of urban agglomerations have started to experience according to the evidence available. Shanghai, Jakarta, Calcutta, Bombay, Mexico City and Beijing are rapidly approaching it.

Under the two differential equations suggested here there are three dynamic equilibria possible. They are located at points of intersection. One is found at the intersection of the two isoclines; it is a quasi-saddle-type equilibrium point sending trajectories upwardly to its left and downwardly to its right. Urban dynamic paths initially converge and then diverge in their motion from it. The other two dynamic equilibria are found at the two intersections of the w-axis with the second degree isocline. The tip of the second degree isocline, where x_i^* is maximum, is thus a quasi attractor.

Also unstable is the lower level equilibrium at the intersection of the second degree isocline with the w-axis, whereas the upper level intersection point is a stable equilibrium. This point is a strong attractor of trajectories in the phase portraits shown in Figures 3.4(a) and (b). It represents a state of small population size cities with relatively high per capita incomes. Trajectories toward this point may be construed as an 'optimistic' future state of affairs in global urban development, although in the long run it may be catastrophic as trajectories eventually lead to competitive exclusion and death for all urban areas attracted by it.

At the lower right-hand side of the phase portrait looms a large area of instability, a field of motion where paths move toward states of progressively lower per capita incomes and increasing relative population sizes. Trajectories in this section of the field point to a potentially 'pessimistic' future state of affairs in urban development. The outlook for a pronounced dualism seems to emerge from the implicit dynamics of the code.

Empirical evidence for each metropolitan area analyzed, with the possible exception of Jakarta and Cairo, seems to confirm the existence and validity of the broad qualitative features of the ADC. All LDC cities inside the arch (Mexico City, São Paulo, Seoul, Rio de Janeiro, Teheran) show growth in both x and y, whereas those outside the arch and below the $\bar{y} = 7.4$ threshold (Shanghai, Beijing, Bombay, Calcutta) show an increase in x and a decline in y. On the other hand, all MDC urban agglomerations (the three US MSAs and Paris) outside the arch (with the possible exception of Tokyo) show decline in both x and y. London and Moscow show decline when outside the arch but inside the arch exhibit growth in y, as expected by their location in the phase portrait.

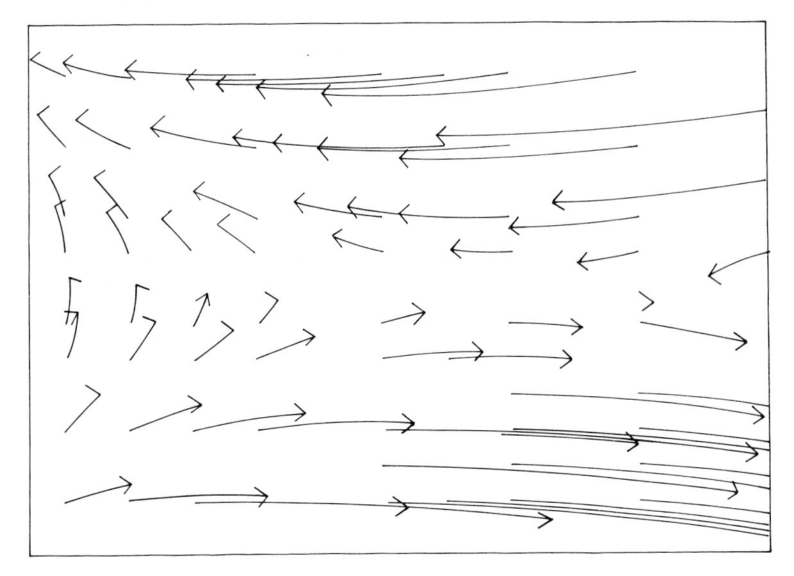

Figure 3.4(a) The phase portrait and field of motion in urban macrodynamics: case when the speed is $\alpha = 0.05$; the horizontal axis (x) varies from 1 to 60, while the vertical axis (w) varies from 0.1 to 4

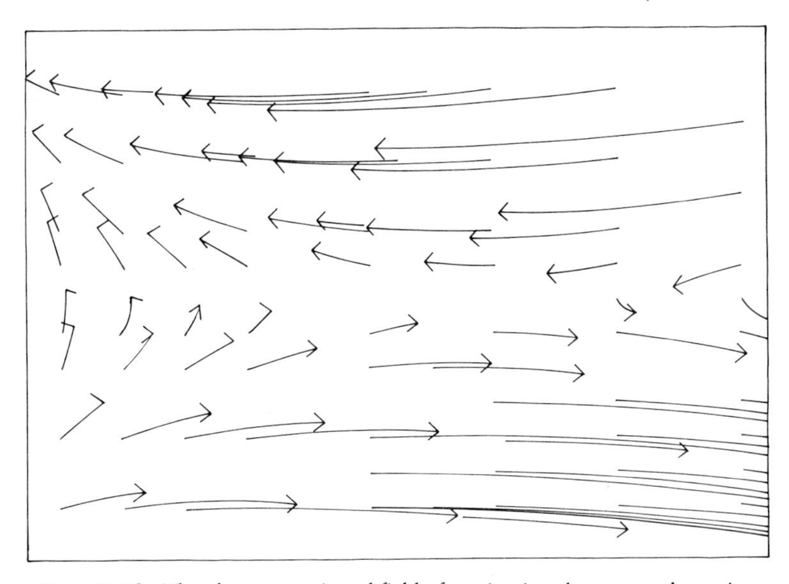

Figure 3.4(b) The phase portrait and field of motion in urban macrodynamics: case when the speed is $\alpha = 0.06$; the horizontal axis (x) varies from 1 to 60, while the vertical axis (w) varies from 0.1 to 4

Tokyo's signature may be very close to the upper (stable) part of the arch. All MDC metropolitan areas show declines in x, as expected by the field dynamics at this part of the phase portrait.

The existence of the code is further confirmed when the case of Teheran is closely examined. Around 1975, Teheran crossed the $\dot{x}_i = 0$ isocline, where $\bar{w}_i = \beta = 2$, indicating an extraordinary speed of development along the per capita income variable. Subsequent events in Iran resulted in a drastic change. The event clearly demonstrates that there is a safe speed for development in both LDC and MDC settings, and attempts to exceed such speeds and use the window of growth in per capita income by crossing rather steeply the $\bar{y} = 7.4$ isocline not only do not succeed in bringing about evolutionary change, but instead may result in significant social disruption. There are no UN data available to trace the post-1976 Teheran path, but it is rather safe at present to assume that it moved south-easterly, and below the $\dot{x} = 0$ isocline.

The maximum in relative population size (x) of the second isocline, is estimated to be at about $\bar{x} = 0.004$. It is likely that Mexico City may reach the boundary by the mid to late 1980s and if so it is expected to follow a drastic decline in y thereafter, unless it crosses the $\ln \bar{y} = 2$ isocline first.

Each metropolitan area has a speed of movement inside the phase portrait. Preliminary experiments with speeds seem to point to a much higher speed for the population kinetic equation than for the income kinetic equation, as $\alpha \gg \gamma$. That the speed of adjustment for population far exceeds that for per capita income must imply that *it is easier to have changes in population than changes in wealth*. In the computer simulation case indicating the field of motion in the phase portraits shown in Figures 3.4(a) and (b) the speeds were: Figure 3.4(a) $\alpha = 0.05$, $\gamma = 0.0005$; Figure 3.4(b) $\alpha = 0.06$, $\gamma = 0.0005$. Dynamics are severely impacted by changes in these speeds, which are assumed to be very slowly altered in time. The effects of such changes upon the form of the field are evident when the speed for population adjustment is increased by 0.01: the slippery section of the phase portrait is reached faster, whereas the motion toward the upper level equilibrium is not as fast.

The window found in the relative developmental threshold may be used by certain urban agglomerations to escape the income depression. If the appropriate development speeds are applied, not being either too high (Teheran) or too low (Mexico City), then the selected LDC urban centers inside the arc and close enough to the dividing isocline (where $\bar{y} = 7.4$) may eventually escape and approach (in per capita income) those of MDCs. Finally the presence of Seoul and the two Brazilian urban agglomerations inside the arc below the $\bar{y} = 7.4$ line area might indicate the location of areas attracted by this window of development.

Promise for the evolution of urban agglomerations on the globe currently belonging to LDCs could be found in this window. The speed of motion inside this area, and motion toward or away from it, may prove informative in the evolutionary process, collectively and individually. As it now stands the two Brazilian agglomerations plus Seoul and Beijing are much more likely to undergo evolutionary change, i.e., shift areas in the phase portrait, than any other city among the nineteen presented here. Thus, by using relative macrodynamics a much richer picture is painted than that which simple Malthusian trap dynamics would imply. The aggregate development code contains a variety of possible paths metropolitan agglomerations might follow, all critically dependent upon their current location in the phase portrait, a blueprint for evolution.

Now one is poised to ask some more basic questions. Where does this ADC come from? Why does it seem to have such a small number of variables? Why does it have such simple outcomes? How durable is the ADC's form as shown in Figures 3.3 and 3.4(a) and (b) likely to be? What does it imply for policy making? Shifting from the technical discussion of its performance to the broader issues involved in some of these questions, a few qualitative points are made next. By responding to the first question the analysis then takes a closer look at another major thesis of this work, namely the relative parity principle argued as the component giving rise to the code.

The existence of this code over the study period leads one to the following two conclusions: first, the possibly enormous number of variables involved in and defining urban development – viewed either in a partial (sociological, economic, demographic, environmental, political, historical, anthropological, cultural, psychological, etc.) or general (ecological) framework, or at a very fine level of spatial disaggregation – have acted in such a manner that persistent non-random patterns have been observed and recorded for the two central urban macrovariables in aggregate. Second, that regions, nations, cities, individuals and their multiple social, economic, political and other institutions in these macro socio-spatial units have behaved collectively (interdependently) and individually in a dynamic complex game theoretic manner with outcomes (cumulatively) consistent over time with the postulated code of development, which involves a basic dichotomy between MDCs and LDCs.

Put differently, a process of selection at the macroscale is present in spatial macrodynamics worldwide. It operates in such a manner that it ties the multiplicity of nations and their inhabitants in an interrelated and vast array of variables and a complex choice of social (individual and collective) actions, among a very large number of possible (feasible) alternatives. This complex behavior gives rise to an ADC which in turn governs the relative development of each spatial unit (region, nation, city), its urban macrodynamic fingerprints given by Figure 3.3. No model is quite intelligent enough to replicate this soup of interaction among all agents of all spatial units over the long horizon and then generate macrodynamics from the bottom up.

In biology the governing principle in selection is individual survival. In spatial macrodynamics the principle may be more complex, involving evolution of a collective behavioral pattern in conformance with the ADC. In this case, the unstable dynamics of the individual urban agglomerations must be consistent with the unstable dynamics of the aggregate, i.e., the collective behavior giving rise to the code. This collective dynamic code can be thought of as having successfully competed in time with other macrocodes (or its mutations at any point in time) and that it has been sustained over the study period. In accepting this hypothesis it is not necessary to also accept that this choice has been made consciously and collectively by all agents involved since the starting point of the time horizon considered. It has not.

No collectively conscious optimization principles can be established or identified as having been employed to discriminate among many (presumably feasible) codes. No evidence exists to imply that other competing (distinctly different or mutant) codes have been detected within the quarter century horizon. If there have been or currently are other codes competing with the one depicted here by numerical evidence this is not at all clear. It is of interest to speculate that a number of neighboring, and close in dynamical qualities, codes can be traced, possibly coexisting in a quantum-type state.[14] What is not evident at all, given the analysis presented, is the contention that distant codes, which are different in dynamical qualities (possibly components of a master supercode allowing for lower level distinctly different codes to coexist) have been or currently are present in global urban macrodynamics. Neither is it clear what form these codes or their quantum-type state might have assumed.

Growth is not necessarily related to optimal adjustment, nor is decline in population abundance or income necessarily related to suboptimal adjustment in social (individual and collective) behavior with respect to the ADC. Instead, the ADC dictates collectively what individual spatial units' trace will be at any point in time at their current location in the phase portrait. Normative aspects are absent. Individual urban areas adjust deterministically to this collectively prevailing dynamic pattern. The existence of this ADC allows one to speculate as to the processes underlying socio-spatial dynamics, along the lines of Chapter 1 and Appendix 1. It also points to the limitations of individual urban areas' (or nations) actions in terms of policy and decision making trying to affect such long-term, large-scale forces at work. The dynamic path traced by each urban area is much more strongly influenced by the ADC than by any public or individual action.

Only a snapshot of history (twenty-five years) has been looked at. But the test was taken at a presumably random point in the long history of urban evolution, if one assumes that our current position is not a privileged one compared with past (or future) positions. If so, then this type of ADC must have characterized past urban settings as well, and is likely to characterize future spatial stocks too. No significant perturbations of urban relative dynamic paths were recorded over the past quarter century at that scale of analysis, in spite of the limited chaotic motions traced. One must then presume that no drastic changes are likely to occur, in the next quarter century period at least, to affect in any significant degree these relative dynamics (i.e., effectuate evolutionary rather than developmental change) at that scale.

The likelihood for evolutionary change, a motion which might guide urban areas to either exit the developmental threshold passing through the $\bar{y} = 7.4$ window or to dip below/cross above the unstable growth bent ray (the second degree isocline), is higher for urban areas close to these critical areas of the phase portrait than for those far away from the two isoclines. Accordingly, the socio-economic effort needed to cross this threshold must be inversely related to distance from this line.

ADC also seems to suggest that from the many candidates of LDC urban agglomerations likely to incur such evolutionary change a few are likely to come through in the short run. Distance from the line $\bar{y} = 7.4$ may not be the only factor but is currently a good indicator as to which urban agglomeration is likely to undergo such evolutionary change. Such a specific expectation coming out of the model might be a very informative qualitative future test of the model's validity.

Macrodynamics of metropolitan areas following this code are 'fast.' Changes in the constants – parameters constituting the code – are 'slow.' Inferences on slow dynamics cannot be made at present given the existing database, although in Chapter 4 some speculative scenarios are analyzed. Next, an attempt toward establishing a worldwide urban interdependence is made and the 'relative parity' concept is discussed as a possible source for the ADC.

Relative parity

The reader has now reached the critical point in the book, where the comprehensive linkages in the dynamics of the urban (and rural) settings of the globe are precisely stated. To follow this subsection the reader must clearly distinguish between a participant in and an observer of social dynamics. Their functions and roles are different, as will be more graphically shown in Chapter 4. Such differences extend over the stocks they observe and the lens they use to observe them, the constraints, objectives, and decision-making rules they perceive and obey, as well as the degree of complexity found in these components by both. Ultimately, their roles differ by the extent to which their perceptions and actions affect the performance of the social system at hand. By elaborating on their behavior one can examine the linkages between the disaggregate (micro) and the aggregate (macro) socio-spatial worlds and their *modus operandi.*

The individual participant's behavior in space–time

At any point in time, an urban area i is perceived by a specific participant (a perspective or current resident, or by any collective) j as being in a state $s_{ij}(t)[K_{ijm}(t);\ m \in M_j;\ t \in T]$, where the state s_{ij} is a function of the currently perceived and expressed in *mostly absolute terms* current and future level of a bundle of attributes or stocks (the quantities in $\boldsymbol{K}$). Examples of the diverse variables included in the vector $\boldsymbol{K}$ are: other individuals' state; perceived total population size of the community a family resides in or of the urban area an individual inhabits; individual or perceived average per capita product (or income); quantity and quality of the built capital stock; local and urbanwide air pollution levels; topographical, cultural, and social amenities and activities enjoyed locally; political barriers to inter- and intra-urban movement, as well as possibility for social mobility, etc. These variables, designated by the subscript m, are found in an individual agent-specific bundle M_j.

The above attributes are but a few among the approximately infinite number of variables (monetary and non-monetary, quantifiable and non-quantifiable, material and non-material, complementary, competitive, or substitutory) which in an aggregate or disaggregated manner constitute the vector $\boldsymbol{K}$. They affect in a positive or negative way, and define, the currently perceived present and future state s of urban area i, according to actor j. *These individual or collective perceptions are not necessarily identical or congruent* among individual agents.

Due to their extraordinary complexity, the legal barriers, and other reasons (the degree of fuzziness being one of them), someone outside these participant individual agents cannot know to any significant degree of accuracy any of these functions $\boldsymbol{s}$ and their constituent variables $\boldsymbol{K}$. Some of these components are not even exactly known to the individuals themselves. An observer can only indirectly infer a very limited number of components in such functions, and obtain at best partial and relatively dated information about such states and their variables. At worst an observer can collect erroneous and useless data, possibly only good for very restrictive purposes and time horizons.

These states $\boldsymbol{s}$, none the less, convey a level of utility or composite profit to the corresponding agents: $U_{ij}(t) = f_{ij}[s_{ij}(t)]$ where the notion of utility is here employed in a much broader sense than in microeconomics. Indices $\boldsymbol{U}$ are non-observerables, and only perceived by individuals in a manner largely still unclear in the field of psychology. Further, these functions $\boldsymbol{U}$ need not be continuous (smooth), strictly convex, and well-behaved; indeed, they rarely are, contrary to simplistic neoclassical economic assumptions of consumer behavior. Inherent in it is a set of constraints perceived by the agents. As discussed in Chapter 1, pp. 34–7, and in Appendix 1, these present and future perceived states, utility functions, and constraints are interrelated through, among other things, social speculative behavior. Social, individual and collective action ensues from such functions $\boldsymbol{U}$, and the states $\boldsymbol{s}$. By pursuing a location of maximum $\boldsymbol{U}$, individuals are thought of as agents in pursuit of composite profit or utility, always and everywhere.

The net currently discounted stream of future gains to current or prospective residents of an urban area i includes actions anticipated by agent j, by producers, consumers, markets and governments. Since all these elements are included in the quantity $U_{ij}(\boldsymbol{K})$, the notion is significantly more encompassing than the concept of utility employed in the spatial economic literature.[15] Traditionally, individual consumers and producers are assumed to be either identical, or different types (groups) of agents as assumed in standard microeconomic analysis where each group is represented by a typical agent (a trivial extension of the first case).

In contrast, individual or collective agents are assumed in this study to be heterogeneous speculators with different time horizons and discounting methods. Consequently, they do not act in unison on what is currently occurring (from the point of view of an observer) in an urban area i, but rather on what they perceive the state to be, given their own important bundle of factors $\boldsymbol{K}$, over their own relevant time horizons, all appropriately discounted at its individual present net worth.

Individual locational equilibrium

Individuals and collectives act by altering and responding to factors $\boldsymbol{K}$ as a matter of course and perceptions. Analytically, at the level of an individual, a decision to remain at location i implies that for this individual city i is the perceived best location to be at present. The specific individual perceives to be at least as well off in urban area of residence i as in any other location considered by the agent over a time horizon T_j, an horizon dependent upon the individual j:

$$U_{ij}(t) \geq U_{h \neq i,j}(t) - TC_{ih} \qquad i, h = 1, 2, \ldots, I$$

$$U_{ij}(t) = \int_{t}^{T_i} u_{ij}(\tau)\, e^{-\lambda_j \tau} d\tau$$

where $u_{ij}(t)$ are individually expected levels of utility at each time-period from time t to time horizon T_j, TC_{ih} is a perceived transaction cost (transportation cost being one component) in moving from city i to h not included in the bundle of elements found in $\boldsymbol{U}$, and λ_j is a rate of discount for agent j. Observers must expect that this equilibrium condition may or may not always hold, however, since an individual always has partial, erroneous, or dated information. But this is only a conjecture on the part of the observer, and not relevant for an individual participant. Contrary to standard economic analysis, the above perceived individual equilibrium conditions can only be stated *but not derived* from the original formulation, either by the individual or by an outside observer.

If for two urban settings i and h, for the same individual j, $U_{ij}(t) = U_{hj}(t)$ then this must not necessarily imply that the levels of corresponding stocks perceived by the social agent contained in both locations must be equal. Given the substitution effects embedded in $\boldsymbol{U}$, there must be significant perceived negative differences in some variable(s) between the two settings for an agent to compensate with positive differentials in other stocks. For example, a per capita perceived high economic commodities consumption-based utility level, in an urban setting belonging to an MDC relative to an urban area belonging to a tropical setting LDC, may be counterbalanced by a perceived relatively high per capita environmental-amenities level. Whether, at the margin (and in accordance with marginal utility analysis from microeconomics) for an individual j at the equilibrium location, all factors $\boldsymbol{K}$ in U have equal marginal utility is an open question in view of the discrete nature of locations and non-convexity in the U functions.

It is recalled that U is a comprehensive quantity in which current and future preferences and constraints (in the form of penalty functions) and expected (as perceived in multiple forms) mobility costs are incorporated and appropriately traded off and discounted at present. Individual locational equilibrium conditions might hold even when one is prohibited to migrate, i.e., when mobility (spatial transaction) costs are infinitely high or one possesses partial or erroneous information. Such conditions apply to all urban areas contained in the locational calculus of the individual. Individual locational equilibrium as stated above does not necessarily require absence of migratory restraints and perfect information, although absence of migratory restraints and perfect information do require individual locational equilibrium.

Aggregate behavior in space–time according to an observer

On the other side of the coin is the observer. To observers the behavior of individuals is untractable and impenetrable. To an observer, removed enough from all social agents and aiming at recording spatial interdependencies and interactions at the most aggregate level, individual and even to an extent collective (group) action goes largely undetected. Like individual drops of water in a vast sea, this plethora of social action is unseen from a distance. But at times these individual actions are not insignificant. A very few, by selective individuals in space–time, may be of such scale as to be perceived by an observer as having had a significant effect in aggregate socio-spatial dynamics. Such participant social (individual or collective) actions could be directly observable and recordable. They could be linked to the role that large fluctuations play in non-linear dynamics upon the evolution of socio-spatial systems by shifting them to qualitatively different dynamic states if such states are present in the phase space (i.e., if they are in the cards at all).

Small-scale non-recordable by an observer individual participant action might affect aggregate non-linear dynamics if a chaotic regime characterizes these dynamics. Under such conditions, a very slight perturbation of the system at any point in space–time might produce a quite different end result than if left unperturbed. The conditions are that either chaotic dynamics are already underway, or that stable dynamics operate in a close neighborhood in the phase space where unstable dynamics lurk. The issue is of interest partly because it seems to supply contradictory evidence with regard to the role that varying initial conditions and perturbations have when strange attractors are present in these dynamics. Although this is not the forum in which to address this issue in a great deal of technical detail, a few notes will be made.

It is now apparent from non-linear dynamical analysis that to have any effect in a phase space small fluctuations have to be extremely timely, directionally selective in space, and of an appropriate magnitude if different basins of attraction are present some including non-periodic movement and must be crossed. Chaos theory and structural stability present us with a very rich picture of dynamic qualities, far removed from the simplistic stability or instability dichotomy often analyzed in conventional dynamics. The role that small fluctuations play in non-linear dynamics can easily by exaggerated. The interesting aspect of such small fluctuations is that they can produce widely diverging dynamic paths whilst at the same time being attracted to or confined by a single strange attractor or container. In this sense, their effects are qualitatively unnoticeable. This is the conflicting result which chaos theory aptly demonstrates regarding fluctuations.

But no matter how complex and inexplicable individual behavior might be, infinite in its variety and minute, aggregate social behavior is relatively simple and describable. It is apparently so even when chaotic dynamics prevail, as their chaos is within bounds and orderly. These qualities offer intellectual legitimacy to an observer's role. Hence, for observers to model a socio-spatial system, other properties and elements than those directly affecting individual behavior must apply and enter their calculus.

Whereas it is mostly absolute quantities which enter individual U functions, it is *mostly relative* magnitudes of selected inter-spatially comparable stocks $\boldsymbol{L}$ which are used by non-participant outside observers, deriving observed (as opposed to perceived by participants) states $S_i(t)[L_{in}(t); n \in N; t = 1, 2, \ldots T]$ of the urban areas in question. These magnitudes determine a total (aggregate or average) composite (ecological) net (positive or negative) attraction (benefit or gain). It is here designated as $A_i(\boldsymbol{L})$, $A_i(t) = F_i[S_i(t)]$ and it includes relative per capita income and population in it. Although relative per capita income is included in it, it should not be considered as an income measure and confused with the behavior of the income component in the ADC. Indeed, the index may rise while relative per capita income may fall. Index $\boldsymbol{A}$ is an comprehensive (ecological) reward for population, not confined to economic reward alone.

This attraction index is a quantity not directly observable by the observer, and is attributable to the social aggregate. It is detected by the observer by its effects in space–time. Whereas M_j in the indices $\boldsymbol{U}$ might be a very large number, N is a small number. A very small number of variables enter, that is, in the calculation of $\boldsymbol{A}$ on the part of an observer.

To a participant very close to the system only absolute levels of variables might matter, or at most relative levels within relatively confined environments. On the other hand, to a remotely located observer local absolute levels of variables may not be as important as relative levels in reference to a broadly observed environment. Those varied perceptions need to be accommodated, and the macrodynamic framework adopted here aims at doing just that.

To an observer the attraction index associated with an urban area i acts as a *valve* regulating the size of the city's relative aggregates $\boldsymbol{L}$ as the result of in or outflows. Although it is *per se* independent of individual or even collective perceptions, it does contain enough information for the observer to model inter-spatial and temporal factor mobility. Hence, both $\boldsymbol{S}$ and $\boldsymbol{A}$ do not directly affect the system's dynamics by influencing individual actions. However, they indirectly describe their relative aggregate dynamic behavior.

Whereas at the level of an individual an effort to maximize composite profit or utility is the motivating force, at the aggregate level, the system is viewed by an observer as being driven by a principle according to which aggregates organize in space–time under an effort toward the attainment of parity in net composite attraction among locations. The principle is partly found in collective (from national governments) pronouncements of 'catching up with more advanced societies.' It can be detected in part also in pronouncements by collective agents striving toward 'maintaining their leading edge which is gradually erroding.'

This is the linkage between the $\boldsymbol{U}$ and the $\boldsymbol{A}$, between the micro and macroscopic worlds. $\boldsymbol{A}$ is not simply the sum of the U. It is a linkage that cannot explicitly be modeled. It is a linkage which allows only its discernible aggregate end effects to be described by an observer at the macroscale, whereas individual behavior in its infinite variety and complexity cannot be modeled.

As noted earlier, within the bundle $\boldsymbol{L}$ there are macrovariables, like population and per capita wealth, contributing to the relative attraction of urban area *i*. The behavior of a vast number of individual and a great deal of collective (government) agents in combination produces aggregates in space–time, aggregates in which individually (and collectively) perceived gains and costs as well as constraints and their individual (and collective) decision-making rules in trade offs are discounted. To some individual agents, certain absolute variables in $\boldsymbol{K}$ may be also found in relative terms in the bundle $\boldsymbol{L}$, although the intersection of these two sets may not be extensive to either bundles. Learning on the part of an individual might imply that a greater number of the variables in $\boldsymbol{L}$ enter the bundle $\boldsymbol{K}$.

Contrary to the notion of utility, and separate from it, is the net attraction count $\boldsymbol{A}$, which is *attributed to the spatial setting and not to the individual resident or collective, who may perceive it quite differently if at all.* It is merely an observer's quantity standing opposed to $\boldsymbol{U}$, in contrast to the way a utility is used in microeconomics where no distinction is made between individual and collective participant agents on the one hand and observers on the other.

There are three different adjustment processes in the dynamics of net relative attraction $\boldsymbol{A}$. Individual and collective *fast* actions, based on perceptions, alter the levels of L_{in} in the short or even the very short term – daily, weekly, or monthly – thus altering the currently prevailing level of $\boldsymbol{A}$ according to an observer. These short-term individual participant motions constitute the first type of dynamics.

To an outside observer, there are relative carrying capacities (also referred to by an observer as 'relative norms' or 'targets') in urban area *i* of stocks $\hat{L}$. Relative norms are attributed to levels of comparative locational advantages and are always perceived by an observer in reference to a particular environment. As the observer changes the reference level (environment) carrying capacities change[16]. In a national frame of reference and within relatively long time horizons the two macrovariables are seemingly driven toward carrying capacities following Volterra–Lotka type dynamics. Within a global environment and under a shorter time horizon (quarter century or so) as envisioned by the ADC, the carrying capacities must be close to the current levels for the macrovariables and change possibly as fast as the macrovariables do. The opposite might be the case for much smaller-scale environments.

In absolute terms, and for certain variables in $\boldsymbol{K}$, there are carrying capacities perceived by individuals as well, representing opportunity vacuums. As a result of individual action, a motion toward these relative norms occurs which is of a medium term, maybe at the annual scale.

Aggregate movements toward relative norms constitute the second type of dynamics. Currently observed levels of spatially liquid factors $\boldsymbol{L}$, adjust in an uninterrupted (continuous or abrupt) manner in each urban area i at any t. Empirical evidence seems to suggest that they are in a motion toward their carrying capacity levels or targets, $\hat{L}_{in}$, in urban area i, as dictated by second in magnitude fast adjustment processes found in the relative parity principle, and local conditions. One can expect the motion to be slow in the movement of any factor in $\boldsymbol{L}$ as the difference $[L_{in} - \hat{L}_{in}]$ is small and to decrease further as the difference decreases; conversely, the movement must have a high speed if the difference is great and to pick up in speed as the difference increases. The observer must expect inflow of factor L_n into city i if the difference is negative, whereas an outflow must be expected if the difference is positive.

In summary, these carrying capacities are the result of perceived endogenous and exogenous constraints to urban agglomeration i and to the possibilities present at time-period t for their growth or decline. Whether or not there is a one-to-one correspondence between a level of observed net gain A_i and any vector $\boldsymbol{L}$ is not of direct importance to the analysis here. Contrary to the critical assumption imposed on the form of utility maps in conventional microeconomics (strict convexity), here non-linear functions in $\boldsymbol{L}$ are presumed to exist. Such non-linearities enrich the insights obtained from the analysis, as multiplicity of stable and unstable equilibria can be present, resulting in interesting and surprising dynamics.

To an outside observer (although such an entity does not exist in its pure form) the net gain or attraction of city i in the medium run is specifically:

$$\hat{A}_i(t) = A_i(\hat{\boldsymbol{L}})\ t = 1, 2, \ldots T$$

where $\hat{A}_i(t)$ is the total (currently discounted from a series of future) composite net attraction under a set of carrying capacities. Put differently, the above index of attraction or net benefits is a translation of the currently prevailing conditions from an observer's standpoint, and of those likely to prevail in the relatively close future at urban setting i. The vector of variables $\hat{\boldsymbol{L}}$ depicts the carrying capacity of the various factors included in the composite gain, in reference to the worldwide or nationally prevailing averages. Relative per capita income and land prices may be two of these factors. Economists assume that within them the effect of a great deal of other factors may have been captured (i.e., capitalized or discounted).

Relative per capita income and relative population were the two central macrovariables of the ADC. With reference to the ADC, one should not necessarily infer from the above discussion on relative parity in net attraction that the metropolitan areas of the MDCs have a higher level of $\boldsymbol{A}$ than those of the LDCs; or that population shares ought to shift toward the metropolitan areas of the MDCs (with a presumably higher $\boldsymbol{A}$). A few points are noted. To a removed observer, recording vast shifts in relative population with gains in metropolitan areas of the LDCs and losses in those of the MDCs, the picture might be different from that obtained by a participant agent. A higher relative per capita income does not necessarily imply a higher level of relative composite attractivity.

As it will be discussed later more extensively, the presence of the relative parity principle does not necessarily imply an elimination of dualisms in the relative spatial distribution of factors $\boldsymbol{L}$. Further, instabilities and disequilibrium in the variables comprising indices $\boldsymbol{A}$ may always prevail in spite of the observer's perception of an effort toward relative parity.

Finally, a third adjustment mechanism is built into this framework: levels of the composite quantity $\hat{\boldsymbol{A}}$ are now assumed by an observer to adjust responding to a worldwide prevailing background relative attraction level $\bar{A}$. The speed of this adjustment process is relatively slow when compared with the previous two changes, as such adjustments extend over the relatively longer haul. In this the third type of dynamics involved according to this model, the changes of $\hat{\boldsymbol{A}}$ and the changes in the carrying capacities of the various factors $\hat{\boldsymbol{L}}$ at specific locations are accommodated.

In summary, this framework both contains and distinguishes between perceptions – heterogeneity of and actions by individual agents on the one hand, and perceptions and modeling by observers on the other. Not everybody in the slums of Cairo thinks as western economists do, or as theorists of efficient financial markets would prefer all investors to behave in order to supply the conditions under which financial markets perform efficiently. Diversity in behavior, purpose, perceptions and beliefs must be accommodated in socio-spatial dynamics. One must go even beyond the anthropologists' aggregate notion of 'cultural relativism' and bring the heterogeneity and differences in perception down to the individual level.

Observer equilibrium

To an observer, the socio-spatial system is at a state in which most (but not all) of the time the majority (but not all) of the participant (individual and collective) social agents are in locational equilibrium: that a *spatial* aggregate equilibrium condition does not prevail worldwide or nationwide anywhere with regard to this net composite attraction; that such a composite parity cannot be reached at least in the short or medium run; that spatial equilibrium $\hat{L}$ does not prevail in each and every component $\boldsymbol{L}$ of the attraction bundle.

It does imply that at most points in space–time there is an effort (made by the social system, individually and collectively at a large enough scale as to be perceived by the observer) toward, in the long term: first, the attainment of carrying capacities, and second an equalization of the net attraction in space. These conditions do not require that individual agents operate spatially in accordance with a drive toward global aggregate locational equilibrium, as they always place their own perceived interests first. In so doing they seem to an observer to only obey a global drive collectively in aggregate toward relative parity in net attraction, i.e., toward an observer's spatial equilibrium. The observer's equilibrium condition in net attraction is then given by:

$$\hat{A}_i(t) = \bar{A}(t), \qquad t = 1, 2, \ldots, T; \, i = 1, 2, \ldots, I$$

$$\hat{L}_{in}(t) = \hat{L}^*_{in}, \qquad n = 1, 2, \ldots, N$$

stating that if spatial equilibrium does exist at any time-period t within a time horizon T, at any location i among all locations I considered, the net gain will equal an average or background level $\bar{A}(t)$ prevailing at that time-period t within the relevant environment of I locations. That this condition holds at equilibrium is a result of the bundle of factors $\boldsymbol{L}$ being in the long run at their long-run equilibrium par levels $\hat{\boldsymbol{L}}^*$. If not, in or outflow of these factors will continue to occur, altering in turn the level of $\boldsymbol{A}$. It is recalled that each component in $\boldsymbol{L}$ is expressed in relative terms, so that flows for an observer are shifts in relative shares.

Instead of asking how high $\bar{A}$ is, one instead must ask how low it is likely to be. $\bar{A}(t)$ may be indeed quite low as it is greatly influenced by conditions in the vast rural settings of the globe where the vast majority of the world's population still resides. It is rather sticky, and if varying (increasing or decreasing) in time these variations must be relatively small, remaining constant in the very short run. At the state of an observer's spatial equilibrium it does not vary in space.

Population level and per capita income are central components in index $\boldsymbol{A}$. Collective effort must be seen, among other approaches, as an attempt by the governments of nations or urban areas to retain perceived ecologically effective population and income levels. This might be the substantive answer to the question posed earlier: where does the ADC come from in view of the two underlying fundamental forces? The analytical aspect of the question will be discussed later.

In contrast to conventional spatial equilibrium analysis, aggregate spatial equilibrium $\bar{A}$ is viewed here as being neither unique nor stable. For an observer the social actions taken as a response to a need for attainment of par levels in the bundle of factors found in $\hat{\mathbf{A}}$, and toward an equalization in attraction process among locations, do not guarantee that a unique stable equilibrium state will ever be attained or maintained or that the path toward any equilibrium is smooth, nodal (without oscillations that is), and continuous. For this reason, the presence of a relative parity principle does not necessarily imply elimination of spatial dualisms, instabilities, or dynamic disequilibrium.

An individual city's composite net attractivity may exhibit turbulent and chaotic dynamics. Actions by agents, based on a variety of perceptions and attempting to respond to perceived opportunities, might produce recordable flows of stocks largely congruent and at times incongruous with an observer's expectations based on the relative parity principle. *To an observer, there are always surprises*, although they may be infrequent. A central component in this incongruity might be found in response time lags among all social agents, a central source of chaotic, discrete, and continuous dynamics.

As their dynamic adjustment process takes them below or above their long run at par (ultimate carrying capacity) levels under the relative parity requirement, possibly non-existent spatio-temporal equilibrium requires that net attraction must be spatially and globally uniform (if current perceptions and long-term reality are congruent) for all agents. Two conditions continuously prevail if dynamic equilibrium has been reached in any urban area i and globally:

$$\sum_n \frac{\partial A_i(t)}{\partial L_{in}} \Bigg|_{\hat{A}_i(t) = \bar{A}(t)} = 0, \quad n = 1, 2, \ldots, N; i = 1, 2, \ldots, I$$

a condition which states that at the margin the sum of all changes in factors must not alter the level attained; and:

$$\frac{d\bar{A}(t)}{dt} = 0, \qquad (T_o \leq t \leq T_s),$$

indicating that the average prevailing attraction level globally and locally remains constant if reached in a neighboring time interval T_s although it may vary in the very long term:

$$\frac{d\bar{A}(t)}{dt} \gtrless 0, \qquad t > T_s$$

possibly slowly changing in time through changes in global socio-economic conditions or environmental shocks (i.e., severe climatic events and/or natural catastrophes of a significant scale). Change in the background level of attraction will also result from significant changes in the various conditions and comparative advantages surrounding major regions, nations or urban agglomerations of the globe. Such changes could be smooth or abrupt, continuous or discrete, sharp or extended in time. Like carrying capacities, this background level of attraction could be a moving target for cities to aim at over the very long haul.

Relative parity in net gain does not necessarily imply ultimately equal levels in any or all variables used as arguments in $A_i(t)$ among the various urban settings in question. In case the ultimate carrying capacity level of one variable, a negative component like air pollution for example, is lower between two urban areas then this difference must be appropriately compensated by offsetting differences in at least one other (in this case a lower positive, like amenities) variable's ultimate carrying capacity levels. Similarly, in the medium run carrying capacities of factors in $\boldsymbol{A}$ must not be identical between two cities with equal net attraction. All these current negative or positive differences have been capitalized and weighted by an observer in the current magnitude of $A_i(t)$, or the expected level $\hat{A}_i(t)$.

Growth or decline in any stock size at an urban area at equilibrium does not affect the $\bar{A}$, although it does affect to a certain extent each urban area's $A_i(t)$, and the associated level of other variables in it. As already noted although changes in $\bar{A}$ are relatively slow in time, changes in the current levels of the components of $A_i(t)$ are relatively fast, whereas intermediate speeds characterize changes in $\hat{A}_i(t)$.

Large perceived differentials in net gains (attractivity) are not sustainable. If wide enough differentials in net gain levels occur in the short term, so that there is for example an urban area i such that $A_i(t) \gg \bar{A}(t)$, then significant inflow into i of certain factors and outflow of others must be expected from the observer's standpoint, so that $A_i(t)$ will be expected to decline relatively fast and considerably. Carrying capacities may not be as important under such *highly excited states* with such pronounced differentials in attractivity. Hyper-concentrations are detected under this condition for inflowing stocks into this area, while hypo-concentrations of these stocks at the origin of the movement occur *while the flow is ongoing*. The observer expects the flows to cease only when $\hat{A}_i(t)$ is judged (by the observer) as having reached a close neighborhood of $\bar{A}(t)$. Population is not likely to be a factor to outmigrate from a city with a net attractivity index higher than the prevailing background average.

If on the other hand $\bar{A}_i(t) \ll \bar{A}(t)$, then the observer must expect the flows to be reversed, till again an effort toward reaching a neighborhood close to the equilibrium might lead the system to small enough differences or no differences at all. Inflow or outflow (in relative terms) of factors varies in magnitude depending on the difference $[\hat{A}_i(t) - \bar{A}(t)]$ in the case of significant variations among the $\boldsymbol{A}$, and on the difference between a factor's current level and its carrying capacity (relative norm) at t. It is possible that if such differences become too large the factor flows may be discontinued, and the process might undergo catastrophic change or collapse from the observer's vantage point. In this case, hypo- or hyper-concentrations might trigger evolutionary events. Populations are likely to outmigrate at significant numbers if a city's net composite attraction index falls far below that prevailing as the background.

There is always a condition in which the attractivity of a city (or rural setting) is below that prevailing nationwide or globally. Thus, there is always a drive for populations and other factors to move. Uneven distributions of comparative advantages among locations, and the inherent dynamic instabilities they entail with respect to the macrovariables in the index $\boldsymbol{A}$, perpetuate or even enhance dualisms and differentials in net attraction. In turn, up to a point the spatial comparative advantage differences are maintained, created, or even enhanced by spatial dualisms along the socio-spatial system's macrovariables. In brief, this is the central message from the analysis presented here.

As already noted, the proper way to refer to relative parity in net attraction $A_i(t)$ and its background level $\bar{A}(t)$ is in terms of the expectations and perceptions of an observer who might use flows to detect whether the model constructed is applicable. One must not view these carrying capacities as perceived levels by participant individual or collective socio-economic agents in or outside the urban areas. However, as indicated, social agents express preferences over levels of perceived stocks and constraints, both at present and future expected levels. Expectations imply that socio-economic actors react, and by so doing they either augment or diminish the various stocks at any point in space–time, depending upon the collectively perceived over- or undersupply of these stocks.

To the observer, relative attraction of each location constantly changes (at the margin), subject to intensive or weak social collective spatio-temporal arbitrage under a broadly defined speculative scheme. The observer expects over-supply of a stock at any location to be checked, as the result of social action, whereby some social agents pay a price for having been proved wrong; whereas, when under-supply is filled, then the appropriately reacted agents are rewarded.

Whether the variety of flows at various speeds recorded by an observer in all cities are such that: (a) always enough outflow of excess factor supply (a level above its carrying capacity) and inflow of excess factor demand (a level below its carrying capacity); and (b) changes in carrying capacities (and comparative advantages) occur so as to bring all A_i's close to a background level prevailing worldwide is the key question.

If carrying capacity differentials and very short-term dynamic disequilibrium conditions are not significant enough, then variables with a positive and/or negative net effect will probably react to diminish the disparities. In fact, vacuums in economic factors will be created. Through spatial arbitrage attempts will be made to very quickly fill them by appropriate increases or declines in the other socio-economic (or ecological) variables. The spatial source for this inflow could be the urban area's national or international hinterland or *de facto* extent. Spatial arbitrage will most likely act as if to effectively close those gaps, although it may in the short term under or overfill the voids at the margin. Very rarely, for an observer, are such voids exactly filled.

It is possibly an entirely different story if the system of cities is off equilibrium enough, so that the mechanism described above might break down. In such circumstance, evolution might overtake the system of urban agglomerations and new conditions might arise giving birth to a new set of carrying capacities, comparative advantages, and revised levels of attraction $\hat{A}_i$'s and even a different background attraction level $\bar{A}$ all far away from current levels and not all working in tandem.

Empirical evidence seems to indicate that the major urban agglomerations of the globe are considerably far from such a spatial equilibrium of uniform net attraction. To an observer it looks like, at any time period and in an aggregate effort driving each individual city or collective urban system toward spatial equilibrium, urban areas are linked by unstable dynamics in their macrovariables. These dynamics, identified by the ADC, are not working in concert in the medium term. A quarter century period of observations suggests that differentials in net attractivity are great enough to be detected by significant shifts in the relative size of key stocks worldwide.

The observed shifts are not random, but instead they are selective enough to support the hypothesis of a prevailing relative parity in net attraction principle. Significant differences in prevailing comparative advantages globally, however, and the underlying interdependencies followed by their corresponding flows, drive and hold the system into an unstable state full of spatial dualisms. This contradictory event is of utmost importance in understanding socio-spatial dynamics.

Relative parity in the long run and the individual carrying capacities in the medium term, work as a set of spatio-temporally distributed *different in speed spigots operated in a non-totally random but not fully orchestrated manner either.* Growth or decline in relative terms along many variables in reference to an appropriate environment occurs in a seemingly haphazard way. In combination the limited chaos and order of such changes in space–time provide for the rough and speculative setting, modeling, and studying of the comprehensive linkages among the world's urban agglomerations, particularly among the largest ones and their linkages to their rural hinterlands.

In view of the above, the issue of where the ecological dynamics analytically described in Chapter 2 and the ADC come from can now be addressed. One needs to look for a governing function which can supply these dynamics under different time and spatial scales. Levels of the various variables involved in A_i, two of them being the urban macrovariables used in this study (relative population and per capita product), adjust in two ways: first, under a relatively fast dynamic of the Volterra–Lotka type in reference to national totals; and second, under the ADC-type dynamic when normalized with world totals. As already noted, in a global environment as that portrayed by the ADC, $\hat{\boldsymbol{L}}$ must be close to the actual level of each component in $\boldsymbol{L}$. In $\hat{\boldsymbol{A}}$ they adjust as fast as the components in $\boldsymbol{L}$ do.

Their levels and their interdependent dynamics must, at a local and longer time-frame, be the *first order Euler conditions of a global potential* instigating a local effort toward composite parity in net attraction at all sites. What the form of such potential is, giving rise as it does to the two dynamic equations of the previous section at a global and shorter time-frame, and to the stated earlier spatial equilibrium conditions, remains an open research question. One might use as a guide the potential found to give rise to the logistic Verhulst equation, as derived by Volterra in his classical paper on the calculus of variations and the logistic curve; or the potential derived for the urban Volterra–Lotka relative continuous dynamic equations by Dendrinos and Sonis

(1986). However, this potential must be far more complex, since it must accommodate different time-scales and spatial environments each with a different fast, slow, and very slow motion.

In Chapter 1 (see pp. 18–23), it was alluded to that the individual urban macrodynamic paths could be chaotic within the domains of motion prescribed by the two isoclines in the phase portrait. As already noted in continuous dynamics, it is required that more than two kinetic equations be used for chaotic dynamics to occur, whereas in discrete dynamics this constraint does not apply. Thus, one may be motivated to model the fast macrodynamics of each urban agglomeration in a discrete manner. Further, one might also find it useful to model the inter-spatial flows of stocks within the principle of relative parity in discrete dynamics. These elaborations could considerably extend the findings of this work.

This subsection concludes with a few notes on the effect that interdependencies and their resulting interactions have, through the relative parity principle, on the dynamics of urban hierarchies. Interactions, of course, do not occur in abstract between a particular urban agglomeration and its relevant environment, although they can be effectively modeled this way. In reality, one specific urban area interacts through concrete means and linkages with a set of others in space–time by following the relative parity principle and its dynamics from both an observer's point of view and the standpoint of individuals and collectives. These linkages, i.e., dynamic spatial interactions, give form to dynamic hierarchies. Hierarchies as the result of linkages can be found among the various industrial sectors of a spatial economy as well as among the various urban areas (or regions) of a nation.[17]

The outcome of dynamic hierarchies is that a few cities emerge on the top of the hierarchy, dominating all other spatial units in size, in time. Similarly, a few industrial sectors or firms appear at the top of the industrial hierarchy of a spatial economy at any specific point in time. Urban areas or industries on top are the leading elements in the hierarchy, whereas those industries, firms, or urban areas at the bottom are the lagging ones. Dominance by few areas and sectors is a widely observed and intricately related event.

A central qualitative feature of these dynamic hierarchies is that they are internally unstable. Specifically, the historical record over the long term (which includes the rise and fall of great cities of the past) seems to suggest that the hierarchy does not provide a steady state (unique, dynamically stable equilibrium) to any of the urban areas or industrial sectors or firms it contains. Thus, there is not a single steady state remaining there in the long haul toward which urban areas, sectors, firms, move individually or collectively. This finding is contrary to the classical static central place theory by Lösch and Christaller. Instead, dynamic paths of cities, sectors, firms and their sizes are continuously altered in their respective hierarchies.[18]

Hypo-concentrations of population in the vast rural areas of the globe (in particular in the Asian and African continents), and corresponding hyper-concentrations (particularly in metropolitan agglomerations at the top of the global urban hierarchy), may be the sources of change in the level of $\bar{A}$ either lifting it or dragging it down over long-term horizons. These giant urban agglomerations and their *de facto* spatial extent may unleash large-scale multifaceted inter-spatial inter-temporal flows to fill in possible vacuums created as the result of changes in the environment (small and exogenous to them) and through the globally prevailing effort toward composite relative parity in the net attraction level. Alternatively, they may release excess capacities produced by the changes in $\bar{A}$ as the effects of relative parity propagate in space–time. These changes in turn may cause further changes in $\bar{A}$, thus creating a possibly destabilizing feedback loop, and considerable restructuring in the global urban hierarchy and possibly environment.

It is more likely that sources of such original perturbation, i.e., changes in $\bar{A}$, are found among urban areas at the top of the worldwide urban hierarchy. Nothing prevents however a lower level urban agglomeration from also triggering such dynamics. This is due to the possibly random nature of these start-up changes, with basic technological innovation being at the core of them at that scale of analysis.

THE URBAN SECTOR OF THE WORLD'S FOUR LARGEST NATIONS

The aim in this subsection is to extract the maximum information possible out of existing data sources to estimate the global urban population and the number of urban settlements in the world as of 1980. Use of relative counts is made to attain these aims. The analysis is disaggregated along the LDC–MDC breakdown. A benchmark is set against which future changes in the world's urban hierarchy can be checked.

It is estimated by the Population Reference Bureau (PRB) that approximately 69 percent of the MDCs' population was categorized as urban, whereas in LDCs the portion was about 29 percent in 1980.[19] These shares translate into a total of about 780 million urban residents in MDCs and 953 million in LDCs. Their sum, 1,733 million, indicates that 39 percent of the global population in 1980 lived in urban settings. The definition of 'urban' area varies widely among countries: for example, Bureaus of the Census in the United States of Mexico and the United States of America define as 'urban' any political jurisdiction containing at least 2,500 residents. The UN definition, on the other hand, includes only settlements with a population of at least 20,000.

To what extent then, is the PRB's 69 percent of global urban population estimate accurate? A few observations on a selected part of the urban population distribution curve and for specific nations with extensive data on their urban sector allows for a crude test. According to the PRB, approximately 45 percent of the world's urban population lived in MDCs in 1980.

At the same time, according to UN data, the number of urban agglomerations with a population greater than 4.5 million in 1980 worldwide was twenty-five, which includes six US MSAs (Detroit, Philadelphia and San Francisco, in addition to New York, Los Angeles and Chicago which were included in the previous analysis) and three urban areas, Leningrad (St. Petersburg), Lima and Tianjing. From these, eleven belonged to MDCs, a share of approximately 44 percent in the number of urban areas in the $10.2 \times 10^{-4} \leq x \leq 40.0 \times 10^{-4}$ range.

The population share of all metropolitan centers in this range belonging to MDCs is estimated, using UN sources plus data for the US MSAs from the US Census, to be approximately 46 percent. Thus, the PRB estimate seems to be valid at this range, since both percentages (in number of urban areas and total population in this size category) are very close to the number given by PRB.

However, looking at the range $2.27 \times 10^{-4} \leq x$ category (urban areas with at least one million residents in 1980) one finds a total of 174 worldwide, with 98 in MDCs (56 percent) and 76 in LDCs (44 percent).[20] This does not confirm the PRB share. Although there is no reason to believe that this share holds for all parts of the x-spectrum uniformly, it does seem to suggest that there could be an undercounting of the LDCs' urban areas in this part of the population spectrum. It is of interest to note that the slightly higher population share by MDCs than their share in very large urban centers can be interpreted as an indication that part of their urban size is subsidized by the rural sector of the world and particularly the LDCs' urban sector. Naturally, and in the spirit of the work here, the term 'subsidy' must be taken in a broader ecological perspective, and not looked at in its strict economic meaning.

To gauge the accuracy of the above counts better, next the analysis focuses on four nations that exhibit dominance in either or both of the central variables (x, y): China, India, US and USSR. At the time this book was going to press, events in the Soviet Union were leading toward a breakaway movement by various Soviet Republics. In the light of such events, still unfolding, the data presented next may well require revision – a task left to subsequent work. Some more extensive remarks on China's urban sector are found in Appendix 3, pp. 312–19.

Of the total world population (estimated at 4,414 million in 1980) China and India had a population of 975 and 676 million respectively, with an urban ratio of 26 and 21.[21] These ratios translate into 254 and 142 million urban residents, respectively. The two nations combined accounted for 42 percent of the LDCs' urban population and 23 percent of the global urban population. At the same time, they accounted for 37 percent of the world's total population and 47 percent of its rural size. In fact their rural population was 29 percent of the world's total: about one out of three persons then on the globe was either an Indian or Chinese peasant. Whereas, one out of five persons lived in an Indian or Chinese city, in 1980.

By contrast, the US and USSR shares of the world's population were 5.2 and 6.0 percent respectively, for a total of 11.2. Their urban size was 165 and 169 million respectively, or in combination 43 percent of the MDCs' and 20 percent of the world's urban population. Thus, the two pairs are similar in their share within their corresponding setting, LDCs and MDCs, but quite different when viewed within the global environment. The difference is accentuated when one considers a very large part of the population spectrum. In the $10.2 \times 10^{-4} \leq x \leq 40.0 \times 10^{-4}$ range (three-fourths of the size spectrum), China and India place only five urban agglomerations in 1975, or 20 percent of the metropolitan areas found in this part of the spectrum – slightly less than their share of urban population which stands at 23 percent.

On the other hand, the US places six MSAs, a 24 percent share, while enjoying an urban population share of only 9.8 percent of the world's total. At the same time, the USSR places two, amounting to 8 percent, while having an urban share of 10 percent of the world's total. Clearly, this shows that urban large-scale size in the US, up till 1975, has been supported by the world's rural population and by other nations' urban agglomerations. More recent evidence, in the last ten years, however, indicates that such shares are being drastically altered.[22]

In 1980, with respect to wealth, China and India's combined share of the world's gross product accounted for only 6 percent, approximately one-sixth their population share. On the other hand, the US portion amounted to 24 percent or five times its population share. The USSR's ratio was close to 9 percent, only 3 percentage points above its global population share. It is of interest to note the closeness of the wealth share and the share in the number of urban areas in the very large size category: the US possesses a fourth of the world's total and the USSR one-tenth.

China and India's shares differ significantly in these two counts. Thus it seems that the largest Chinese and Indian metropolitan areas subsidized the United States' and other MDCs' largest cities' populations and wealth, in a relative framework, up at least and around 1975. Meanwhile, the Chinese and Indian economies were being taxed in relative wealth.

Although never precisely measured, it is estimated here that in 1980 there were approximately 130,000 urban settlements worldwide, each with more than 2,500 inhabitants. The estimate is obtained as follows. Assuming that the total urban population was 1,733 million in 1980, and that the average urban settlement of the world contained about 13,000 inhabitants then. Approximately 58,500 towns were in MDCs, assuming that their 45 percent share of urban population applies, and 71,500 in LDCs. From these, about 30,000 were in China and India, assuming that both contained a combined 42 percent of the LDCs' urban population, whereas 25,000 are estimated to exist in the US and USSR, assuming that their 43 percent share of the MDCs' urban population is valid, as the number of urban areas (with a population greater than 2,500 inhabitants) in the US was estimated to be approximately 13,760 in 1980.[23] This provides a strong confirmation on both the estimate on the total number of cities globally, and the percent of urban population in MDCs.

The above analysis demonstrates the dominance of the four nations' urban and rural sectors upon the global developmental process. It also points to interdependencies in the developmental process in each of the metropolitan agglomerations for four of the world's largest nations.

A PHASE PORTRAIT OF GLOBAL HIERARCHICAL DYNAMICS?

Next, the analysis shifts to the absolute lens as it attempts to address a very specific question more fully. The emphasis is no longer on the code of development, or on the largest urban agglomerations, but rather on the relative population size and income distributions of almost all of the world's urban areas and their dynamics. Put differently, the size–income demographic structure in the world's urban hierarchy is examined. The hypothesis is put forward that the *distributions* found in the phase portrait uncovered for the world's nineteen largest metropolitan centers in the domain $[(10.21 \times 10^{-4} \leq x \leq 40.0 \times 10^{-4}), (0 \leq y \leq 55)]$ can be viewed as the result of a maximum principle which extends in the space of all non-negative values for the macrovariables (x, y). Hence, in this section a general analytical approach to the dynamics of all urban areas on the globe is attempted. To ascertain the validity of the suggested hypothesis, certain fundamental assumptions must be made regarding the *static population* size and income ratio distributions for MDCs' and LDCS' urban areas and their *underlying* dynamics. Following these assumptions and available evidence, a deterministic and stochastic model of the world's urban dynamics is outlined.

The relative population size distribution

A particular population size distribution of settlements available in the geographic literature is that by Zipf (1949) and its variations, for example by Simon (1955) who provides an absolute size-rank rule. Observations recorded on relative size distribution for the urban areas in the world with a population of at least 4.5 million in 1980 (covering more than two-thirds of the x spectrum, 10.2×10^{-4} to 33.3×10^{-4}) do not obey this rule. One can show that *any* variation of the rank-size rule would either underestimate the total in the lower part of the spectrum (if precisely depicting the upper part), or underestimate the upper part if correctly replicating the lower part.

This forces the analyst to adopt two different size distributions, seemingly associated with two distinctly different underlying behaviors. *The break apparently occurs in the neighborhood of urban size close to one million (2.27×10^{-4} in relative size) inhabitants* (1980 count). Thresholds in the weights of positive and negative effects of relative agglomeration below and above this size seem to play a key role in this phenomenon.

A function which could be used to replicate the size distribution in the domain of very large urban areas (size varying between 4.5 million and up, i.e., $10.2 \times 10^{-4} \leq x \leq 33.3 \times 10^{-4}$ in relative size in 1980) is the following variation of Zipf's rank-size distribution:

$$\ln N(10.2 \times 10^{-4} \leq x \leq 33.3 \times 10^{-4}) = a(\bar{x} - x)^{\alpha}$$

where N is the cumulative number of cities with size at least x, $\bar{x}$ is a carrying capacity parameter identifying an upper bound in relative urban size, at time t; and α is an exponent parameter. The frontier ($\bar{x}$) moves over time, currently being approximately equal to Mexico City's 1980 relative population size (33.3×10^{-4}).

The values of a and α can be directly computed from the values of the distribution at $x_1 = 10.2 \times 10^{-4}$ where $N(x_1) = 25$, and at $x_2 = 22.7 \times 10^{-4}$ (the 10 million size urban agglomerations) where $N(x_2) = 8$, noting that at the frontier $N(x_3 = 33.3 \times 10^{-4}) = 1$. These values produce: $\alpha = 0.561$, and $a = 97$. Inserting them in the formula, one obtains that at $x_4 = 2.27 \times 10^{-4}$ (1 million in relative 1980 size) $N(x_4) = 45$, a big difference from the actual total of 174. Projected for $N(x = 4.5 \times 10^{-6})$, the relative size of a 20,000 inhabitants town, the distribution produces about 55 settlements, a gross underestimate. In fact, in the range $5.0 \times 10^{-7} \leq x \leq 2.27 \times 10^{-4}$ no realistic parameter values for $\bar{x}$, a or α can replicate the size distribution. Obviously, another distribution rule must apply.

The distribution which seems to depict the size rule in the lower part of the size spectrum is the following:

$$N(5.0 \times 10^{-7} \leq x \leq 2.27 \times 10^{-4}) = e^{b}(-\ln x)^{\beta}$$

or in logarithmic expression:

$$\begin{aligned} \ln N(x) &= b + \beta \ln(-\ln x) \\ b &= -20.51 \\ \beta &= 12.07 \end{aligned}$$

a distribution calibrated with values at $x = 2.27 \times 10^{-4}$ where $N(x) = 174$, and $x = 5.0 \times 10^{-7}$ where $N(x) = 130{,}000$. Outside its domain of relevance, this distribution generates about 3.5 settlements at the point of $x = 2.7 \times 10^{-4}$ (10 million in 1980), which is a gross underestimate.

The derivative of N with respect to x, $dN/dx = n(x)$ identifies the number of settlements at size x in range dx. They are, in the two distributions, one up to 1 million inhabitants, and the second beyond 5 million inhabitants:

$$\begin{aligned} -n_1(x) &= \frac{d}{dx} N(5.0 \times 10^{-7} \leq x \leq 2.27 \times 10^{-4}) \\ &= \beta e^{\beta}\, x^{-1}\, (-\ln x)^{\beta-1} \end{aligned}$$

$$\begin{aligned} -n_2(x) &= \frac{d}{dx} N(10.2 \times 10^{-4} \leq x \leq 33.3 \times 10^{-4}) \\ &= \alpha a(\hat{x} - x)^{\alpha-1} \end{aligned}$$

whereas, in the range $2.27 \times 10^{-4} \leq x \leq 10.2 \times 10^{-4}$ there is an $\hat{x}$ to the left of which n_1 applies and to its right n_2 does. The total relative population, ν, under these two distributions is:

$$\nu(x = 4.5 \times 10^{-6}) = \int_{5.0 \times 10^{-7}}^{\hat{x}} \beta e^{b}(-\ln x)^{\beta-1}\, dx$$

$$+ \int_{\hat{x}}^{33.3 \times 10_{-4}} \alpha a(\bar{x} - x)^{\alpha-1} x\, dx$$

which must equal the urban share of the globe's population, about 29 percent. This integration allows for estimating $\hat{x}$ which is found to lie close to 5×10^{-4}, or approximately 2.5 million. In all points of the x spectrum the number of urban areas found within an interval of x in LDCs must be higher (as empirical evidence indicates) than that of the MDCs up to 1975. This is shown, for the two relative population distributions, in Figure 3.5(a).

The per capita income distribution

A reasonable assumption to make is that in the two settings (MDCs and LDCs) there is a normal distribution of urban areas with respect to per capita income, at each time-period t (see Figure 3.5 (b):

$$p_l(\ln y_i) = \frac{1}{2\pi\, \sigma_l} \exp\,[-(\ln y_i - \ln \bar{y}_l)^2/2\, \sigma_l^2]$$

where: $\ln \bar{y}_l$ is the natural logarithm of the per capita income ratio of the LDCs mean average to the world's average (in our case $\bar{y}_l = 1$); and σ_l is the standard deviation of the distribution for less developed nations – both at 1980. For more developed economies.

$$p_m(\ln y_i) = \frac{1}{2\pi\, \sigma_m} \exp\,[-(\ln y_i - \ln \bar{y}_m)^2/2\, \sigma_m^2]$$

where: $\ln \bar{y}_m = 3$, and σ_m is the standard deviation of the distribution for the more developed economies in 1980. Since data on per capita urban incomes are not always available, the national ones are employed for all countries, except for the US. Evidence seems to suggest that $\sigma_l > \sigma_m$ as the variation in incomes in MDCs is less than that found in LDCs (see Figure 3.5(b).

The possible micro and macrodynamics of the two distributions

Both relative population size and per capita income ratio distributions, as given in the two previous subsections, are static. Over time, however, the model parameters change. Possible changes are shown by the arrows in Figure 3.5. Whereas the LDCs' number of urban agglomerations increase significantly (shown by the upward and to the right movement of longer arrows in Figure 3.5(a)), the number of MDCs' urban areas may increase somewhat or decline (indicated by the two directional smaller arrows in Figure 3.5(a)).

The speeds and directions of motion cannot be precisely estimated currently, due to lack of data. Crude inferences can be made regarding these changes, however, so that if they produce *individual* urban dynamics consistent with the observed distributions over the study period, then they may be accepted with qualifications. Lack of data on such initial distribution makes precise testing currently impossible and requires one to go very far back at the beginning of urban settings formation.

Limited current evidence seems to suggest that the speeds of motion in the parameters of these distributions must vary: the upward motion of $N_{2,1}$ far exceeds in velocity the downward motion of $N_{2,m}$. Further, the east–west movements in $\ln \bar{y}$ for both LDCs' and MDCs' urban areas must have been infinitesimally small during the 30-year period of observations. Of particular interest is the effect of the combined distributions in x and y, over very long time horizons (over a century or so), upon the movement of the isocline $\bar{y} = 7.4$. The downward movement (meaning that the world is getting collectively richer and more even in the distribution of wealth) and upward shift (the world's poorest economies and their urban agglomerations dragging down its aggregate wealth and producing more uneven income distributions) implications will be addressed later, in Chapter 4.

A maximum principle possibly underlying the code of the world's unstable urban hierarchical distribution in per capita income and relative population is derived. Its existence can only be inferred at this stage and future work is needed to confirm or modify it. The size and income aggregate distributions are the result of individual urban area behavior in the (x, y) space. At each time-period, there is a probability that at the neighborhood (dx, dy) of any particular point in the (x, y) space there is an urban setting to be found. Obviously this probability $P(x, y)$ depends on the individual metropolitan dynamics and their starting point at some original time-period t_o.

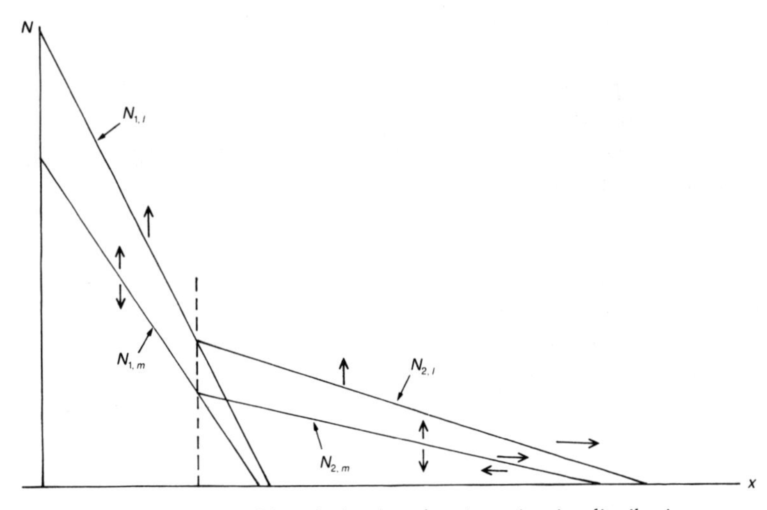

(a) The more () and less (l) developed nations city size distributions

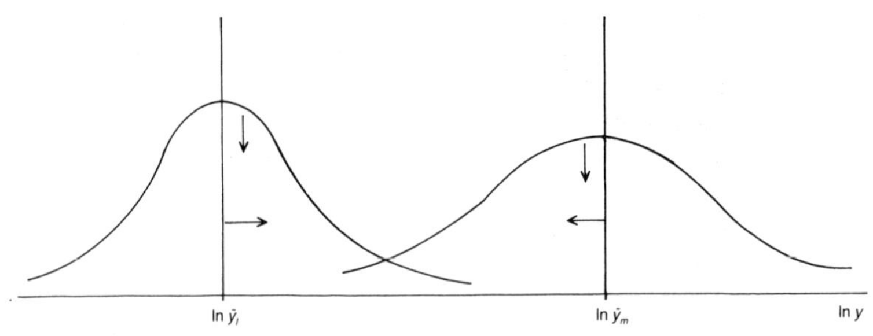

(b) The per capita income ratio distribution for more (m) and less (l) developed nations

Figure 3.5 (a) The relative urban population, x, and (b) relative per capita income, ln y, distributions. Arrows indicate possible slow motions

Evidence discussed earlier suggested that the individual urban dynamics *in the 4.5 million to 14 million part* of the x-spectrum are such that the two isoclines, $dy/dt = 0$, $dx/dt = \infty$ (or $d \ln y/dt = 0$, $dx/dt = \infty$) and $dx/dt = 0$, $dy/dt = \infty$ (or $dx/dt = 0$, $d \ln y/dt = \infty$), identify the individual urban dynamics in the phase portrait. The $d\ln y/dt$ ln y isocline (the straight line ln $\bar{y} = 2$) seems to be the local minimum of a two-peak (maxima) probability distribution in ln y for each x, whereas the $dx/dt = 0$ isocline (the second degree arch) seems to define the points of minima in the two probability distributions for the MDC and

LDC metropolitan agglomerations. The peaks' heights decline as x decreases.

Thus the presence of metropolitan agglomerations at particular relative population size x, of particular time-period t, is linked to a maximum principle associated with a probability distribution scheme, at each level x, for relative wealth y (or $\ln y$). What is the interpretation of this grand probability distribution? Is it equivalent to a master equation in non-linear stochastic dynamics,[24] or is it a set of distributions obeying normalization conditions and specifications different from those of a stochastic master equation formulation? This is still an open research topic. It may be due to a maximum stochastic entropy principle manifested in a deterministic, mean value equation form.

This scenario seems to work well for ranges of urban areas up to about 1 million inhabitants. It is noted that at this point the cumulative urban size distribution rule changes in form. Thus, it is speculated that at this point the probability distributions in $\ln y$ for each value of x reach a point of doubling (or multiplying) the number of peaks (see Figure 3.6). At lower sizes of x, and as x decreases, peak-doubling might be occurring of the exponential form 2^k, $k \to \infty$ (a geometrically increasing accumulation of peaks at exponentially declining sizes of increments in x). Each of these peaks identifies the maximum point in the probability distribution, $p^{max}(x)$ that a city exists of size x, with a per capita income level y or $\ln y(x)$ in the corresponding setting (see Figure 3.7). Possibly a chaotic regime appears, where no regular increments in the number of peaks takes place which may exist till $x = 5.0 \times 10^{-7}$, the relative size of 2,500 inhabitants in 1980. The precise formulation of this probability function $p(x, y)$, giving rise in its deterministic form to multiple peaks accumulation, is still to be derived so that the available empirical evidence can be satisfactorily replicated.

CONCLUSIONS

The sharpest possible focus has been placed in this subsection (pp. 175–92) on the available empirical evidence to document the fingerprints of the current aggregate development code (ADC) of the world's urban macrodynamics. In doing so the signatures of the nineteen largest metropolitan agglomerations on the globe, as of 1975, were traced and the code was precisely estimated for approximately a 25-year period. First, the theoretical foundations of the spatial agglomeration phenomenon were laid out and the notion of an agglomeration gradient was elaborated. Its dynamics were tied to locational comparative advantages within an environment of interacting locations. The urban developmental code is a function of spatial relative agglomeration gradients.

By intense examination of the evidence, it was found that the code contains unstable individual and collective urban macrodynamics. It also contains two distinctly different paths for the LDC and MDC metropolitan agglomerations. A developmental threshold was detected at a per capita product level equal to about 74 percent of the world's average. In accordance with arguments presented in Chapter 2 (pp. 18–23), one could also detect in the macrodynamics of Figure 3.3 aggregate national economic growth or decline patterns.

Lines were drawn between micro and macrobehavior. For an observer, the disaggregate microworld is too obscure and complex to deal with directly, but the aggregate macroworld is relatively clear and simple. Linkages between the two worlds can only be picked up and modeled by macroscale modeling containing macrovariables. The dynamics of these macrovariables for the largest urban areas of the globe were shown to obey bounded and ordered chaos. Macrodynamics were attributed to the presence of a governing principle. A worldwide and nationwide relative parity in net attraction and the carrying capacities embedded in it were suggested to act as spatio-temporal valves for inflow or outflow of various stocks in or out of individual urban areas through the presence of a ubiquitous arbitrage process.

To an observer, this composite relative parity principle allows for modeling the comprehensive interdependencies among the world's largest urban agglomerations. The observer can test it against the spatial flows of a variety of stocks. Relative-parity-driven dynamics provides a source for global dynamic instability. Vividly put, among other settings it links within a global perspective the dynamically unstable present and future states of the very large Los Angeles metropolitan region to the present and future state of the small Mehrana village near New Delhi, to the medium-size urban agglomeration of Tashkent.

At the macroscale, globally relative parity in net attraction among metropolitan and rural settings and its inner dynamics necessarily work toward equalization of attraction and uniformity. The problem is that there is not a unique path to such equalization, as will be further seen in Chapter 4. For an observer, massive actions by individuals and collectives are used as evidence for detecting the relative parity in attraction principle at work, most often by recording significant enough locational shifts in stocks as those detected by the ADC. It is on the other side of the spectrum – at the microscale and individual level – that

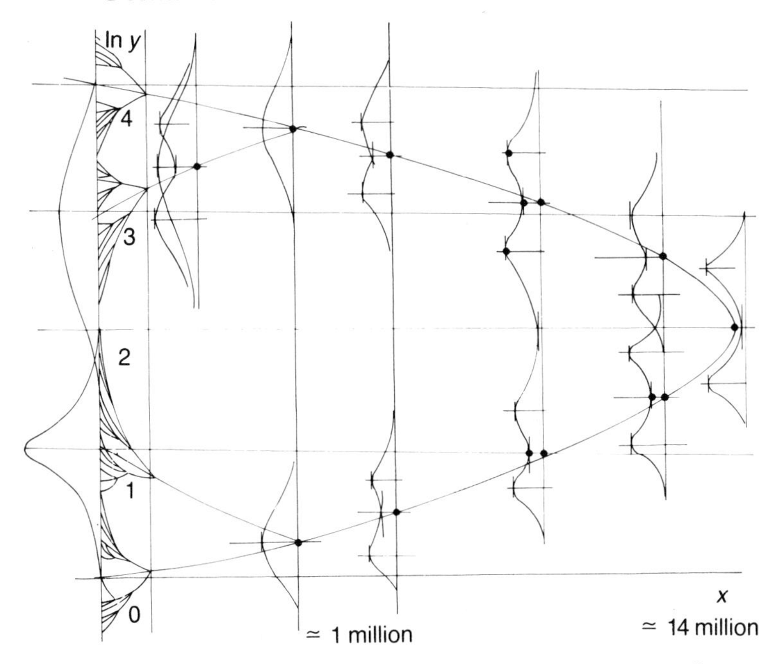

Figure 3.6 The world's urban hierarchy: the system is schematically constructed as a decomposed bifurcation tree

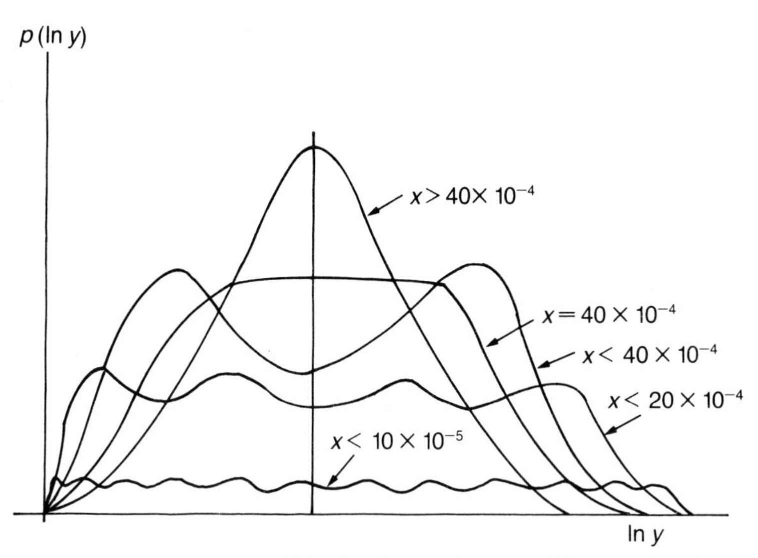

Figure 3.7 Schematic probability distribution for ln y at different urban sizes

we can assume that locational equilibrium occurs most often in space–time.

Focus was then placed on the urban sectors of four dominant nations: US, the (former) USSR, China and India. It assisted in providing an estimate of the total number of urban areas on the globe as of 1980, approximately 130,000 (with a population of at least 2,500), with the average urban size at about 13,800 then. Next, the population size and income (product) urban hierarchies were set for the globe. Relative population and per capita product size distributions were proposed, and a disciplined speculative excursion into the dynamics of these distributions was carried out. It was argued that each urban area's stochastic dynamics must produce a deterministically unstable worldwide hierarchical structure. The form of the world's urban hierarchy was sketched as constantly undergoing change. Its composition, by specific urban agglomerations at each level of the hierarchy, was viewed as being constantly in a state of flux. Changes were found to be more pronounced at the top of the hierarchy. It is of interest to note that all these findings were based on data from the last quarter century. Future time series may enrich these conclusions further.

4

EPILOGUE

If two more persons had been alive in the year 10,000 BC the earth's population would be greater today by about 2,000 people, under an average annual exponential growth rate close to 0.000576 over that period. If two more humans had escaped death at infancy in the year 1,000,000 BC there would be one additional city of approximately 85,000 people on the globe in 1989. Or, would it be so? From the mathematical theory of chaos we now understand that slight changes in the initial conditions may have quite different end results, a phenomenon known as the 'butterfly effect' in weather predictions.

The phenomenon is due to E. Lorenz,[1] who stipulated that a butterfly flipping its wings in Brazil now may alter the weather in New York in the next week or so. Whether the action of a vendor in Kalemie today can affect the economy and ecology of Paris tomorrow is a similar question we have come a full circle to ask. A butterfly effect touches issues of neighboring or distant future paths, the presence of basins of attraction in the phase space of global macrodynamics, of extreme sensitivity to initial conditions, of determinism and stochasticity. At the end, this issue begs the question whether what has taken place so far over the historical period in the socio-spatial world is any indication at all of things to come.

Perturbations are of extreme importance in chaos theory. The future state of affairs under chaotic regimes is extremely sensitive to the initial conditions (perturbation). However, the 'butterfly effect' may be an exaggeration. It is highly unlikely that the current action of a small entity somewhere on the globe can significantly affect a much larger entity at a faraway distance in the relatively distant future. To be of such significance, the timing, importance of actor, and magnitude of action must be neither random nor minor. The 'butterfly effect' is in all likelihood not satisfying these requirements. Thus, a modified principle

– that of a 'timely and important actor effect' – may be a more appropriate way of stating the importance of perturbations in socio-spatial dynamics.

A question which keeps cropping up is that of a 'deterministic' history versus a 'scenario-based' history. Chaos theory (and quantum theory to an extent) sheds some additional light as to what an observer (and participant) might assume about social history and its state(s). Although to an observer it might seem that a determinism might govern the course of history, its state might be fuzzy as demonstrated in part by the sizes of urban metropolitan regions (see Table N.1 in note 12 of Chapter 1). To participants, individuals and collectives alike, social history might also be fuzzy, i.e., neighboring histories might have coexisted at least in interpretation and perception. As discussed in Chapter 3 (p. 190), distant historical paths coexisting in a quantum-type state are not evident or required by an observer to be assumed in order to describe what has occurred in the relative macrodynamics of the globe's largest cities during the past quarter century.

In a stochastic context, a quantum probabilistic state[2] might be a new and very productive manner in which to view socio-spatial dynamics (histories and futures). It lends itself to dealing in an efficient and insightful way with fuzzy social outcomes under one code, but this will not be further explored here. Next, some speculative statements are made, still disciplined by the bounds of the previous analysis, addressing issues of worldwide urban growth in absolute population and wealth, based on a deterministic environment. Globally prevailing socio-spatial heterogeneity and related relative deterministic growth issues are also viewed in their prospects for enhancing or eluding dynamical stability over the long haul.

The point is repeated that these findings discount social actions, ethical stands, or policies under a macro spatio-temporal ecological determinism detected by an outside observer in the macrodynamics as recorded now and currently perceived as possible. How these macro-dynamics and its inner and multiple local microdynamics unfold is nobody's (and thus maybe everyone's) guess, and of course bet.

SOME FUTURE SCENARIOS

One cannot avoid discussing global population and income growth when addressing the more confined topic of urban population and wealth. In doing so, one cannot fail to note the variety of perspectives regarding these issues, either from the doomsday or optimistic side.[3]

The view from ecological determinism is that these, and all other viewpoints in between, are made to serve particular purposes and in so doing their effects are discounted in the deterministic dynamics recorded.

An ancient Greek belief had it that 'in the days before the Trojan War the world was too full of people!'[4] Throughout the millennia, similar beliefs have periodically appeared when perceived population-carrying capacities had been reached, fueling fears of overpopulation, or when perceived population voids created fears of underpopulation. Equivalently, periods have witnessed strong popular impressions of too little wealth to go around and moments in history where excessive wealth prevailed, although the former is much more frequently and strongly felt than the latter. Significant social upheaval often followed such perceptions, either feeding or checking these fears.

Apparently, the only sustained (stable) state is that of instability. Many socio-spatial conditions offer grounds for unstable dynamics to occur and be observed. Dominant among them, notwithstanding conflicting conclusions drawn from changes in income distribution curves of certain spatial economies, are the phenomenon of inter- and intra-regional dualism manifested in relative poverty; the presence of very few relatively very large spatial centers associated with the dominance of a few production sectors; the actions by very few social actors dominating the economic, social, and cultural milieu. These phenomena are encountered at all points in space–time, no matter what the ecological conditions present. Parallel to these phenomena, the record demonstrates exponential growth in the absolute levels of the total human population (see Appendix 3, pp. 319–23) in spite of any or all Malthusian and other checks. The exponential growth rates have been declining in magnitude, however, as the time horizon stretches out.

Fast and slow dynamics have been fueling and are fueled by the increases in global absolute population and wealth. Are these increases and their associated dynamics sustainable? It is clear that there are still increasing marginal social returns, in terms of the composite wealth index, to the world's population. This must be the main force behind the sustained and so far positive, albeit decreasing, exponential growth rate in absolute human population abundance. Whether a decline in either stock is associated with growth in the other, as a second phase of a cycle, is still unknown. Also unclear is whether decreases in any of the two stocks are imminent, or even if they are to occur at all in the foreseeable future.

The human population has been increasing for almost a million years. This suggests that it will keep increasing for a long time to come, unless either an unforeseen catastrophe begets the human race, or we happen to be at a privileged position close to a different state than that which has persisted for so long. Is a period of aggregate human population decline foreseeable? Can wealth increase under such a condition?

Whether a basic periodicity, similar to the spatial Volterra–Lotka continuous dynamics found to hold for urban settings, drives these stocks at their worldwide aggregates is one hypothesis. Another has the empirical evidence suggesting that the speed of change ($\dot{X}$) in the current population level (X) is levelling off. On this premise, demographers currently estimate world population size to reach a steady state, what ecologists refer to as 'carrying capacity,' which is defined as the level at which a crude birth-rate equals the crude death-rate, and to stay at about 8 to 10 billion in about a century or so, over a relatively prolonged time-period.[5]

According to prevailing demographic theory, the population abundance in all nations is on a course toward a (utopian) steady state, although that steady state is not to be reached at the same future time-period for all nations. These future horizons are not far apart, as they range from 2025 to 2100, depending on the demographic study. From a longer-term viewpoint, say a millennium, these are very close time-periods. That such a narrow band is so close to the current time-period, the time these estimates are made, makes this event highly unlikely even if one goes along with assuming a drive toward a steady state unless one assumes again that this generation is a privileged one, when compared with the past and the ones to come, for being so close to a state of 'nirvana.' Further, demographers are mute on how long such states are expected to last, if ever reached. Apparently, they expect the underlying dynamics responsible for these 'steady states' to go on *ad infinitum.*

Whether the absolute current population size worldwide is at an isolated logistic or oscillatory motion over a very long time-period is an open question. Given the available historical record one might find it more profitable to study $\dot{X}$ rather than X, or even $\int X(t)dt$, the cumulative human life. Cumulative life significantly preoccupied Volterra in his classical paper on mathematical ecology.[6] The rate of change, $\dot{X}$, in all likelihood obeys an oscillatory chaotic motion.

Parenthetically, this is a very general approach to viewing stocks in a non-spatial context, and it applies equally to wealth (its first time derivative being income), capital (its first time derivative being investment), information, natural resources, and commodities. In all cases, the current addition (first time derivative) is the birth (replenishment) minus death (depletion) and thus net accumulation (or decumulation) rate, which accounts for new production, conversion, or appreciation rates minus elimination, destruction, or depreciation rates. In a spatial context, net immigration rates have to be added.

Demographic analysis suggests that given the relatively young age of the current population stock (the Population Reference Bureau estimates about one-third of the world's population to be below the age of 15, in 1989) one must expect a significant increase in $\dot{X}$ within a generation's time as the current young female population reaches its age of fertility, barring global large-scale social shocks of extraordinary proportions. There are, on the other hand, cases where the current female fertility rate is below replenishment levels as for example in a number of medium-size European countries.

Are these simply temporal events in the fertility cycle instigated by the current local population stocks? Are they to be followed by still another fertility rate transition, possibly induced by migrating populations into these countries? Is all this a part of a cycle toward an eventual steady state? Is stability at the most aggregate level consistent with the constancy of the destabilizing elements mentioned earlier? Could it be that this stability is elusive because human populations are not homogeneous, and neither is the space in which they are distributed nor the time-periods they live in? Is dualism (in either its geographic or non-geographic version) a ubiquitous event because it is stable? Must all dualisms be eliminated? Can they be eliminated? Is there a real effort made to eliminate them?

What is clear is the fact that population densities are not uniform and neither is density of wealth in space. And that heterogeneities do not seem to be evening-out in time but instead they become more pronounced, enhancing various dualisms and hyper-concentrations. From the snapshot of a quarter century, data available seemingly support the hypothesis that, no matter the effects of the absolute worldwide and urban specific population and per capita product growth dynamics, their inter-urban relative distributions are locked into a 'fast' motion.

Urban macrodynamic motions were argued to be outcomes of an aggregate code of development, and a prevailing relative parity in net attraction principle among metropolitan agglomerations in a worldwide environment. Simultaneous urban macrodynamics of relative population and wealth seem to be determined by this code and by relative parity in the time span of a quarter century.

This code is depicted by some constants in form and parameters, which all metropolitan areas' dynamics obey. Seemingly these constants identify the discounted collective effect of a very large number of variables, what was identified in previous chapters as the 'composite' or 'soup' of interaction. They are testimony to underlying ecological (relative) forces at work which guide the relative development of nations and their main urban agglomerations within a community of urban areas, regions and nations.

Constants depict long-term (of about a century), large-scale (worldwide) conditions. Part of the code is a set of constituent elements in a drive toward the equalization of a net attraction index operating for an observer under the principle of a relative parity among locations. Inequality (off par) in particular factors and the presence of opportunity vacuums by participants result in social, collective, or individual action. Equalization in net attraction among locations may never be achieved and the underlying motion of stocks in space–time may indeed be unstable.

According to inferences made from empirical evidence individual urban areas do not seem to affect the code's constants over short time-spans (about a decade or even a quarter century), no matter what their current relative stock's (population, wealth) size. Urban macrodynamic paths depend on their current position in the phase portrait of the code, which dictates their eventual state.

Dynamics embedded in this code are associated with fast changes in the two central variables, relative current population and per capita product – the fundamental economic–demographic force. Dynamics associated with changes in the constants themselves are slow dynamics in the interplay between the current relative wealth and population of metropolitan agglomerations. What are the underlying dynamics of the slow motion? Available evidence cannot address this question with vigor. One might speculate systematically, in view of the long-term (about half a century) effects of the fast motions built into the code, by looking at neighboring futures accommodated by the code.

Were one to assume a scenario according to which the code remains unchanged over a prolonged time horizon, say an additional quarter of a century, then the following might occur: the very large Chinese and Indian subcontinent urban agglomerations, and possibly Mexico City, São Paulo, Cairo, Lagos and Jakarta, will move toward extremely low relative per capita wealth and very large population sizes.[7] Beyond a threshold *and* a specific time-period, the constant of the code $\bar{y} = 7.4$ will be affected (see Figure 4.1).

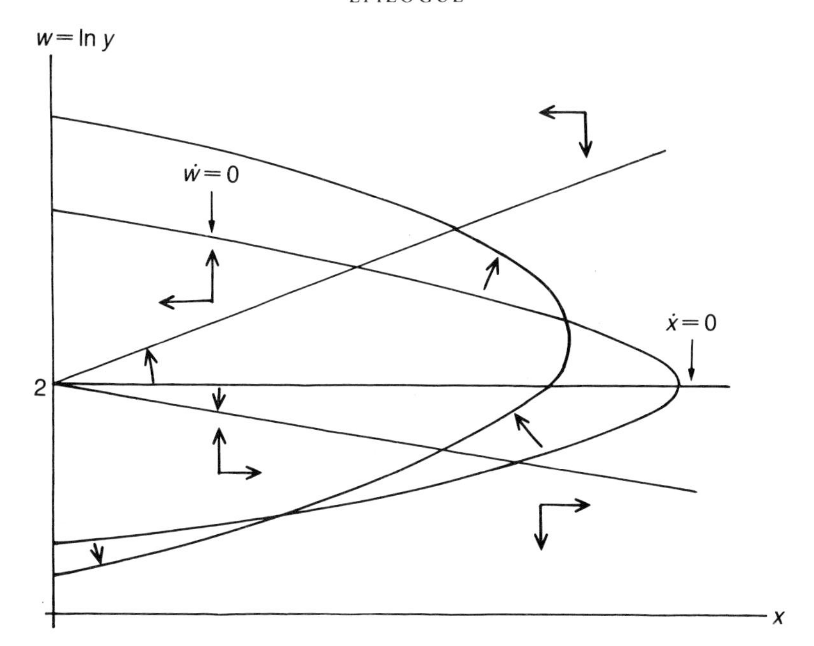

Figure 4.1 A possible set of slow motions in the aggregate developmental code. The cases of a downward and upward sloping $\dot{x} = 0$ isocline (developmental threshold) are analyzed in the text

As these urban concentrations move in the lower right-hand corner (the very slippery part of the unstable domain) in the phase portrait, the $\ln \bar{y} = 2$ line could become *upwardly* sloping as a greater disparity between the average per capita wealth in MDCs' and LDCs' large urban agglomerations materializes. The case is depicted in Figure 4.2(a), where the fields of motion in the phase portrait of the urban macro-dynamics is shown under the specifications of the code as in Figure 3.4(a) of Chapter 3, with the population adjustment speed having been reduced to 0.03 from 0.05.[8]

The slope of the linear isocline has been assumed at +0.0333 (in magnitude comparable to that of x, where x is the relative urban area's share of the world's population); the qualitative properties of the phase portrait are not affected by the magnitude of the slope which depicts in this case the positive relative population response to increases in per capita income. The unstable section of the phase portrait leading toward a state of very pronounced relative population levels and extremely low per capita incomes has expanded considerably. At the

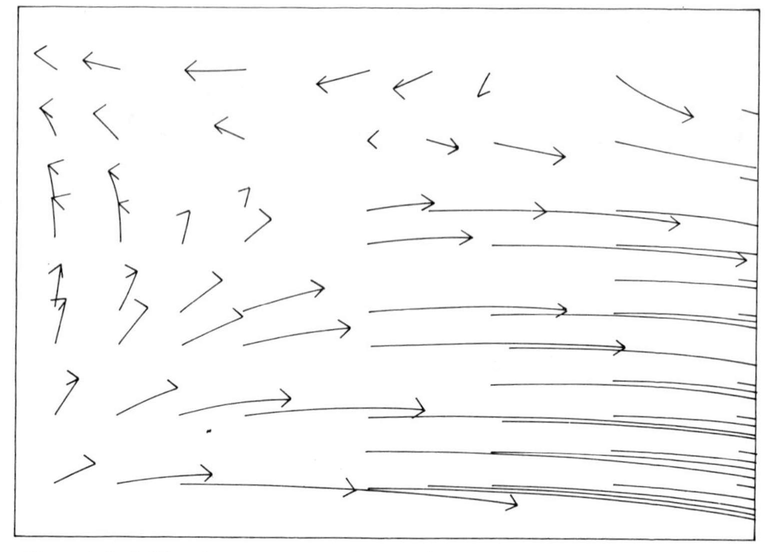

Figure 4.2(a) The phase portrait and field of motion in urban macrodynamics under an upwardly sloping developmental threshold ($\alpha = 0.03$)

same time the progress in the stable region of the phase portrait, guiding paths toward the upper level stable equilibrium, has slowed down appreciably.

Alternatively or concomitantly, the arch (in the form of the second degree equation) could become flatter, pushing urban areas below the $\bar{y}$ line out of the income growth region and possibly further accelerating the upward movement of the $\bar{y}$ line. A squeeze of the economic growth bent ray, the arch of the code, may have significant consequences: the maintenance of a few urban areas, mostly in MDCs, of relatively very high wealth coexisting in a world with very many urban areas, mostly in LDCs, of relatively very low wealth. This tenuous coexistence could be shortlived, in terms of human time-scale, as it is recalled that the fast urban macrodynamics embedded in the aggregate developmental code are unstable by the very form of the ADC.

Evidence from a 25-year time-span of urban–national development points to a world becoming more wealthy on a per capita basis. However, the less developed nations absorb a fast increasing share of the world's population with a sharply decreasing share of its wealth. It is evident from the dynamics of the code that the LDCs might pull down the world's per capita wealth eventually, if these trends continue unchecked. In accordance with the relative parity principle, if spatial equilibrium prevails, things will probably be very miserable for all. This constitutes the *global misery scenario*.

Even if this scenario on the population–per capita income space does materialize, the spatial disequilibrium in net attraction may imply different enough levels of utility to different cities' dwellers, possibly drastically different when compared with today's levels. As was argued in Chapter 3, when differentials in net attraction levels become too wide, the effort toward equalization might break down. Social measures to keep at bay the residents of the cities with net attraction levels significantly below those of their counterparts in the more developed states might have to involve coercion on a scale unthinkable at present.

There is an alternative scenario, still involving neighboring futures. The arch of the code could move outwardly and to the right in Figure 4.1, thus becoming sharper. This slow motion will put the growth area of the very large urban agglomerations of the LDCs inside it. According to this scenario, the world will become more uniform in terms of per capita wealth as a larger number of the fast-growing urban agglomerations may now go through an enlarged window of per capita income growth and move toward the stable upper level (low population size, high per capita income) equilibrium. Such a change will be associated with a change in the magnitude of the $[b/a]$ ratio in the second differential equation's non-linear isocline. It would allow for a higher than the present maximum relative population size (a position currently held by Mexico City) to positively affect change in relative per capita income levels.

As was the case with the previously examined set of alternative scenarios, the threshold line might become downwardly sloping independently or in combination with changes in the arch. The case of the downward sloping linear isocline is shown in Figure 4.2(b), where the slope is assumed at -0.0333 depicting a negative relative population response to rises in per capita income. According to this alternative vision of the future, the section of the unstable part of the phase portrait shrinks considerably, enabling many urban areas to move through the per capita income growth and population decline window. This is the *convergence to uniform austerity scenario*. The scenario needs, however, direct outside interference or perturbation, some exogenous (albeit slight) change of the code in order to materialize. It does not depend exclusively on the individual urban areas' current motion as did the previous scenario, this relying on *continuous* increases in the size of currently large and relatively poor urban regions.

It is not clear what will propel the latter, slow movement scenario,

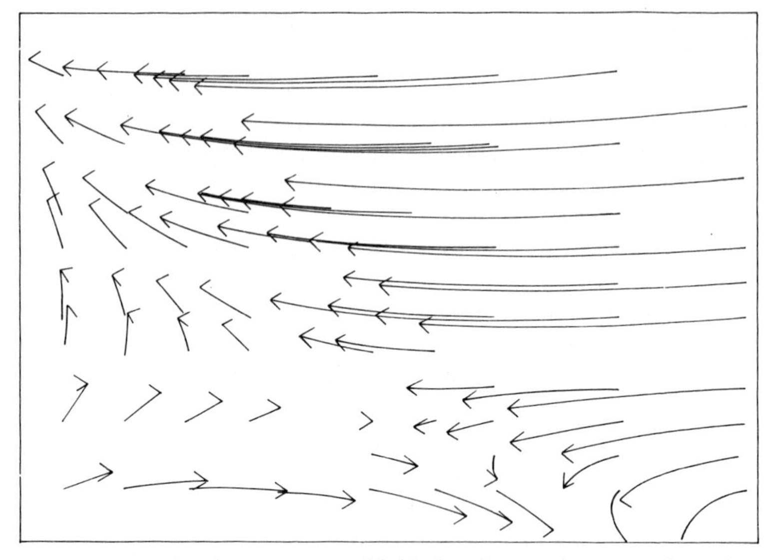

Figure 4.2(b) The phase portrait and field of motion in urban macrodynamics under a downwardly sloping developmental threshold ($\alpha = 0.03$)

given the current fast dynamics. As the first scenario requires a less unexpected change and a continuation of current trends, it might be thought of as more likely to occur at present than the second hypothesis. Current fast dynamics smoothly bringing about a change in the slow dynamics, rather than slow dynamics being affected by forces other than the current fast dynamics, render the likelihood of the uniform austerity scenario rather small. There is no evidence, as already mentioned, to suggest that historically incomes of various locations worldwide have become more uniform in the past or are doing so currently. From an ecological standpoint uniformity is only stable at the phase of competitive exclusion, a state associated with the extinction of most stocks from most locations – a final state of utmost instability.

Glancing at the last few millennia of human history, from what one can infer from a very incomplete and filtered record, uniformity in either wealth or population has never been experienced. This could imply either or both of two things. Either that uniformity is an unstable part of the phase portrait (code) containing stable or unstable fast dynamics. Or that the code, in its present form or close to it, containing unstable fast dynamics may have been very resilient and not likely to change or to have been changed. A quarter-century history of individual urban areas (and nations) seems to attest to the *preservation* of an instability producing code: spatial units may be continuously or discontinuously perturbed from their dynamic paths, but the qualitative features of their signatures and the code they obey remain basically unaltered.

Code mutation presents an alternative perspective to the two previously mentioned scenarios. Slight variants of the present code might be competing all the time, albeit for a short time-period each. Such code mutations, the result of economic, social, political, environmental, or other individual (including changes in preferences) or collective decision-making processes (including such diverse policies as family planning, environmental policies, or restrictions on migration into core urban regions), are either unsuccessful in replacing the original code or they do not last long.

At times some mutations could have successfully competed in their attempts to marginally or mildly alter the prevailing code. Such mutations must not be viewed within the framework of a master code containing possible alternative sharply different codes. As the subject is totally outside the confines of this work, it is unnecessary to make such an assumption as the evidence supplied can be adequately described with the help of a single code. If there is a master code with significantly different scenarios present, then uniformity in incomes or population densities must have been unstable events in all of the possible codes.

Over human evolutionary time the very definition of the environment, in which the code or its competitors (mutations) appear(ed), has also been changing through regional discoveries, technological innovation, and topographical change. The historical record is full of regional expansions, accompanied by redistributions of both wealth and population. The rise and fall of dominant civilizations, nations, cities, and industries at any time or space, can be seen either as direct testimony to these redistributions or as indirect evidence of the workings of the code on the grand scale.

Cycles in socio-economic spatio-temporal stocks have taken place on a lesser scale. Large scale inter-regional population migration movements are examples associated with the expansion of spatial horizons and the discovery of new natural resource stocks. Spatial changes and expansions were associated with bursts in the production and destruction of wealth, built capital and population stocks, innovation, and changes in sociological, political and cultural attributes.

All these changes take place motivated by individually and collectively perceived and negotiated ecological composite profit. Composite profit is accrued to individuals when firms, industries, cities, regions, or nations are created and rise. Profit is made when they fall and are dismantled. Individuals move in space–time in perpetual search for composite profit. Changes in the multifaceted ecological attributes of spatio-temporal attractivity may impact the slow dynamics affecting the code. As is the case with fast motion one must not expect stability in the slow motion of the code.

A picture in the above analysis is painted by a snapshot of history which shows the world's heterogeneous social stocks distributed over heterogeneous locations in dynamically unstable paths periodically being shocked. Disequilibrium (or unstable) dynamics stemming from interdependencies and interactions among these stocks and locations are always likely to be present.

Similar to fast dynamics, slow dynamics must also have a mechanism which favors instability. Could it be that the source fueling such slow dynamics is due partly to the increasing difference in wealth between the poor and richest settings, worldwide? Certainly this difference fuel's expectations of an increasing world population. Dualism in relative wealth and its external effect on stimulating expectations from the more to the less wealthy nations may underlie forces toward competitive exclusion. *Increases* in relative wealth disparities may tend to accelerate the eventual extinction of certain populations and other stocks. Dynamics producing dualism are fueled in a positive feedback by the formation of dualism and heterogeneity. All this in spite of (and indeed due to) a drive toward a global relative parity in the composite net attraction principle evident to an observer.

These two processes, dualism and heterogeneity, seem to be on the increase at most points in space–time and in retreat at a few others. Whereas inter-national heterogeneity in population may be lessening through migration, intra-national social heterogeneity and dualism in MDCs is increasing. Through inter-national migration and differential population growth rates among different ethnic groups the recently arrived stock share of an MDC's nation's population is altered.

The US presents a case in point in the mid-1980s. The 1940 Census of Population reports that the European-Americans accounted for 89.8 percent of the US population, with African-Americans (blacks) amounting to 9.8 percent and Asian-Americans (including native Americans) totaling 0.4 percent.[9] In 1980 these shares have been changed to: European-Americans (non-Hispanic whites) amounting to 76.7 percent, Hispanic-Americans (mainly Latin Americans) 6.5 percent, Asian-Americans (including native Americans) 2.3 percent, and African-Americans (blacks) 11.8 percent.[10]

Similarly, with an influx of Asian and African populations into Europe, Western Europe's ethnic composition is undergoing changes, as well. An influx of foreign workers into Japan from various origins, particularly from South Korea, the Philippines and Taiwan, is also recorded. This evidence could signal some locational redistribution of the current population stocks from LDCs to MDCs, so that ethnic differences among populations on the globe may be under constant change. One must distinguish, however, between ethnic and locational differences in the dynamics of the population and wealth stocks. The code's dynamics, reported in this book, seem to apply to population groups classified by *location* not by ethnicity, race or any other factor of social disaggregation.

Two other forces may be part of the bundle of causes in slow dynamics: a rise in the absolute level of worldwide population and a change in its collective wealth. Their two rates of change, or their absolute levels or both, may impact upon the form of the code and its parameter values. Which one of the above hypotheses proves to have been most appropriate will have to await the accumulation of the time series of at least one century to verify.

The underlying motive of many public policies has frequently been to 'bring about stability,' in a world perceived as 'becoming more unstable.' Dynamic instability may indeed be the collectively selected macropattern because it brings the highest composite profit to the current or future dominant social stocks or spatial units. Successful speculators find periods of price volatility in stock exchanges the most opportune time to acquire profit. The social system, collectively, may find periods (or spaces) of instability the most appropriate time or space for its successful social speculators–agents to incur the maximum comprehensive (bundle-like) ecological net gain.

Broadly, price volatility is due to differences in expectations among agents on risk and payoffs associated with the future performance of stocks, financial or tangible. It can also be due to surprises when expectations and events differ. Individuals or collectives can exploit this volatility.[11] Primary and secondary financial markets are examples of forums where price volatility can be desirable to some. Broader ecological markets allow for similar opportunities in the trading of options or futures contracts on various scenarios or states by a nation's socio-economic stocks. Socio-economic, spatio-temporal volatility may be a collectively and individually sought-after condition, and thus a collectively selected (albeit unplanned) macrofactor.

OBSERVERS OF AND PARTICIPANTS IN SOCIAL DYNAMICS

Throughout the text the terms 'observer' and 'participant' type models of socio-spatial dynamics have been used, and a distinction between participants and observers has been made. In this subsection a few comments are made on the nature, purpose, and use of these two species of models. The mathematical models presented previously are not directly relevant for policy making. As already noted, they contain a few macrovariables which are not subject directly to public intervention as policy instruments, and well beyond individual action. Their macrodynamics are removed from any intentional social, individual or collective, action. They are of the utmost possible 'observer' descriptive type; utmost in the sense that 'pure' observer-type models do not exist.

In a social science field like urban and regional economics, geography, or planning, where policy-sensitive models of a behavioral 'participant' type are revered, this admission could be construed as a severe limitation. Participant-type social science models are 'structurally biased': they are developed with financial or other immediate gains to their developers, financial sponsors, and users in mind; they are funded because they choose or are expected to adopt a viewpoint acceptable to their sponsor; they address the constituency the sponsor deems it desirable to address. Their bias may be severe at times. Universities and faculty members actively pursue them, because of and largely due to their immediate monetary and other rewards. Some of these social science models become the basis for public policy, or they are employed directly as public policy instruments; these instruments are means by which collectives (governments) effect social change along lines they desire. The objective here, however, *was not to provide another instrument to manipulate socio-spatial systems*. Rather, it was to more fully understand them by finding a model which would possibly discount for the effects of any feasible social action. Models of a participant type respond to a market determinism for their existence. There is demand for, supply of, and prevailing prices for their inputs and outputs to their creators and adopters. A formal or informal market determines the type and quantity of output they produce.

Various expectations which models generate are subject to similar market determinism. They deal with large-scale economic, social, and political resources (like large-scale national econometric models do, for example). There are only a few, due to the very high fixed costs to build their infrastructure and maintain them. Consequently, their output impacts the dynamics of the social system they model, at least in the very short term. They have a self-fulfillment/defeat built-in mechanism of feedback with the social system they pertain to model their behavior on. Thus, their presence is serving specific interests of specific social groups betting on the outcome or the social responses they generate.

Insights into the variety of 'observer' and 'participant' type models of social systems can be obtained by references to real games, claims placed on their outcome, their determinism, and their agents.[12] There are limitations to this analogy, as a number of caveats apply: social events do not always represent zero-sum, purely competitive games; they are continuous and not discrete in time, with a beginning and an end; they are not repetitive; observers enter and exit continuously rather than in discrete intervals; they are played under constantly shifting rather than constant rules; they have fuzzy outcomes rather than the sharply defined outcomes that real games have. As is the case with all analogies, they do not constitute proof (if proof exists at all in the social sciences), only sources for insight.

At the beginning of any game of American football or baseball, for example, every side starts from an equal position: a state of uniform distribution. It is rather rare, however, for a game to end in a tie: uniformity is an unlikely outcome. At any point in the game, and while a team or player is losing, to an observer it seems that the effort by the losing team (or player) is to drive toward an equalization and then toward going ahead, while the winning team (or player) is attempting to maintain the winning margin or to even increase it. To an observer, the effort by the losing agent toward equilization is stronger than the effort by the agent ahead in the game toward increasing the winning margin assuming that there is enough capacity to support such effort. At the end, the result is likely to be uneven and at times quite lopsided. Dualism prevails while at all times (at least from the side of the currently losing team) the effort is toward parity.

Similar to real socio-spatial systems, in a real ball game there are very few players, a much smaller number of coaches and a much larger number of spectators. All are participants, but their degree and importance of participation varies considerably. The diverse agents involved directly with the game carry out certain specific tasks at particular time-periods. Preceding their action, they employ a variety of strategies for decision making and implementation. Clearly, not all construct and use the same model to perform these tasks, even in the case of agents belonging to the same category (players or spectators).

Very often the same agent does not employ the same model during the course of the game. Although they may have models of the relevant variables they are affected by and can affect, they do not always act according to the model's recommendations. If they did, then they would be at a disadvantage in a competitive game for others could learn their model and act accordingly. Thus, they either do not want to act so as to reveal the model, or do not act consistently with the model's recommendations.

There is an interesting location theoretic aspect to the analogy, associated with the notion of 'viewpoint.' Spectators are there to obtain pleasure out of watching the game. They pay an admission fee to do so. Once admitted, spectators are distributed over the space of the stadium so that no two spectators have identical views. There are neighborhoods of (qualitatively) similar, but not identical, viewpoints. Positions closer to the center of the stage are fewer in number, offer a more clear view than those further away and their seats are more expensive than those further out. A game creates a variety of markets, the location to view it from being only one of them. There are many claims placed on its resources by a variety of agents and these take a variety of forms. The outcome of the game is by far the most important resource claims are placed on. Competition, cooperation and predation are dominant interactions within and among the various agents (players, spectators, coaches, etc.). Agents have memories. These memories extend either over a game's observations and actions (marginal), or to previous observations and actions on previous games (cumulative).

At the starting point, there are a few inquisitive and presumably ignorant spectators whose main objective is to figure out the game's rules and code. They are not as much interested in the outcome of the game, but instead are eager to learn more about its strategic and tactical aspects. Understanding is their main source of utility. Further, there are speculative spectators who place bets on the game's outcome. The first kind corresponds to a pure academic, whereas the second type includes the involved (quasi) academic and the enterpreneur. Players and coaches are some of the game's active agents, others being the announcers, reporters, photographers, owners of the teams, umpires, etc.

There is a very relevant quote attributed to the physicist John Wheeler (see Bernstein 1991: 96), which links in particular the role of umpires to the fuzziness (quanta) of social outcomes; three umpires discuss their job:

Umpire No. 1: 'I calls 'em like I see 'em.'
Umpire No. 2: 'I calls 'em the way they are.'

Umpire No. 3: 'They ain't nothin' till I calls 'em.'

Obviously the game's diverse agents do not employ the same calculus throughout the game. Their computational requirements differ. The quantity of time involved in computing, the timing of the computations, their nature, and the subsequent actions the agents are engaged in, differ significantly. Second guessing the game's outcome is worthless to speculators and is left to the unsophisticated spectator or the analyst.

Speculative and inquisitive spectators only consider a few key factors in a timely fashion during their computing. In their computations, observers (both speculative and inquisitive) do not attempt to replicate the individual behavior of all active agents involved in the game and their decision-making calculus to figure out the contest or predict its course. It would be nonsensical to try to simulate the game in its entirety with all its complexity, as it would be ludicrous to attempt to predict its outcome by trying to foretrace the ball's trajectory throughout the game. In fact, the speculative spectator does not have either the time or the interest in doing so. Further, their bets may affect its outcome, through a variety of (allowable or prohibited) means.

As the inquisitive spectator does not have access to the book with the game's instructions, the effort is to make the basic, possibly minimum, necessary observations which are not violated during its course, and enjoy it up to an acceptable level. The spectator has all game long to reflect upon these rules, and test alternative schemes. Not all spectators who are of the inquisitive kind acquire the same knowledge of the rules during the game. What input information this type of spectator uses, and the output of the computations, differs from what the speculative spectator considers. The latter is preoccupied only with a very few factors relevant to the bet, in a very short time-period to take or give odds. This spectator's memory depends on the time-period the bet is placed in. Although partly based on some information about the game, the players and the conditions surrounding them, the gambler's reacion to the current odds are mostly instinctive in imperfect options markets. In an interactive manner, odds affect the betting behavior by speculators and at the same time betting affects the odds.

On the other hand, a player in the game is faced with a very different set of conditions. A player is constantly under a fast action mode, having a very short time span to make a series of decisions and take a series of actions. There is simply not time for reflection on actions during the game, a factor which obviously depends on the overall speed of the game, but certainly there is never enough time to reflect. Faced with a complexity of factors to cope with during each play, the player's decision making is very dense timewise. Not many alternatives can be considered at any point in time during the game. Depending on the overall speed of the game, most of the time the 'do nothing' alternative prevails as either the player's environment proves overwhelming, i.e., the player is overly constrained, or the action occurs away from the player.

Coaches are the closest one comes to pure policy makers, to use this analogy. Coaches are mission-oriented, but their mission is different than the player's objectives. Whereas the players' aims are first to put up a good show for themselves and then to win the game for the team, the coaches' objectives are first to win the game and then to satisfy the crowd and through them their employers, the team owners. Coaches are the agents of owners, the interests of which they simply serve.

Their coaching strategies are also informative. They enter into a speculative interaction with the other team's coaches, trying to outguess and to an extent deceive one another. Their mode is less reflective than the non-participant observer but more so than the players. They continuously attempt to figure out the other coach's instructions; by the time they do so, thinking that they have gathered enough information, the other coach changes instructions. Much like agents' behavior in stock exchanges, this analogy is quite generic and informative in theoretical social dynamics.

Some alternatives are evaluated during the contest, particularly when tactical planning coaches carry out. A game plan, or strategic planning, presumably took place before the start or during the half-ime break. Rarely is it adhered to through the whole course of the game, particularly by the losing team. Being very close to it and given their mission, coaches often have a very distorted (i.e., partial, partisan, erroneous and certainly normative) view of the game. Although there are alternatives considered before the game, the game and its conclusion are unique but are nevertheless subjected to multiple reporting.

Finally, a few remarks must be added regarding the code. Its formal rules are only part of it. The social, economic, cultural, and other conditions associated with the game are also part of the code. Among the rules there are some which are essential in order to follow the basics of the contest at all time-period.[13]

Coaches may fully know the rules and to an extent the code of the game. But they do not win all games they coach. Indeed the vast majority of coaches do not last long, as they cannot keep having winning teams continuously. Observers in aggregate may not have a full knowledge of the rules or its code. They attempt to enjoy the game and marginally influence its course by voicing their pleasure or displeasure. Their level of utility is weakly tied to the perceived outcome of the game while it is in progress. Inquisitive observers may never have a full knowledge of the rules or code, but they enjoy a more detached and clear view in their attempts to learn more about it. Players, coaches, and the speculative spectator may have a better knowledge of the rules than the average observer, but as their enjoyment is strongly linked to the current likelihood of the outcome they have their bets on, this enjoyment is certainly less than that of a casual bystander.

The above comments strongly suggest that academics (economists, sociologists and political scientists alike) involved in participant-type modeling efforts attempt to fulfill two very different roles. Their efforts, as a result, may be falling between two stools and thus accomplishing neither role's objectives. Macroscale modeling affords the academic a forum to derive considerable insights into the socio-economic spatio-temporal structure of society, detached from any allegiance to any specific interest, no matter how noble this interest may currently sound. So far, as social scientists, we have not fully availed ourselves of this vantage point. The manner in which society rations out the very few spots it affords for observers might be a reason why.

APPENDICES

1

TOWARD AN ECOLOGICAL DETERMINISM

Introduction

In this appendix, a few additional remarks are made to outline the principle of a macro spatio-temporal ecological determinism and to place it in a theoretical perspective. This principle is the theoretical backdrop to the empirical findings of this work. It is partially revealed by the relative population and per capita product growth or decline paths of. the world's largest urban agglomerations, although its full extent goes far beyond the domain of these two variables or the urban focus of the study.

Ecological determinism, in a capsule, is a broad market-based determinism containing the components of demand, supply, and prices. The markets are predominantly informal, although formal markets of exchange are accommodated as well. All elements of the social system, which is comprised of physical and non-physical stocks (economic, social, environmental, cultural, religious, ethical, etc.) are subjected to this broad market determinism. Demand, supply, and prices are multifaceted in this market structure, and they are not exclusively limited to their economic attributes.

Economic and ecological competition, cooperation, isolation or predation occurs in these markets. A variety of forces (economic, sociological, psychological, cultural, environmental, legal, etc.) affect market determinism (i.e., the elements of demand, supply and prices) of a particular stock. Conflicting, cooperative, and predatory interests are resolved within an ecological (i.e. comprehensive) framework. The principle of macro ecological determinism can be stated in more detail as follows:

1 All stocks or components of a social system (physical quantities like final products, wealth, natural resources, etc, as well as non-

physical entities like public policies, religious beliefs, modes of production, or governmental structure) are exchanged in multifaceted formal and informal markets and are ruled by a market determinism. Subject to a bundle of forces defining a supply function, the expected quantity supplied of a particular stock is simply determined by the composite price currently perceived and expected by the supplier(s) in the exchange. Under a bundle of forces defining demand function, the expected quantity demanded by the buyer(s) is determined by the composite price currently perceived and expected in the exchange. The two expected composite prices are rarely identical. They usually differ, as in the case of informal markets, where under a barter system one type of commodity is exchanged for another. Whereas in formal markets, when another commodity (most frequently money) is used as an intermediate (*numéraire*), the money-based exchange price must be the same to buyers and sellers for exchange to occur. Otherwise a rationing scheme will prevail, or a continuum of disequilibria will occur with excess demand succeeding excess supply conditions (as in the well-known cobweb problem). However, in the vector of components constituting the 'composite price' to the buyer or seller there could be differences, at times considerable.

2 As with stocks traded in the market place, the underlying multifaceted behavioral attributes of the social agents (preferences, constraints, production, and regulatory conditions giving rise to the bundle-type demand, supply and market functions, and to governing rules and regulations as well as decision-making functions at the individual and collective levels) are also exchanged in multifaceted markets and are subject to a similar determinism. Prices, market, and individual as well as public decision-making conditions are the outcome of ecological (multifaced) interactions. These interactions are partly fragmented (isolated), party integrated (interactive) and future-oriented, through the agents' ability to act speculatively. In individual and collective action, perceived external effects from interaction have been internalized.

3 Speculative interactions among all agents and markets in space–time result in aggregate macro ecological interactions (chief among them being competition, cooperation, predation, isolation, amensalism, commensalism). The prevailing aggregate end result is the outcome of macro spatio-temporal ecological determinism, i.e., the composite outcome of all complex, very large in number, and relevant underlying ecological interdependencies.

How this ecological determinism works for the factors involved in the problem at hand, and the extent to which statements 2 and 3 apply, is demonstrated in Chapters 1–3. In this appendix, it is mainly the first of the above statements which is elaborated. Some extensions to the analysis along statements 2 and 3 are also provided. Reference is also made to the connectance between micro and macrolevel determinism. At the individual level micro ecological-economic determinism occurs in which possible macrolevel outcomes have been discounted. By a process of internalization inherent in social (individual and collective) action, all external effects of an action to other social actions perceived by the actor, and the perceived effects of all other actions by all other actors to the actor have been discounted at the moment of action. The link between micro and macro aggregate outcomes is far too nebulous and complex to model, leaing macrolevel ecological interaction and macro ecological determinism as the only interesting areas for an observer to study.

Within this ecological approach, relevant economic factors or sectors must be considered including labor, capital, information, natural resources, technology, output, prices and profits. These factors are traded and determined in composited multifacted markets and are subject to the economic market determinism. There are political, economic, and social factors operating within these markets as well. A central feature of the composite markets in ecological determinism is that they are never efficient (when the term 'efficient' is used in its pure economic definition).

Economic market failures, trade and other spatial barriers, the exercise of political power, multiplicity (rather than uniqueness) in the rules for social (individual and collective) choice, the variety of cultural (aggregate) and behavioral (disaggregate) norms (defining a cultural *and* individual relativism), a variety or morals, and ethical as well as legal standards (moral relativism), the structure and inner connection of the vast variety of social institutions (social relativism), etc., are examples of non-economic factors affecting market conditions.

To the variety of individuals as well as collectives who inhabit this planet, preferences, constraints, decision-making rules and their associated demand functions are not static, smooth, continuous, or strictly convex. The participants in markets are far from being homogeneous, identical (or very similar) consumers. Instead, they are not smoothly heterogeneous, but rather sharply different in their preferences and behavior. Overly simplified assumptions regarding consumer behavior are used for convenience in micro and macroeconomics, where smooth downward-sloping individual or aggregate demand curves and simple additions of individual functions abound in the analysis. Instead, preference, constraint and demand functions are highly non-linear, discontinuous, and fast-changing within an environment characterized by a high degree of heterogeneity and interaction among consumers. Contrary to common belief, it is now quite clear that any two residents of any suburb in any US or European metropolitan area differ a bit from one another. Specialty retailing firms and market analysis have to an extent realized this much sooner than economists.

Equivalently, production, resource constraints, and the associated supply functions of the vast variety of producers involved in diversified production on the globe, are neither smooth, continuous, nor well-behaving in a manner suitable for teaching freshman students in economics at colleges of the west. Demand conditions are always unknown, whereas supply conditions and prices by other firms may be better known to observers than to an individual producer. Analysis at the individual level is never possible, whereas analysis at the aggregate level is more promising. All these specialized markets and their elements are highly interactive. Interactions produce and resolve conflicts, and produce cooperation or predation among socio-spatial stocks and their markets. These interactions force socio-spatial stocks to evolve within deterministic ecological-economic markets.

By addressing 'demand,' 'supply,' and 'prices,' in any of these various factors, the analysis does not adopt an economic perspective. Nor does it rely on the definitions of exactly what 'value,' 'price,' or 'profit' are, or on what their precise economic connectance is. What Marx or Hicks define as value is simply the Marx and Hicks definition of value and of those who decide for one reason or another to adopt them for analysis or action. It is presumptuous to infer that these definitions are the perceptions of 'value,' 'price,' or 'profit' shared by a resident of Westchester County in New York State, by a slum dweller of south Chicago in the US or north-west Lagos in Nigeria, or by the approximately 1.4 billion Indian and Chinese peasants of the 1980s.

The present analysis does not rely on the definition proper or the role of economic variables to model macrobehavior. To the extent that per capita income is concerned, it is used in the empirical part of the work as a proxy for the much broader notion of 'income.' The theoretical part of the work is considerably broader in scope than dealing exclusively with economic relationships and particularly with economic competition, since other forms of interaction are also recognized. It adopts, however, the deterministic mechanism of economic analysis. Similarly, by adopting the six ecological interactions, the analysis does not adopt the animal and plant species ecological perspective or any substantive analogy to it. Rather, it adopts the deterministic mathematical mechanism of ecology in linking one social stock to another. This is the foundation of economic-based ecological determinism.

As social events are the outcome of individual or collective action, the economic determinism in these social actions must be analyzed. Next, determinism of individual actions is presented through an analysis of individual preferences, constraints and their market determinism. Then, determinism of collective actions and public policies is elaborated upon. Emphasis in this exposition is placed on dynamic aspects of these markets and particularly on the role of expectations and social, individual and collective, speculation. Then, two broad phenomena produced by macro spatio-temporal ecological determinism are addressed. They are central in this book, found throughout its exposition. The two phenomena are: (i) dualism, a basic heterogeneity observed in the relative accumulation of social stocks, provided in a spatio-temporal framework; and (ii) dynamic instability in the ecological interaction among socio-spatial stocks, viewed as the cause for dualism.

INDIVIDUAL PREFERENCES AND PUBLIC POLICY MARKETS

In this subsection, the market components of preferences, constraints and decision-making rules are discussed as giving rise to an economic market determinism in social (individual and collective) action. Market determinism is detected in the choice of preferences among social agents, the imposition of relevant binding constraints upon their actions, and of decision-making rules leading to actions. Of especial interest to planners, the adopted speculative framework is expanded to examine the process of planning, and plans in particular, in a speculative context. Numerous insights and broad principles emerge in understanding planning and plans through this lens, at the individual and collective (public policy) fronts, at any spatio-temporal context. Initially, the subject of informal market determinism in preferences for individuals is presented: this is followed by the public policy informal market determinism.

Individual preferences

In accordance with the principles of psychology adopted by behaviorists in neoclassical economics, individual agents are characterized by preference functions (maps) which translate their self-interest into a map of evaluated individual and social states of affairs. In accordance with standard microeconomics, consumers do this through a composite utility function defined over substitutable commodities. Such a function may contain material and non-material (spiritual, moral) arguments (or components), with only physical quantities being recognized in micro-economics. Producers have a composite profit function on the basis of which they rank alternative production states.

The perception to action connectance for both producers or consumers is not so simple, however. In a complex interplay, perceptions affect actions in individual behavior, but in an interactive manner actions affect perceptions, too. This interplay between action and perception leads to a quantum mechanical structure of human behavior which may permit one to observe either but not both simultaneously (Dendrinos 1991). Often, and within various time-scales, this two-way linkage between perceptions and actions exhibit fast and slow motions. It is within such a framework, and when individuals interact with other individuals and collectives, that individual and collective speculative behavior is formed. Some of these social behavior arguments will be further explored later in this appendix.

Under these preference and profit or cost functions, and subject to the relevant constraints and initial endowments they possess as well as a set of prices, individual consumers or producers act in an 'economically rational' manner whereby, by applying a decision-making rule from the theory of optimization, they endeavour to maximize (or minimize) their level of satisfaction, utility or profit, (or disutility, pain) subject to the prices and constraints present. The optimization process is the rational decision-making rule and the deterministic component in the process. Thus, rational action is determined by the psychology of a particular form of individual self-interest, constraints, and the appropriate decision-making rule.

Where do these preference functions, constraints and decision rules come from? Economic analysis has very little to say about all this. Economists only address the point when they deal with social equity issues in deriving the socially optimum distribution of resources (constraints) among individuals in a social group. Then, the position is taken that the adoption of a particular set of aggregate preferences, constraints, and social choice rules is the outcome of some social choice mechanism. This is clearly inadequate in describing the source of social (individual and collective) preferences, constraints, and decision-making rules.

Preference functions are demanded for and supplied in informal preference markets. Adoption of a particular preference function depends on the multifaceted price of holding such a function as currently perceived by an individual. On the supply side, the generation of specific baskets of preference functions is tied to the multifacted price of offering such a basket depending on how this is currently perceived by a collective unit. On what basis are choices among preference functions made by individuals? One cannot assume the existence of a preference function on preference functions, because this would be leading to an infinite regression and thus to logical inconsistencies. A force operating at a more fundamental level must be searched for. Indeed, a *biological* drive must be sought in determining the psychological preferences. This biological force is the drive for survival and socio-spatial dominance, as argued by biologists for animal and plant species. Self-interest is the result of this biological drive which also leads to efforts toward reproduction and access to a multiplicity of resources. Each individual evaluates, instinctively or otherwise, alternative preference functions as to their currently perceived expected likelihood to contribute to survival and dominance. Emotions may be more dominant than reason in these evaluations, although this is not a subject that comes within the bounds of this book.

Biological drives may undergo transitions. For example, drive for survival at a point in space–time may be substituted by a drive toward extinction. Then, the individual will adopt, in accordance with the psychological force for self-interest, the most likely preference function to effectuate self-extinction among the ones available. In a given social group, in space–time, a mix of the two may possibly be found. Presence of suicidal individuals in a pool might be attributed to such a thesis. Hence, three levels of determinism operate at the individual social agent scale: the biological drive, the psychological self-interest force, and the decision-making process underlying rationality at the individual level. It will be argued later, at the aggregate scale, that as a result of a very complex set of interactions among a multiplicity of individual agents, a macro ecological determinism prevails void from these attributes.

A distinction can be made between the 'preference immobile' individuals and those who can freely commute among preference functions. The first category contains those individuals whose preferences are very strongly (culturally) bound, being inherited and transmitted almost unaltered from one generation to the next. In this case, the biological drive to survival might be collectively and hereditarily selected. On the other hand, in the second group containing 'preference mobile' individuals, discrimination occurs among preference functions by each member of the group. The biological drive to survive is individually carried out by a selection process. Each individual adapts to the survival and dominance drive differently. Causes for such differences are not among the topics of this book.

There is also a hybrid group, found between the preference immobile and preference mobile individuals. Individuals who belong to such a social group, through either environmental conditions (war, famine, etc.) or through coercive action by the State, are forced to adopt preference functions. Under different environmental conditions or state of affairs, or by choice, they would not have adopted the preference functions they carry. Supply of preference functions is determined by the multifaceted and currently perceived economic, religious, sociological, cultural and political profits going to the collective unit offering the opportunity for different preference functions to be traded in preference markets.

That there is an informal market for preferences, in space–time, does not necessarily imply that it is always cleared, i.e., that it is always found to be in static equilibrium. Excess demand for, or supply of, preference functions may occur and then a rationing scheme allocates them to individuals – certain individuals are allowed to exercise their choice of opting for particular preferences, whereas other individuals are suppressed from doing so. For it to survive in the preference markets, a preference function may require a minimum number of individuals to adopt it. Moreover, excess demand for a number of preference functions may prevail; or excess supply in the number of preference functions provided may exist in an almost perpetual disequilibrium state with one such state succeeding another. As a result of these conditions, the currently perceived informal market composite price of specific preferences may fall or rise in time, unless perfectly inelastic conditions prevail. Many other economic aspects of these static markets can be inferred from standard theory. They are all left to the interested reader. But a few dynamic aspects will be addressed.

At first, the speculative aspects of preference function holding will be briefly discussed. In its broader definition, speculative action is simply defined as an action by an individual or collective so that when subject to foresight, i.e., when this social entity acts under a spatio-temporal horizon, it discounts currently expected future events for the present. Next, the notion will be used in its stricter definition, whereby speculative action is defined as an action by agent(s) with foresight purchasing futures or options contracts in markets of assets (commodities, stocks, or any other asset) made available by another entity which wishes to diminish its exposure to risk (i.e., hedge) against possible adverse future movement in the value of the assets that it holds.

The entity which holds the valuable resource is a collective, often identified by the State, but not necessarily confined only to this entity in this analysis. However, the State is a dominant component in it, with religious institutions and the family being other relevant entities. Here, a broader ecological entity is thus assumed to be the holder of such assets, defined by cultural, economic, sociological and political components. It is the collective that an individual belongs to, ecologically. The state of this collective is the valuable asset which the collective possesses. There are many possible future states of this collective. Each of these possible states is the outcome of specific social (individual and collective) actions based on the preferences, endowments and constraints the collective's agents possess at any relevant time-period resulting in this expected state.

Supply of options on alternative states as it regards preferences, endowments and constraints goes far beyond mere risk management by the collective in this case, and it involves a broad array of ecological factors. Individuals usually belong to more than one collective, possibly in conflict with each other; the options for preferences and constraints an individual is exposed to may be viewed as risk management from a broader perspective (environment) to which these in-conflict collectives may belong. Both individual buyers and collective sellers of preferences enter these exchanges with stakes in the future state of the social system in terms of individual or collective expected net profit. Individuals and collectives develop conflicts by having different expectations about the future state of the social system or different degrees of net composite profit. Since the future is unknown to social agents there are risks involved in holding expectations.

To hedge against these risks, the State (or, more broadly, the relevant collective) allows for options on these future states to be traded in preference, constraints, and associated expected actions options markets. The adoption of a specific preference function by an individual (and its supply by the relevant collective) represents the purchase of an option (and the supply of that option) on the future state of the social system. Thus, both the preference function selecting individuals and the preference bundle offering collectives act as speculators or hedgers on the future state of affairs in the socio-spatial system in question. The individual and collective expected utility to be enjoyed at that future state of affairs, a variable equivalent to the striking price of the option at its expiration date, has a current price. This is either the composite price that an individual incurs at present for holding onto, over a time horizon, a particular preference function, or it is the currently perceived composite social price offered for that basket of preferences in the preference markets by the collective.

Individuals may drop the specific preference function they hold at any time and acquire another one. However, in preference function markets, there is no resale value as resale of preference functions is not possible. An individual decides to drop the current preference function it holds because it perceives its composite price to have collapsed. Getting closer to the option's expiration date – i.e., the date the expected state of affairs for the individual materializes – the premium on time takes its toll on the composite price of the option unless the individual has perceived this composite price to have risen enough to compensate for the time premium drop. Typically, these events are associated with individuals changing their 'life styles.' All individuals hold, at all time-periods, a preference function. All collectives supply (offer options, i.e., access to) a preference function bundle, at all points in space–time. The bundles may be different, but the underlying process is the same. No individual (observer or participant) or collective is immune to holding such preference functions or bundles, and neither are collectives in supplying such bundles.

Individuals do not speculate at will. During the course of speculation for preference functions they expend resources, both biological and psychological in kind. As a result, they can only afford a few speculative trades. For one, they spend their biological time on these speculative enterprises, during their lifespan. Ineffective speculators, having run out of resources to speculate, stop chasing after many options. They only get involved in a few transactions. Thus, the process allows only for the most successful speculators to persist in trading in options, enhancing the overall chance of the system to move and adapt to changing conditions by means of having it driven by the most successful of its speculators.

This process can be construed, in turn, as representing the conditions whereby the choice of the state of a social system is carried out by means of natural selection based on speculative behavior by its most apt social agents. The fact that ineffective speculators may be limited in their participation in the preference functions, constraints, and market conditions does not necessarily mean that ineffectiveness, as a characteristic for speculators, vanishes too. Its survivability is guaranteed by the fact that it is needed, since the successful individual or collective speculators need the unsuccessful collective or individual ones to profit in the exchange.

This may be why the misery of some may be the source of happiness for others.

Preference markets, as all informal and most formal markets, are inherently inefficient. They are subject to many imperfections and distortions due to their prevailing characteristics. Contrary to efficient commodity markets in various exchanges (where a rather large number of buyers and sellers of the commodities exist, where reselling of options is possible, where a high degree of homogeneity in these commodity markets is found), preference markets are deficient on many counts. They are highly heterogeneous, fragmented, disjointed, localized, and thin.

Close to the subject of this book, typical markets for exchange of preferences among other forums are conferences where, among other things, issues relevant to cities are raised and discussed. An example directly related to the topic of this book was the May 1986 Conference on 'Population and the Urban Future' held in Barcelona and sponsored by the United Nations Fund for Population Activities. It was one in a series of such meetings (not all sponsored by a UN agency) over the period 1974–87. It produced a book on the predicaments of the world's largest cities. The preoccupation of the Conference was to cite unavoidable 'big numbers,' and to accompany them with a recitation of the usual list of predicaments associated with urban conditions in developing (and some developed) countries (poverty, slums, lack of sanitation and health facilities, contaminated water, air pollution, congestion and the like) (see Rusinow 1986). A selling pitch, among others, was for governments to react to these conditions as many public officials were present there. International (and national) development agencies walk a tightrope: they exist because of poverty (the main source for their being in demand) and they survive on the hope of being able to do something about it (the main source of their supply of program and funding functions). Thus, they have to sell to the public the existence of poverty, but not too much of it or its hopelessness for this would render their *raison d'être* obsolete.

Forums where preferences on similar issues are supplied are provided by the vast medium of published material. For the subject at hand, an illustrative case is the 1987 book in the series by L. Brown *et al.* on the *State of the World*, and its section on 'the future of urbanization.' As is usually the case, the *State of the World* reports supply numerous 'consciousness raising' problems along many dimensions ranging from environmental to economic and demographic. By citing many examples of 'impending catastrophes' and by drawing from various areas of the globe, various calls to action are made. Another such outlet is the annual *World Development Reports* by the World Bank.

In all cases, the aim of these opinion-making (supply) forums is to enter some of these components (particularly what is in the current political agenda, and especially those associated with the earth's environment) and various social conditions into the preference functions of influential individuals and policy makers. All such forums apply, as well as advocate, positions on preferences. They do so under the assumption that preferences are acquired. As mentioned earlier, short of accepting that individuals are born with and carry hereditarily transmitted preferences, one must expect that most of the social agents (individually or collectively) accumulate preferences through a 'socialization' or 'acculturation' process, as sociologists put it.

It is of interest to look at the mechanics of this accumulation process, to place it in its ecological perspective. What has been overlooked in the process of acquisition of preferences is that the holding onto or discarding of preferences responds to the movement in the currently perceived composite maintenance cost to an individual of the option to carry a specific preference function at a point in space–time. Since there is no salvage value and no trade of used preferences, the relevant price to an individual holding a preference function is its perceived maintenance cost. Individuals have the option to act upon such functions at any time in the future or in any location in space, responding to individually perceived movements in these maintenance costs in space–time. A feature of particular interest in such perceived costs is that they fluctuate, at times sharply.

What most of the forums in which such preferences are supplied fail to do is to provide the expected dynamics (and associated risks), as well as the exact nature of these composite prices and costs to all relevant social groups. This may not always be by design, but out of mere inability to do so. In any case, individuals are left guessing and consequently their only recourse is to take bets on such functions in these markets. That is the case for those who have access to these markets. But the vast majority of the people on the globe do not and cannot afford such access. In this case, particular preferences become sticky in time, transmitted through socialization almost intact from one generation to the next, as in traditional (or primitive) societies. They are, in this instance, highly localized, belonging to preference clubs.

In other instances, as already mentioned, at times individuals are subjected through coercion to a State-induced change in preferences by various preference-changing procedures, or through a variety of economic and other (individual or collective) inducements or incentives/ disincentives. In general, because they are highly fragmented and thin, preference markets operate like exclusive economic clubs with a very high access cost at times. Despite the occasionally sharp fluctuations in the perceived prices of preference options, there is very low mobility among preference functions, in space–time, particularly at the low (and very high) end of the social hierarchy. Commuting costs among preference functions are high, particularly the psychological and social components of the cost structure. As a result, at times of drastic price fluctuations, there is considerable disequilibrium observed in the preference markets.

An example can illustrate the importance of access costs to preference markets quite vividly in terms of purchasing power parity (see R.J. Gordon 1978: 557). A book published by a press in the US at a price of US$25 in 1985 had an equivalent cost, given current per capita income differentials, of US$1,500 in India. This mere transfer in price equivalent terms highlights not only the economic inaccessibility of intellectual markets in low income areas, but also another force governing trade flows (their type and intensity) among rich and poor nations.

Naturally, there are other barriers to the exchange in preferences, not economic but broadly defined as socio-cultural. They may be far more difficult for various preferences to penetrate. As an example, one may cite the fact that certain slum areas in the large urban agglomerations of the Third World (as well as some in MDCs, the South Bronx of New York City and East Los Angeles being cases in point), together with certain rural regions of many nations, are inaccessible to outside observers or ideas. The problems cited by Brown *et al.* would have a hard time finding their way into such communities in isolation.

These observations reinforce the point that what concerns some social groups may or may not concern others. The existence of fragmented (spatio-temporally) and heterogeneous preference markets may be the single most important factor in the presence of conflict among social groups or individual agents. Further, the presence of these conflicts is exactly why speculative behavior (speculating or hedging) may underlie markets of preference functions.

Some may argue, O. Lewis (in *La Vida*) (1966) and E. Banfield (in *The Unheavenly City*) (1970) for example, that the informal class does not speculate, because they lack foresight. This is an erroneous view stemming from the fact that these authors were not aware of the principles of speculation in dynamic markets or ecologies. That time-spans for speculative action by individuals belonging to the informal class are short can be attributed to the conditions facing these groups. As societies at war do not find it appropriate to speculate over long-term horizons (for one, they do not know when the war will end), so do members of the underclass. It is not efficient for members of the underclass to do so, not because they are (biologically, psychologically, economically, or socially) incapable of speculative action as Lewis and Banfield imply. They do speculate over very short ranges.

But speculation requires resources. Individuals of the underclass do not have as plentiful resources as the other members of society to speculate extensively. As they turn up losing, they seem to rapidly deplete their small reserves. Thus, they do not speculate a lot because they lack resources, and not the other way around as Lewis and Banfield argue. These arguments hold not only on the subject of speculation on preferences, but also on the subject of speculative economic behavior given a set of preferences by the members of the underclass.

One can posit similar arguments on the subject of constraints, and not much will be repeated. Inherited, environmentally or socially imposed or individually perceived bundles of constraints are conditions affecting the constraint basket markets of individuals. Given these preferences and constraints and the manner they are being adopted by social agents, the focus now turns briefly to the subject of decision-making rules.

Choice rules are indeed acquired in a manner similar to the one discussed earlier for choice of preference functions. An optimization (problem solving) process always underlies an individual's action, but the form of such an optimizing function is unknown to an observer, and possibly to the individual too. It must be highly complex, and directly affected by matters of individual and collective emotion, intelligence, knowledge, ability to generate and evaluate alternatives, learning, and the quantity of information it is capable of absorbing, etc. It must contain dynamic non-linearities in its interaction with other agents' decision-making rules. An individual chooses a bundle which simultaneously involves preferences, constraints, decision-making rules and discounted interactions from other agents' choice of bundle.

Such a process is in fact the process by which an individual or group is 'socialized' at any point in time within the broader socio-spatial system in which it is to function. To an observer whose interest goes beyond any single individual (or group) and extends to the study of the socio-spatial system, an effort to detect the aggregate behavior might reap more reward than the effort to isolate components and study them separately and then attempt to link them. The second path, so extensively followed in contemporary social science, may be a labyrinth from which the social scientists will find it difficult to escape.

Public policies, public actions, and their markets

An area where the hypothesis of ecological determinism may be questioned is that of public action. The question may arise as to whether a public action may change the deterministic course of the socio-spatial system's dynamic. This question may be answered negatively on two grounds. First, that if a large enough socio-spatial system is considered (like the system at hand) then there is no specific public agency with enough potential action to significantly affect it, the UN notwithstanding. Second, and most important, is the point that any public body's action (and policy) must be grounded in some form of public opinion, i.e., it must be demand driven. Thus, what enters the public decision-making calculus, and therefore affects policy and public action markets, indicates that there is little left of a non-deterministic nature.

Public actions are traded in informal inefficient public action markets, similar to public policies being traded in informal inefficient public policy markets. Whether these policies and actions are congruent it is not of direct interest here, as is the case regarding the question as to whether public actions and policies are sharp or fuzzy at any point in space–time. The quality of these attributes is itself subject to a market determinism. Next, the analysis focuses on public policy markets; it is equivalent to the analysis one could carry out regarding public action markets.

Policies are traded in policy markets where individual and collective conflict resolution, cooperation, etc., occur. There are many types of policies at any given point in space–time, each having its own policy submarket. For example, housing policies for low income families and mass transit policies are two cases of sectorally, spatially, and social-group targeted, partly fragmented, but ecologically interacting (in a competitive, cooperative, predatory, isolative manner) policy markets. Within each policy submarket different policies may be supplied at any particular point in time. For each policy in this submarket there is demand by different socio-spatial groups affected by it. Demand for a particular policy and supply of it are directly linked to the complex, composite perceived price to its adopters and supplier(s). Market clearing mechanisms are present, but they are now different than in the case of preference or constraint bundles.

Market clearing mechanisms, or conflict resolution guidelines, are referred to as 'social choice rules' in policy sciences. The outcome of these rules for mutually exclusive policies is for one policy, rather than a mix of policies, to be adopted at any point in time. Over time, however, public bodies may experiment with alternative policies. They may adopt a 'pendulum'-type mix of strategies whereby one policy is substituted for another from one period to the next. At any given moment the winning policy prevails, having survived among a number of mutually exclusive alternative competing policies in the particular policy submarket.

The choice of a particular policy in the policy submarket results in a social state of redistributed resources among competing social groups. Further, the social choice rule responsible for the choice is thus directly linked to the new expected future social state. This may be the reason why the procedural aspects of public policy making may be much more important than the substantive.

Suppliers of policies (and the collective interests they represent) and adopters of policies have expectations regarding their level of satisfaction from the envisioned future state of affairs. These expectations may be partly or fully congruent; or, they could be incongruous. The composite price they incur by participating in policy submarkets and by adopting or supplying a particular policy reflects this expected future payoff. As expectations about this future state change in time, so do the composite current prices of policies leading to such a future state of affairs.

Thus, policies are nothing but options governments and individuals trade in policy option markets. One of the composite costs involved in a policy option is the cost of planning for that option. Plans are not implemented (they get shelved) when their option composite price collapses in the options market as it is no longer desirable to exercise the option (i.e., implement the plan). Policy option prices fluctuate considerably in time. The likelihood of executing specific public policies, i.e., their adoption and/or supply, largely depends on the magnitude of movements under these fluctuations.

There are linkages between policy submarkets. In urban areas, for example, transportation, housing, public utilities, health services, and environmental policy submarkets interact to a great extent. Under this perspective one could structure the above arguments to an individual policy within an ecological framework, in reference to a bundle of policies – traded, supplied and demanded – and a social choice rule resulting in the choice of a surviving bundle of policies, rather than to an individual policy within a policy submarket. As numerous policies interact within a spatial unit, the possibility for instability is strong there.

Another interaction of interest is that between the individual choice of preference function and constraint bundle and the adoption of a bundle of policies, and the supply of preference function and constraint bundles and the supply of policy bundles. These interactions now occur in an ecological deterministic framework simultaneously as one directly impacts the other. This interaction takes place in an ecologically complex manner, through a process of discounting one for the other.

Social groups, collectives and individuals, react to the existence of preference, constraint, and policy bundle markets. Each social agent discounts in a current choice (leading to individual action) all other relevant social agent choices (and expected actions). Consequently, the choice of particular bundles of preferences, constraints, and policies is the end result of a complex ecological game. Its outcome is in turn discounted in the type of ecological interaction prevailing among the many socio-spatial groups in time. Choices regarding residence by individuals are indeed choices among spatially competing bundles of this type. So, the statement (see point 3, p. 244) of the ecological-economic based determinism materializes, as aggregate macro ecological determinism governs social dynamic behavior.

At times in 'policy science' the argument is made that aggregate studies are not policy-sensitive (which is true), and that disaggregate analyses, because they allegedly contain extensive lists of social groups and variables including 'policy' ones, are useful for policy making (which is erroneous). It is still to be demonstrated either that, in any policy science field (micro and macroeconomics, demography, political science and public policy, urban planning, or any other field), modeling of social systems alone (divorced that is from a patron) can make or have made a difference in what is or has been an ecological (i.e., culturally, economically, and politically bound) decision. Or, that enough disaggregation can be attained in any policy-oriented model to be close enough to individual preferences or constraints so that an otherwise unanticipated public action becomes possible and thus what is considered as the true finding from a policy-making model becomes effective.

Despite these failures, the efforts are not over, and social scientists still seem to pursue the elusive goal of 'detached' disaggregate behavioral modeling and policy making. It is not so in the case of 'attached' modeling efforts where motivation for such efforts mostly comes from government sources which need 'policy-related' justifications in order to either fund these studies or justify programs they wish to see implemented. How strongly justified this effort is on scientific grounds, however, remains an open question. Policy making and planning are simply manifestations of broader sociological, economic, political, cultural, and other conditions prevailing in a social system in space–time. To make any claims to the opposite is simply preposterous.

DUALISMS

This subsection will address the general phenomenon of dualism, a major and widespread socio-economic spatio-temporal event, and its ecological determinism foundation will be laid out. It will be argued that sustained socio-spatial dualisms are due to the dynamically unstable multiple stocks, multiple locations ecological interactions.

In the field of economics, dualism has been studied from a rather narrow and production standpoint and in reference to agricultural versus manufacturing economies by N. Kaldor (1967) and P.J. Verdoorn (1949). Kaldor's two-section growth model asserts that the propeller to growth in the (leading) advanced (manufacturing) sector is its ability to draw (surplus) labor from the lagging (agricultural) sector. Maturity (and thus an end to growth) occurs when the leading and lagging sectors' wages equalize and thus no labor flows take place. A key element in Kaldor's model is Verdoorn's law derived from empirical evidence, which states that productivity increases as output in the (leading) sector is increased too. Casetti (1984b) analyzed Verdoorn law in reference to the Sunbelt–Snowbelt shifts in the US regions during the 1958–76 time-period.

A particular version of regional development dualism was analyzed by Dendrinos (1984a, b), in a region–antiregion framework. The dynamic regional ecological (population–income) structure of the US was shown to be leading toward a spatial dualism. A configuration involving two stable and one unstable equilibria was detected. The two stable equilibria represent an upper (developed) and a lower (underdeveloped) state, and the nine US geographical divisions seem to have gravitated toward these two states during the 1929–79 period. Limited work by Tabuchi (1988) regarding the Japanese regional structure turned up qualitatively similar results to the extent that population shares are concerned. Dualism has also been observed and documented in the regional population structure of the People's Republic of China (Dendrinos 1989, Dendrinos and Zhang 1988).

Spatio-temporal dualism is much broader than economic production shifts. It covers a broad array of stocks, locations, and their behavior in various time-periods. Broad ecological forces are at work. In general, ecological determinism and the associated developmental or evolutionary events are strongly linked to the heterogeneity of social stocks in space–time. Heterogeneity feeds ecological socio-spatial interactions, and in turn socio-spatial interactions contribute to socio-spatial heterogeneity.

In a capsule, as socio-spatial heterogeneity increases and interaction becomes more extensive, but also thinner, instability sets in whereby some socio-spatial stocks grow or decline exponentially due to their own strong rates of change and as the effect of interaction diminishes among these stocks. This is the developmental aspect of the heterogeneity and stability link. The evolutionary aspect is that, due to a change in parameters, the nature of the stock interaction may undergo phase transitions. Either stability or (most likely) instability occurs, shifting the ecosystem to a new dynamically stable (or most likely unstable) equilibrium.

There are many ways one might approach and record heterogeneity in social stocks. One can disaggregate social groups by income, ethnic composition, age, religion, education or any other socio-economic factor. In spite of the possibly infinite number in dimensions and measurements one can use to group human micro and macrosocieties, one in particular stands out. It often evokes emotional reactions. This form of social segregation is refered to as the phenomenon of social dualism.

Specifically, heterogeneous socio-spatial stocks, i.e., groups like the underclass, middle class, or the elites that differ along a number of socio-economic dimensions (like income, or education, or at times ethnicity), have a self-growth component and an interacton (stock transformation) component at any point in space–time. The stock dynamics of these socio-spatial groups is directly related to the number of heterogeneous groups interacting in space–time, the magnitude (size) of the (heterogeneous) groups interacting in space–time, the magnitude of their interaction coefficient, and finally, each group's self-growth or decline rate. Environmental changes may affect these groups' dynamics by altering the qualitative properties of the prevailing dynamic equilibrium. If one group's self-growth rate is relatively large compared with other groups' rates, then this group will tend to dominate the dynamics of the community of interacting groups in the ecology either through its explosive growth or by its extinction.

Socio-spatial aspects of dualism

Large-in-scale urban dynamical events, as described in this book, reveal the presence of a macrolevel 'dualism' in the evolution of large urban agglomerations. Distinctly different dynamic paths are traced by the giant urban areas of less developed nations than those of more developed ones. This dichotomy is not unique at the inter-urban dynamics presented here. It has been identified also in intra-urban dynamics.

Dualism in its intra-urban form has been associated with 'informal sector' activity. It is multifaceted. On the production side it is used to classify economic petty production or distribution activity or the very small-scale enterprise widespread among the non-formally employed residents of cities in the Third World – 'non-formally employed' being a group which various censuses classify as the officially unemployed or underemployed, or 'other,' or avoid classifying altogether. Of course, any one who is not formally employed, and survives, must be considered informally employed, no matter the means employed to secure food.

On the consumption side, the term is used to classify (legal or illegal) street or slum dwellers. Many studies have described the phenomenon of dualism in the urban scene of both the less, as well as the more, developed nations. Depending on the authors' ideological bias, dualism is at times referred to as 'marginality,' 'informal' sector activity, or 'underclass.' What role this informal sector plays in the urban development process is of course subject to dispute, see Perlman (1976) and Linn (1983) among many others.

The topic of dualism has attracted considerable attention over the past quarter century, mainly because – depending on the angle one takes on it – justification is obtained, or the impetus for value-laden urban and national policy making and intervention is ignited. Inter-urban migration constraints, applied in many developed and developing nations, form one of these policies aimed at limiting the population abundance of the lowest class and its ability to colonize space, in particular that of the core region or primate cities. At the same time, another implicit or explicit objective of these policies is to increase, at least in relative terms, the population abundance of the formal sector.

In the 1970s and 1980s, among many other places, such policies were explicit in Singapore, Jakarta, and the four largest urban regions in China. Population control coupled with strict migration policies were attempted during that time-period in the People's Republic of China to limit colonization of space by the rural (or small urban) areas' poor. Transmigration policies out of the island of Java and Jakarta were implemented in Indonesia to this effect.

Central to these debates has been the issue of market versus optimum city size and the role that *externalities* play in this debate, a role which has been extensively addressed in the economic (non-spatial and spatial) literature and advocated as one of the main reasons for having governments and to accept their regulatory actions. Arguments for a market-city size rest on standard neoclassical microeconomic efficiency principles: if the *marginal city dweller* perceives a higher utility level in a city compared with any other location within a nation, then city size will (and ought to) grow. Obviously there is an externality present, and the perceived individual marginal utility is not the same as the marginal utility from a social standpoint.

Arguments for population control are basically based on this external effect where the argument is laid out. An individual's action, according to the argument, has an effect upon the rest of the community, with these effects not taken into account by the acting individual at the time of his action (i.e., they have not been internalized).

On social equity grounds, another *raison d'être* for governments, the accent is on externalities. If by the marginal individual entering the city non-internalized net negative external effects of agglomeration occur, for the city or the nation as a whole, then governmental intervention is deemed warranted to prevent growth, and at times strict governmental control follows.

What exactly is meant by an 'externality' is not at all clear, and neither is the notion of a 'non-internalized' external effect. Every action has a bundle of effects; some of them can be positive whereas others can be considered to be negative at particular time-periods and by particular individuals or social groups for a variety of reasons which might not be durable in space–time. Thus, it is a relative notion. An individual might cause congestion in a road by going to work, thereby possibly creating a negative (and, if not perceived or tolled, non-internalized) external effect. By pricing-out some of this activity, through the imposition of a toll to handle the congestion externality, one eliminates some of the product of this worker's labor which could be of benefit to society in a manner not fully accounted for by the wage paid to the worker.

But most importantly, the notion of non-internalized externalities is subject to serious doubt on the following grounds: on what count can an observer conclude that an individual trip-maker, incurring (un-compensated) as well as causing (non-compensating) congesting costs in the daily traffic conditions for the trip to work (for instance), has not taken these conditions into account when deciding to participate in the congestion game? From a speculative standpoint, the trip-maker can be viewed as having very likely taken both costs into account in a repetitive game played daily, where position in the road is close to random each day. On these grounds, it is not clear what is the regulatory authority of objectives of a transport agency charged with regulating (pricing) congestion.

Externalities imply a prespecified spatial and temporal extent over the boundaries of which they 'spill.' Redefining these spatio-temporal boundaries makes these externalities disappear. The setting of political jurisdictions over which these externalities occur must not be regarded as being divorced from the current or eventual appearance of 'downstream' or 'downwind' effects. In the definition of spatio-temporal jurisdictions the so-called external effects have indeed been discounted.

Similarly, arguments can be suggested for the question of externalities in producing offspring or receiving migrants, the two factors which produce increases in a setting's population. In the decision to have children the parents may have discounted for the costs to be created and the costs to be occurred by the offspring to themselves, the offspring, and the society at large. More importantly, the rest of the individuals in the group to be impacted on by this action must have discounted that possible action by their very presence in that setting. Equivalent are arguments for the migrating populations.

To have considered and temporarily accepted the possible (at least reasonable and legal possibly even illegal) actions by others it does not necessarily imply that an individual or collective might not act on these perceived 'non-internalized' effects. Oftentimes this reaction takes the form of attempts to change the rules of the game, for example the laws and regulations impacting on actions which may carry perceived non-internalized effects. The example of workers in nuclear arms or reactor plants and their families residing nearby who seek legal action for effects incurred while working there is a case in point. Even malicious intent by employers knowing about the danger and not informing their employees must be considered as having been discounted in an employee's willingness to work there. No matter how unreasonable a group of one's peers is likely to find this argument to be, to an observer an agent could well be assumed to act as if this is the case.

One might find the premise that individuals must internalize in their actions all external effects directed toward them as too much of a responsibility imposed on anyone, and pushing the limits of what is considered as negligence a bit too far. Furthermore, to assume that the one causing the externality without paying for it might not be carrying as much legal burden as the individual incurring it (and presumably expected to take steps to avoid it), could be considered ethically appalling or even illegal under particular systems of jurisprudence. But this value judgement does not violate the basic premise of the argument which holds agents responsible for internalizing all possible perceived effects from their, and all other, relevant actions and thus not being negligent. Those that have not successfully speculated on such external effects, or have been too negligent, pay the social and individual cost imposed on them.

There are urban economic agents (producers, consumers, or governments) that do not internalize (i.e., perceive and in turn account in their actions for) the full net costs resulting from their actions. They must fall into two broad categories: those who decide to do so out of rational choice and those who are unaware of such externalities. Those who incur these negative effects may have already discounted them in their decision to participate in the game.

Let us look closer into the externality question from the point of view of those who cause it (who may also be those who incur it, as well). The first group contains agents who, although cognizant of these effects, are not induced or coerced to internalize their (negative) externalities. This group contains economic agents with utility or production functions explicitly containing components such as congestion, pollution, and other negative external effects. Individuals in this group hold a utility or production function that places relatively low weights on negative externalities when compared with other components in their utility or production function. In the second group are individuals indifferent to or ignorant of such negative effects, i.e., agents whose utility or production functions do not contain such components.

Individuals in the first category, in the absence of appropriate incentives, opt to not fully internalize the negative consequences of their action, whereas in the second case individuals are lacking the social consciousness or necessary information to make choices relevant to these issues. One must know that by participating in a social group where both of these types of individuals exist they run a risk. In other words, one has internalized in the choice to locate there for (a) the possible negative and positive actions of all other groups, (b) the appropriate odds regarding the expected actions by these groups, and (c) for the benefits and costs one expects to enjoy or suffer by one's own actions of being member of the group. It does not mean that there are no surprises, but these can be discounted altogether as they may occur anywhere.

But who are these individuals that are either motivated or coerced not to reveal the external effects of their action (group one), although they are aware of them? And who are those that are unaware of the external effects of their action (group two)? Whose interests are governments serving? To answer these questions one must resort to a social disaggregation of society into groups.

Agents coming under the first category include individuals who belong to middle and upper income classes, to the social, political and cultural 'elites' of various urban areas, who are upper-echelon public officials, or medium or large-scale producers. On the other hand, the lower income and the underclass of the urban agglomerations worldwide are comprised of, to a great extent, the second group of agents. It includes the slum dwellers, the petty producers or traders, the lower-level public servants. The marginal city dweller – the individual entering the city at the margin – is a member of this group.

In view of this antithesis (the recognized 'dualism' in the consumer and producer sectors of an urban economy) it is of interest to go one step further and ask the question as to who defines what 'negative' effects and the associated notions of 'optimum city size' are. Of course, such definitions come out of the cultural, social, economic, legal, religious, ethical, and political norms of the formal class. Concerns of the various elites within this class form the social agenda of problems addressed by public policy. Their agents, the governmental bodies who represent the dominant and prevailing interests within these urban areas, regulate and implement these policies.

The underclass, in this basic dichotomy, antithesis, or disaggregation, does not articulate such concerns in written or verbal form. It does not explicitly express its preference structure, which is only represented or communicated through the lenses of representatives or observers from the formal class. No members of the underclass participate in conferences or write books to tell us how they think and act.

To put it vividly: that Jakarta's population was reaching 12 million people in the mid-1980s was of little concern to the newcomer in the slums of the western outskirts of Jakarta in 1985; and the fact that the world's urban population had risen from 20 percent in 1950 to 43 percent in 1986 was certainly not the cause of sleepless night to such a dweller. The fact that rain-forest ecology was being depleted at accelerating rates in Borneo, the Amazon region, or along the eastern coastal area of Madagascar during the 1980s has and will have little meaning to a street dweller in Bombay. As lung cancer mortality rates are much higher in Mexico City, Seoul, or Shanghai than their nations' average in the mid-1980s was, is, and will continue to be of little concern to the members of these cities' immigrants or underclass, or to a resident of

Cairo's City of the Dead, when compared with other factors keeping or attracting them there. In fact all these events might tell different things to these dwellers than they convey to planners or policy makers.

Problem definition does not occur in a vacuum. Different expectations and perceptions by a number of social groups in space–time are involved. Problem definition implies that there are specific well-defined or fuzzy social groups which currently define and perceive a problem; that certain social groups are the current main cost-bearing entities; that certain social groups are the current main beneficiaries of the problem; that certain social groups are most likely to be the beneficiaries *after* the problem has been 'addressed and/or resolved'; that certain social groups are most likely to be the cost-bearing entities after the intervention; and that certain social groups must be involved in carrying out the intervention and incur the associated, as currently perceived, costs and benefits to carry out the public action to 'correct' the problem.

One might argue that there is significant overlap in the perception and assessment of 'problems' among all these social groups. Urban problems, it may be argued, cross class barriers. The very presence of these 'problems' is proof to the contrary, however. It may be argued that 'altruistic' behaviour by those that 'know better,' substitutes, or should substitute, for the lack of knowledge of those that 'do not know.' Of course all this might also be viewed as gratuitous paternalism.

Existence of altruistic behavior, whereby an individual's utility level contains the utility level of others, is a means through which economists account for charitable behavior. But altruistic behavior contributes to the welfare of the donor. It is consistent with the principle of self-interest, and thus consistent with market determinism.

While the subject of market versus optimum city size was looked at from the viewpoint of the marginal city dweller, and the basis for intervention and policy making by urban and national governments was made on the grounds of net negative external effects of agglomeration to the particular city or nation, one can transfer these theoretical considerations to a worldwide environment.

The dualism between the formal and informal urban sectors now crosses national boundaries and becomes the dualism between the formal and informal sectors worldwide. Whereas the focus was previously on urban-related problems, now the attention is drawn to global problems. Urban growth policies are now global population policies, and inter-urban migration controls now become equivalent to national immigration rules.

Intra-urban net negative (or positive) external effects of agglomeration now become net external effects of global population growth. Internalization of external effects now crosses national boundaries, as do externalities of all kinds (not merely environmental, but also economic, political, ideological, cultural, and other). Thus, one could substitute 'world' for 'urban' in all prior statements made.

Ecological interactions and instability

Now the analytical aspects of dualism are addressed. Each of the diverse socio-economic groups found in urban agglomerations interacts in space–time with all other groups. The socio-spatial interaction is very complex but its ecological classifications are small in number and quite specific. Mathematical ecology provides a comprehensive framework for addressing and modeling such interactions and their effects upon the size of the urban socio-spatial stocks in time.

If each group's pair-wise interactions with any other group are looked at, then there are in all six possible ecological interactions feasible: competitive, cooperative, predatory, isolative, commensal and amensal. The various social stock sizes, on the other hand, exhibit one among a small number of alternative dynamic paths: they can grow or decline exponentially but slowly, follow a (periodic or non-periodic) cyclical movement, remain at a steady state, follow an exponentially fast growth path and become dominant over all other social groups, or follow an exponentially fast decline path and become extinct. Thus, coexistence or competitive exclusion of these different social populations can be studied. These are the two qualitatively different end results one can obtain by modeling socio-spatial group interaction in an ecological framework.

One can consider a number of disaggregations in reference to an urban population by recognizing the existence of four main social groups: an extremely small-in-size elite, a modest middle class, a sizeable working class, and a vast underclass. This is the conventional manner in which sociologists carry out their analyses with an eye on developing nations.

Possibly one can be a little more aggregate and recognize only three groups: the upper class, the middle class, and the lower class. Or, one can be very aggregate and consider only a two-group breakdown: a formal and an informal class. As it will be seen later, a three or more stocks disaggregation does not alter the qualitative dynamics. These dynamics are significantly different when one moves from a two-group to the three-group breakdown.

Under these social specifications, one is interested in looking at the constituent elements in the dynamics of population abundance for each social group. In the case of the underclass, for example, two rates of change are of interest. First, the rate of replenishment of the underclass, which is the rate of transformation into the informal class by downwardly mobile individuals belonging to the upper classes, and the birth-rate in the underclass. Second, the rate of depletion of the underclass, which is the rate of transformation into the formal class by upwardly mobile individuals belonging to the lowest class, and its death-rate. Upward or downward mobility are the interaction coefficients; the net growth rate is the difference between the birth- and death-rates.

A theorem in dynamic mathematical ecology, due to May (1974), is instrumental in classifying these social dynamics. The full exposition of this theorem is provided in Chapter 2 (see pp. 78–82) for the case of international trade, and it is discussed throughout Chapter 3 in reference to inter-urban interactions. An associated and more basic theorem on dynamical stability is due to S. Smale (1966). More recent numerical simulation on discrete and continuous dynamic models also provide new information of the stability properties of socio-spatial dynamics (see Dendrinos 1985, Dendrinos and Sonis 1990).

In summary, these findings seem to support the following conclusions: dynamic systems are overwhelmingly unstable, the more so the larger the number of interacting stocks or locations. Only when two populations interact, and then under very restrictive conditions, can they coexist in the long run. In all other cases the likelihood is very high that one will grow exponentially large and dominate all others. At the same time, at least one (and possible more) of the other stocks will become extinct.

In a multiple social groups interaction, one must expect that the transformation (interaction) coefficients are relatively small, that the self-growth or decline coefficients are dominant, and that the growth rate of the lower classes is much higher than that of the upper classes. One must expect under these conditions that the underclass, having the highest self-growth coefficient, will dominate the community of interactive social groups. The reasons why the underclass has the highest growth rate were touched upon in Chapter 1 (particularly on pp. 48–54) and in Appendix 2, on theories of economic growth. Why are the upper classes so interested in controlling the lowest classes' population growth rate is very evident from these qualitative properties of the postulated dynamics.

Under the three groups breakdown, and given the internally unstable dynamics of three species interaction, one might expect the middle class to be only a temporary phenomenon with population stock gravitating to two poles, the upper class or the underclass. A different scenario may lead to the extinction of both the upper class and the underclass, and the emergence of a single egalitarian social middle class. The latter is also the outcome of unstable dynamics. Under the two groups breakdown of an upper and underclass, the dynamics can be stable, i.e., the two groups can coexist in the long term if their association is of the predator–prey type.

The qualitative statements of ecological determinism do not necessarily select among these alternative community associations, and thus futures. They only classify the possible outcomes. Although the dynamics at the original perturbation, i.e., the starting point, are unstable they may lead to eventually stable and sustainable end states.

This analysis presents a source of potentially conflicting conclusions to be drawn from current evidence. Is what we currently observe in social group dynamics the outcome of continuous stable dynamics since the original perturbation or is it the tail-end of originally unstable dynamics? Is, in other words, the presence of a strong underclass coexisting with elites evidence of stable dynamics, or is it evidence of the tail-end of continuing unstable dynamics which have not played themselves out yet? Some speculative statements are provided in Chapter 4 in answer to this question.

In principle, these dynamics lend support to the argument that intrinsic analytical properties of ecological systems underlie the strong presence of an underclass. They do not support the hypothesis that some socially conscious decision, whatever its source (government, individuals, or specific social groups) produced this dualism. The term 'conscious decision' is used to distinguish it from 'collective decision' as used in the ecological determinism framework.

2

A BRIEF REVIEW OF MAJOR THEORIES OF DEVELOPMENT

PURE ECONOMIC THEORIES OF DEVELOPMENT

Theories of (absolute) growth in population and wealth and their interaction are found in classical and modern economic growth theory and in the theory of international trade, both in the field of macroeconomics. The classical work took place in the past two centuries by A. Smith, D. Ricardo, R. Malthus and J.S. Mill. Their work will not be reviewed here, but it provides the basis for the currently available purely macroeconomic theories of development, modern economic growth theory, and the theory of international trade to be reviewed next.

Macroeconomics deals with economic aggregates and its subareas are diverse, depending on the particular school and time-period considered. The field of macroeconomics is as varied as that of demography. Many subjects are covered and numerous interpretations and schools exist on these topics. To a great extent any abstraction of the available literature is subjective, depending on who is doing the survey or extension of the original works these fields are based on. It is, however, a fair statement to say that the major emphasis is in the role that capital plays in the function of an economy.

For an ongoing debate in economics as to the definition and measurement of capital, prices, value and (rate of) profits, see the discussion of J. Hicks (1965), P. Sraffa (1960), I. Steedman (1977, 1984), and E. Mandel and A. Freeman (1984) among others. The manner in which dynamics are viewed within capital theory is of limited interest here. Thus, the currently central themes from macroeconomics are only tangentially relevant to the matters discussed in this book. A few books that can supply slightly or significantly different surveys of macroeconomics are (chronologically): G. Ackley (1961), where the subjects of macroeconomics are split on the basis of the time-period they address (medium or long term); R.G.D. Allen (1968); M.H. Miller and C.W. Upton (1974); B.N. Siepel (1974); C.L.F. Artfield, D. Demery and N.W. Duck (1985), and others.

All seem to have a certain overlap, as far as the subject of this book is concerned, regarding the topics covered. Namely, economic growth theory, international trade, and technological progress are always topics addressed by these and other works on macroeconomics. All seem to draw from the classics, A. Smith's (1776) *The Wealth of Nations*, D. Ricardo's (1891) *Principles of Political Economy and Taxation*, R. Malthus's (1798) *An Essay on the Principle of Population* (first edition), and J.S. Mill's (1848) *Principles of Political Economy*. The modern macroeconomic theory is largely based on the work by J.M. Keynes which appeared in his *General Theory of Employment, Interest and Money*, first published in 1936.

Briefly, macroeconomics can be broadly classified under classical macroeconomics (involving Say's law, the theory of full employment and equilibrium, and Wicksell's model), Keynesian macroeconomics (with the question of non-full employment, and speculative aspects of demand being dominant elements in it), and neo-Keynesian macro-economics (addressing issues of business cycles, neoclassical economic growth, technological change, and international trade): see G. Ackley (1961) and a survey by Choi (1983). Surveys also include those by Hahn and Matthews (1964), Nelson and Winter (1974), Harris (1975) and Todaro (1982) among many others.

In more recent textbooks in the field there seems to be a breakdown in terms of fiscal and monetary issues, as population subjects retreat when compared with earlier texts. For example, in R.J. Gordon (1978), the role of government budget deficits is emphasized, together with the flows of goods and capital among open economies with fluctuating exchange rates. In J.W. Elliott (1979), the concern is for revelance and realism in addressing cyclical behavior in macroeconomic indicators; the classical, drawing from the work by L. Walras (1874), W.S. Jevons (1871), and A. Marchall (1922), and Keynesian macromodels are discussed: the classical model implies that most markets are efficient, that prices in all markets adjust to equilibrium levels all or most of the time, and in the labor market in particular the demand for and supply of labor together set the wage rate (at a level equal to its marginal product), p. 132; in the Keynesian model, an increase of government expenditures, through a multiplier effect, brings about an increase in real output, p. 146; and a comparison between the two according to which 'shifts in aggregate expenditures (such as government spending) do not produce changes in output and jobs in the Keynesian model if one assumes wages continuously adjusting to product price changes,' p. 156.

In W.L. Peterson (1980), the Keynesian fiscal approach (government expenditures, taxes and their multipliers) and the monetarists' approach (money supply and interest rate policies) are juxtaposed; a good introductory discussion to the IS (interest equals savings at equilibrium) and LM (liquidity preference equals supply of money at equilibrium) curves is given: their interaction, in the interest rate–net national product space, defines an overall equilibrium for the economy. 'Although the model is originally due to Keynes, Hicks' (1937) formulation of the IS and LM curves is the basis for contemporary work' (Peterson 1980: 388). How much of all that is brilliant fiction or reality is still an open question.

The choice of topics included in this brief review reflects the fact that they bear more directly on the two macro ecological variables' role in development worldwide than other macroeconomic subjects or aggregates (for example, models about inflation or money supply) that are not addressed here in any detail.

These major theories in macroeconomics were and are being used to commission large-scale intervention in economies of both the industrialized and developing nations of the world. Among other public fiscal or monetary policies and actions, massive lending practices are stimulated by these theoretical constructs, and significant trade policies are carried out in their name. Their use, however, in either studying or broadly affecting development leaves a bit to be desired, from both an insider's or outsider's standpoint. Their heavy reliance on prices, ephemeral in their stability ratios of stocks, makes these models and theories pernicious for policy making. Use of unrealistic aggregate demand and supply functions limits their insights. Emphasis on the peculiar assumptions regarding the behavior of economic agents, and the necessary and sufficient conditions evoked for efficiency, clouds their theoretical basis.

These might be analytically convenient conditions and assumptions to structure an edifice of economic orthodoxy, but they are not the means on which to model a social systems behavior. They might provide the grounds from which to launch government economic policy shaped by the orthodoxy's disciples, but this does not mean that these assumptions and efficiency conditions educate us on the behavior of the social system. Indeed, at times a reading of these texts on macroeconomics gives an impression of how these economists would like the system to be (without calling their theories normative), and its individuals to behave.

These texts are laden with value judgements in a manner similar to those of various Marxist theories, although they differ on the exact value judgements. Imposing these theories through the form of public policy or social action to societies (the variety of societies that such systems have been imposed on their individuals, at a variety of locations on the globe, and during different time-periods) carries, at times, significant social costs. Moreover, by reading these theories one can learn about these authors to a much greater extent than one can learn about society.

Modern economic growth theory

This branch of economic theory includes Keynesian macroeconomics and its derivatives, the Harrod–Domar formalism on economic growth and neoclassical economic growth theory. For the purposes of this book the most pertinent is the Harrod–Domar model (see Allen 1968:197–202); the Harrod–Domar model is a name given to a combination of the original work by R.F. Harrod (1939) and by E. Domar (1946). The reader must also consider its close neighbor, neoclassical economic growth theory, which will be briefly outlined.

Two central variables are argued in economic growth theory as being at the core of national growth – namely, savings and capital investment (income foregone from current consumption and transformed into future investment, i.e., capital accumulation) and technological change. The first variable figures prominently in the works by A. Smith and D. Ricardo. The Harrod–Domar growth models and their extensions originate in the work by Keynes as they attempt to identify the relationship between investment and income (output) and employment.

In the 'observer-type' positive (equilibrium) aggregate macromodel by Harrod and Domar there is no explicit population growth present except that the labor force is assumed to be growing at an exogenous (natural) exponential rate which drives the system together with a savings (dynamic, kinetic) condition. Capital accumulation occurs in the absence of technological progress in this one-sector model, with no substitution allowed between capital and labor and with an aggregate production function. There are two basic versions of this model: a fixed-coefficient production function, and a multiplier–accelerator version related to the capital/output ratio. Both versions assume a constant propensity to save. In both cases, an equilibrium growth is produced, a steady state in which output must grow at the natural (labor force) growth rate and in which full capacity and employment must occur.

Investment equals savings in this 'golden age' phase of economic development. The ensued discussion focused on the restriction imposed by the Harrod–Domar model on the capital to output ratio.

Neoclassical growth models appeared as a response to the Harrod–Domar model, and are mostly due to Solow. Substitution was allowed into the production function among the two input factors in production (capital and labor), diminishing marginal products to each factor were assumed in a homogenous (Cobb–Douglas) production function of degree one. Modelers assume that there are many production processes employing various capital to output ratios, with a distribution of ratios making it reasonable to approximate the actual aggregate by a single ratio. The outcome of the neoclassical model is that at a steady state all factors in production (capital and labor), as well as output, grow at the exogenous and fixed (exponential) 'natural' growth rate. This is in variance with the original Harrod–Domar model which includes the possibility of output growing at a rate different from the 'warranted' rate.

The original Harrod–Domar and neoclassical models are interesting because of their simplicity and insights into economic dynamics, by standards of that era. Among others, N. Kaldor and J. Mirrlees extended the neoclassical model by introducing the classical savings function (Allen 1968: 217); J. Hicks and P.A. Samuelson extended it by formulating the two-sector model, (Allen 1968: 220–2); R. Solow, J. Hicks and N. Kaldor introduced technical progress (Allen, 1968: 236–56). Of particular interest in that evolutionary path in the economic growth theory is the work by Solow (1956), Samuelson (1959), and also by Mead (1961). However, it is at this point (in the mid-1950s) that the mathematical modeling efforts in economic growth theory switched to a path which has since led to what some economists characterize as the current impasse and malcontent. See for example, the work by F. von Hayek (1975), and in general the Austrian School critique of modern neoclassical theory. A parable, a simple observer-type model, was taken and extended by a host of subsequent economists in an attempt to transform it into a 'participant-type' model, an instrument for public policy.

Economists criticized the original Harrod–Domar model and the neoclassical extensions to it over the next twenty years for not being appropriate enough to handle a host of factors: for being too simple to account for the complexities involved among an economy's multiplicity of sectors; for having too many exogenous variables; for prices not being included in the original Harrod–Domar model; for being based on too many unrealistic assumptions; and finally for not being justified on empirical grounds.

These criticisms were labeled against the observer-type Harrod–Domar or the neo-classical growth model as a result of institutional support for various views congruous or incongruous with these early efforts. It is very difficult, if not impossible, to pinpoint the supporters or critics of each model and its derivatives and interpretations in space–time among academic circles, funding agencies, research and development and planning divisions of major corporations, and the financial or governmental agencies in MDCs and LDCs. It is also very difficult to ascertain the thickness for such support or criticism. In any event, large-scale capital investment and betting took place, based on the outcome of some of these models, by major private and public actors. An attempt to identify the specific actors in these markets and their actions in the past quarter century falls outside the intent of this book.

Some insider and outsider's criticism of these models will be alluded to in the analysis which follows. From our vantage point, however, and in retrospect, only the last criticism seems to be justified: the empirical grounding of these models was and is extremely weak, particularly when medium to long-term horizons are considered. This shortcoming is only rivaled by their emphasis in proclaiming the existence of a 'golden age' type steady state. The manner in which they viewed dynamics as being equilibrium-bound is a topic this book takes issue with. Their emphasis on the role of savings and investment also makes their use, in the context of LDCs, at least open to question. In LDCs, financial and particularly credit markets are very weak and under-developed if they exist at all.

The subsequent plethora of mathematical models of economic growth fall under the positive (market equilibrium) and normative (welfare optimum) type. A 'turnpike' kind of dynamic was frequently located or sought after. Output and capital per unit labor stocks were found to adjust in an efficient manner (close to the turnpike) toward a planned or targeted steady state. In normative models, under an optimization framework, often a social objective function was defined over aggregate (per capita) consumption levels appropriately discounted over time, and subject to foresight. Under natural (either exhaustible or replenishable) resource constraints and labor-saving technological progress the optimization was carried out, so that the inter-temporal aggregate social welfare function was maximized. Their boundary (traversality) conditions involved either a prespecified time horizon or a targeted per capita consumption level. In a market framework, the conditions for a steady state were analyzed under some decentralized investment condition. But it was the optimum formulation and its associated policy making (through mathematical control theory) aspects which attracted most of the attention in the years to come (see Morishimo 1964, Cass 1965, Phelps 1966, Shell 1967, Arrow 1968).

The main thrust of these models was to state the necessary and sufficient (first and second order) conditions for the postulated unique and stable dynamic states to emerge and be maintained. Central to this was the outcome that under a market (decentralized) or a normative (centralized) decision-making process, input factor prices always reflect and should be equal to the factors' marginal worth. For a more detailed view of this type of model the reader is addressed to Sen (1970), Mirrlees and Stern (1973), and Jones (1976) among others. Of particular interest is the treatment of capital in neoclassical economics, since its very essence (let alone its measurement) is a subject of considerable controversy; see for example the analysis by P. Sraffa (1960), and the discussion in the Introduction of D.M. Hausman (1984) where it is argued that Sraffa supplies a neo-Ricardian version of capital with strong influences from Marx.

As more and more variables and parameters were gradually inserted into these models, the analysis became more complicated but the strength of insights gained diminished. Questions of empirical validation became more severe. Many variables used were unobservable; others, like capital, had definitional and measurement difficulties. As more and more complicated functional forms were considered, issues of relevance were brought up. The mathematical sophistication was countered with the simplistic and utopian view of uniqueness and the existence of long-run equilibria that economic policy may steer the system toward. Quite recently, however, some of these restrictions on unique and stable dynamic equilibria has been lifted as it became apparent that these and associated models are capable of reproducing chaotic dynamics. This undoubtedly will be the subject of considerable future work in the field of economics.

Technological change

A second major force considered by economists to fuel economic growth is technological change. Emphasizing this factor, Schumpeter, in his classical 1939 treatise on business cycles, presented a theory of technological change as the central element to business cycles – then a topic of current concern. Schumpeter was writing in the midst of the economic depression, the same time-period that Keynes also presented his *General Theory*. Kuznets, a quarter of a century later and in the midst of a broad economic upturn of the world's economies, also viewed technology as the driving force in economic growth. The work by Kuznets (1966) was a contribution to the theory of economic growth, as was the work by W.W. Rostow (1960) on stages of economic development and the role technology plays in them.

In the theory of technological progress there are two types of innovation studied: process innovation and product innovation. It is generally accepted that in the short term process innovation is labor saving, whereas product innovation is labor augmenting. It is unclear how these activities affect demand for labor or capital or any other input factor in production in the long term, as it is unclear if they positively or negatively affect total output or income. In neoclassical economic growth theory technological progress is assumed to be exogenous to the economic production and consumption process. It is also assumed to be continuous. No matter how realistic these assumptions are the qualifiers do not resolve with any certainty the definitional and measurement problems associated with technological change.

An interesting notion that has emerged from the economic literature on technological progress and economic growth is that of a 'catch-up' process. According to this principle, the technologically less advanced a nation is in reference to other nations, the faster it must be expected to grow. A corollary to this notion is the presence of what geographers refer to as a logistic technological 'diffusion' process, which distributes technological innovation in space–time.

Economists and geographers, following the pioneering work by Schumpeter (1939), have started to explore the spreading of innovation in the past thirty years or so in a voluminous literature which is largely based on simple Malthusian or S-shape logistical growth models of temporal (and spatial) diffusion (adoption) of innovations; see for example the papers by J.C. Fisher and R.H. Pry (1970), E. Mansfield (1968), F.M. Bass (1969) and, from a geographic standpoint, T. Hagerstrand (1969). More recently a number of authors have analyzed the economic and geographical aspects of technological diffusion; see for example C. Marchetti (1976), S. Davis (1979), R.R. Nelson and S.G. Winter (1982), H.B. Stewart (1982), E.M. Rogers (1983), C. Freeman (1983), V. Mahajan and R.A. Peterson (1985), T. Vasko (1987) and, among geographers, L.A. Brown (1981) and Sonis (1983, 1989) where a number of papers on the subject were presented.

In P. Maitra (1986), the issue of endogenous and exogenous (introduced) technological change is addressed from a sociological perspective; the author discusses Boserup's theory and also that of Malthus and

Marx on the linkages between technological change and development, and he also reviews the contributions made on this issue by Mao and Ghandi. He advocates a preference for endogenous technological innovation. Overall, without the infusion of non-linear (discrete or continuous) dynamics this seems to be a saturated field. The process identifies a saturation level in the share of potential adopters of an innovation closely following an S-shape curve of the Verhulst or Pearl–Reed type. All these approaches can well be captured and significantly expanded by applying the universal map of discrete relative dynamics (Dendrinos and Sonis 1990).

Flowing from relatively advanced technologically regions, innovation spreads to technologically inferior areas as an equalization process of technological know-how in space–time. Severely hampered by difficulties in empirically testing these hypotheses, the above two principles have remained at an underdeveloped stage. There is so much uncertainty regarding this construct, as a result of the measurement problems it faces, that analysts have difficulties agreeing on whether Japan has narrowed or widened its technological gap with the US over the post-war period (see, for example, Choi 1983: 87).

A number of questions remain unresolved. Under what conditions is a technological innovation labor saving or labor augmenting? Must technological innovation be treated as a continuous or a discrete process, occurring randomly or subject to expectations? Is technological progress indeed exogenous or does it depend on the other factors which are associated with development, i.e., is it an endogenous event? Does technological change ignite, at times, evolutionary events?

More importantly, since technological progress – to the extent that it is process innovation – results in lowering average costs it is not clear how it contributes to economic growth, at least in the short run. As it results in lower prices, higher level of competition, and overall less labor for the particular industry impacted by the innovation, one wonders how it can fuel growth. It may not be coincidental that periods of economic depression are also periods of relatively high levels of innovation. But is technological progress the cause or the effect of these economic upturns or downturns?

The deterministic ecological framework of this study may shed some light on these basic questions. Technological innovation is a dynamic process, itself an element of the broader ecological framework within which a social system operates. In the short term, there is demand for and supply of alternative technologies in space–time when *marginal* technological changes are contemplated. Adopters and suppliers of technological innovation interact in ecologically complex markets, where multifaceted social economic and environmental costs and benefits are considered.

Obeying this ecological–economic determinism, technological innovation is a cause–effect factor in the unstable socio-economic spatio-temporal dynamics, and one element of the underlying multiple interdependencies and interactions among various sectors of an economy. In that sense, technological innovation is not only the cause of cycles or instability, it is also the effect of the cycles and instability which emerge under the prevailing, and sustained in time, interdependencies and interactions among a nation's or region's diverse industries.

Interactive adopters and suppliers of marginal technological innovation respond in space–time by purchasing future options contracts in technological innovation speculative markets. These contracts are part of a diversified portfolio of the many and multifaceted contracts they hold. By doing so, they have made technological change another element in the complex web of ecological determinism. Models of discrete or continuous dynamics capable of reproducing periodic and non-periodic motion seem to be most appropriate to capture the technological factor in economic development processes. S-shape simple logistic models are not powerful enough to depict the forces underlying technological progress and its effects.

When in those rare moments of *basic* scientific innovation fundamental (random and discrete) technological changes have occurred, they can be construed as exogenous shocks. In effect, evolution in the dynamics of socio-spatial stocks may have been triggered. Developmental dynamic paths are radically shifted under these conditions and new qualitative features emerge in their spatio-temporal dynamics. These topics are further discussed in Chapter 2, pp. 131–41, as they relate to socio-spatial cycles.

International trade theory

The role of international (or inter-regional) trade in the overall economic growth of a nation or a region has also been debatable. There are those arguing that foreign trade plays a significant positive and negative role in the process of national development, while there are others arguing that its role is minor, see among others Todaro (1982), Ch. 12; and Choi (1983), Ch. 5.

In the first category, opinions which assert that national or regional economic growth is export-driven are dominant. These analysts cite certain instances where the portion of export valuation to a nation's gross national (or domestic) product can reach 90 percent (for small oil exporting nations, for example, during the last quarter century). However, a more comprehensive view of export/import valuations to a nation's gross domestic product reveals that this is not always the case, particularly for nations large in population and/or gross domestic product. In Table A.2.1, the 1950, 1960, 1975 and 1983 exports and imports in ratios to gross domestic product for selected developed economies are shown. In most cases they do not exceed half of GDP, and in the case of Japan and the US their ratio is close to 10 per cent, although rising over the period.

Were one to compute the net difference (exports minus imports) share of GDP it would be a very small ratio indeed. The US merchandise trade deficit in 1986 amounted to about 3.6 percent of the GNP. Two informative surveys on international trade are those by Bhagwats (1964) and Findlay (1979). The conclusion seems to be that economic growth is mainly domestic-markets-driven. This does not necessarily mean that foreign trade deficits cannot be used as social issues for political or other aims. In accordance with the principle of ecological determinism, trade deficits have multifaceted interpretations, with diverse potential costs and benefits.

Table A.2.1 Exports and imports as a percentage of gross domestic product for selected developed countries, 1950–83

	1950[1]		*1960*[1]		*1975*[1]		*1983*[2]	
	E	*I*	*E*	*I*	*E*	*I*	*E*	*I*
Australia	23.0	15.9	14.9	17.8	15.3	16.0	22.1	18.0
Belgium	23.7	28.1	32.9	33.9	46.2	47.9	59.1	62.9
Canada	21.8	19.7	17.5	18.7	23.3	24.0	24.6	20.5
France	13.4	13.1	15.0	12.9	19.6	21.7	16.2	18.7
FR Germany	13.6	10.6	19.0	16.5	24.8	23.5	24.2	21.6
Japan	7.2	7.1	11.0	10.5	12.7	11.4	12.9	11.0
UK	20.1	17.3	21.3	22.6	26.7	30.5	18.1	19.7
US	4.3	4.7	5.1	4.4	8.6	9.5	6.1	7.8

Sources: [1] World Bank, *1980 World Tables*, 2nd edn, pp. 228–69, 189, Johns Hopkins University Press.
[2] United Nations, *International Trade Statistics Yearbook*, 1985 Table 178 (on imports and exports by country); on GDP by nation: US Dept of Commerce, Bureau of the Census, *Statistical Abstract of the U.S.*, 1986, Table 1472 and Table 718 for the US GDP.

International trade theory has its origins primarily in the *Principles* by D. Ricardo and in J.S. Mill's work, although A. Smith's *The Wealth of Nations* did touch on it as well (see Isard 1965). Basically, it draws from the notion of comparative advantages, a key socio-spatial concept, see B. Olin (1933) and its transformation by P.A. Samuelson (1947), who introduced other factors in the model (land and capital) with substitution effects allowing for variable factor proportions and spatial adjustment.

The literature on international and inter-regional trade has coined the term 'international division of labor.' According to this construct, spatial (national, regional, urban) economies specialize in the production of basic (export-oriented) output for which they enjoy comparative advantages over competing locations, as well as competing commodities (natural resources or final products). This principle implies that spatial economies specialize in sectors where minimum minimorum (or close to it) average cost per unit output is attained, or that they concentrate in sectors where maximum maximorum (or close to it) unit profit is achieved. Thus, by specializing in these production and extraction processes they can outbid all other competitors, spatially and sectorally.

The classical work by Ricardo and Mill was a static, one-factor (labor) approach to trade theory. Modern international trade theory is based on the Hecksher–Olin factor price equalization theorem. In the spatial sciences, locations are viewed as competing with one another for exports markets, economic production activity being footloose. The work by C. Tiebout (1956) was central in the development of these notions. Employing such models, one could trace the effects of changes in input and output prices, and/or exogenous demand for the exported commodities upon a nation's economic growth scheme. In the long term, nations with 'locational advantages,' i.e., able to quote lower cost, insurance, freight (c.i.f.) prices at the various points of destination, under predatory pricing, are successful in establishing viable market areas and export the largest quantities of the particular products imported.

Embedded in this approach are yet unresolved issues surrounding (spatial) monopolostic or oligopolistic competition. The topic of monopolistic or oligopolistic competition was first addressed by H. Hotelling (1929) in a spatial context, and later by E.H. Chamberlin (1938). International or inter-regional trade is the basis of spatial sciences. The original work by the German school, von Thünen (1826), A. Weber (1909), A. Lösch (1944) and W. Christaller (1966), as extended later by Isard (1956), M. Beckmann (1968), W. Alonso (1964) and E.S. Mills (1972), addresses the efficient and optimum allocation of economic activity in space; it also deals with efficient and optimum spatial flows in stocks.

Efficient production, distribution, and input factor combinations toward spatial (inter-national, inter-regional, intra-national, intra-regional) specialization are attempted to be modeled using such constructs. Most are formulated in a static manner, and rarely empirically tested. These, however, are not the most serious objections to these models. Economic considerations are not the only factors accounting for production and distribution of commodities and resources, and their trade, in space–time. An observer-type model of direct national (regional, urban) linkages exclusively based on economic forces of prices and exchange is bound to be at least partial and at times irrelevant. Economic forces are only part of the broader ecological forces of competition, cooperation, and predation, affecting international flow of commodities and/or natural resources.

In reality, political, military, sociological, economic, ideological, legal, and cultural elements enter the decision-making calculus both in affecting foreign as well as domestic (inter-regional) trade. At the economic front, politically motivated protectionist import quotas and the manipulation of prices through either direct subsidies or tariffs are examples of the presence of non-competitive and non-economic forces at work. Thus, economic efficiency in international trade becomes elusive. Market failures and imperfections due to various external effects of interaction, multifaceted transaction costs, and the existence of oligopolies or oligopsonies further impede mere economic efficiency in trade. The long-term pure economic efficiency conditions, in all likelihood, never materialize.

There is no evidence further to suggest that at any time-period the system as a whole moves toward more efficient outcomes. International trade decisions are affected by more than economic and ecological factors – they are shocked by technological change which includes innovations production and consumption, by the constant entry and exit of products, and by natural environmental disturbances. All these conditions severely impede various tests to detect a long-term economic efficiency drive.

Fluctuations in the prices of the commodities and final products exported or imported in international commodity markets, and fluctuations in the relative value in the currency of the importing or exporting nation, contribute to an inherent instability in international flows of natural resources, agricultural products, manufactured goods and services. This instability, as it will be seen later in more detail, fuels the unstable interaction patterns among spatial economies. All nations act under the objective of increasing exports. Some also act on objectives of decreasing imports, either through import substitution and/or decreasing demand for imports through imposition of tariffs and/or quotas.

All nations prefer a state in which each runs a positive trade balance. Clearly, it is not feasible for all nations to simultaneously experience a positive foreign trade account at any point in time. In terms of merchandise trade accounts, but not in terms of inter-regional social welfare, inter-regional trade is a zero-sum game. P. Samuelson puts it differently: to have all nations pursuing positive trade balances 'is as sensible as everybody trying to be taller than everybody else,' (Samuelson 1970: 670). Not only do all nations wish to have positive trade balances, but they also tend to be selective with reference to the nations with which they wish to trade. The latter is partly due to preferences by the exporting nations for particular currencies.

The framework of ecological interactions is of some interest with this realization in mind. Nations may not sustain exploitive trade relations for long. Instead, they might be willing to shift roles in time in either of the following two ways: by switching an initially predatory connectance in terms of trade to a cooperative one or by alternating between the roles of 'predator' and 'prey' over time.

Trade is clearly a subject which demonstrates that economic forces are neither dominant nor always unique in spatial interaction. This makes international trade a good forum in which to place the major arguments of this book, these being that the existence of bundle-type forces are continuously influencing international interdependencies and interaction. Further, due to its internally unstable nature and its spatio-temporal pattern of linkages, foreign trade is shown to be a prime contributor to inter-regional spatio-temporal instability, another major thesis here (see Chapter 2).

Two conflicting bundles of forces affect spatial trade at any point in space–time. One bundle is associated with the factors determining comparative advantages of locations in terms of proximity to natural resources, markets, major transport routes, and topographical features. All these factors are durable and are characterized by locational fixity. They are, that is, very slow-moving variables constituting the elements of this slowly changing bundle of forces. Fixed costs in industrial production are typically associated with these factors. On the other hand, there is another bundle of forces affecting spatial trade in space–time. In this bundle relatively fast-moving variables are contained. They include changes in prices of input and output factors in production, international currency fluctuations, changes in tastes, marginal technological innovation, socio-political changes within nations, and international realignments.

Thus, nations entering the international trade activity take significant risks. They act speculatively in potentially highly volatile markets. The term 'speculatively' is used here in its broad meaning. It implies that nations act under risk when at some time-period they either decide to enter international markets as sellers of a product or natural resource or to avoid doing so. Nations have to invest (or forgo investment) in the extraction and production costs of the commodity. They have to invest in transportation infrastructure for shipping the particular products. There is a time lag between the decision to invest (or not to invest) and the time the commodity to be exported (or imported) is sold (bought). Unanticipated price fluctuations may occur in the international markets of these commodities, possibly rendering the investment decision *ex ante* erroneous.

Naturally, no agency exists at that level to underwrite the risk of operation. The only recourse individual nations have is their own, and at times of collusion (cartels), collective leverage to force a desired level of prices in the marketplace. That leverage, in many instances, might be very limited and short lived. All these points demonstrate the highly complex and multifaceted aspects of international trade, and the need for a comprehensive ecological approach to the subject.

DEPENDENCE THEORY

Another family of socio-spatial growth and decline theories, much broader in scope than merely economic, is (the various versions of) 'dependence theory.' A basically Marxist construct, found mainly in the non-economic (political science and sociological) literature, it encompasses many economic, sociological, political and historic notions. The origins of this theory are found in the work by P. Baran (1962), and are further elaborated in B.J. Cohen (1973), I. Adelman (1975) and G. Palma (1978). See also D. Seers (1981) and R.H. Chilcote (1982). This is a theory which is severely criticized for lack of rigor even by its own proponents – see the Introduction in Seers (1981). It is a theory of spatial interdependencies, rather than interaction, in spite of its emphasis on trade.

According to the dependence hypothesis, the wealth in one nation is attributed to exploitation from trade with other nations. The basic 'dependence' related elements of this model are the existence of certain 'core nations' each with a 'periphery,' and the existence of 'surplus value' flowing from the margins to the metropolis so that the centers grow/grow faster/decline slower than the periphery which declines/grows slower/declines faster than the core as a result of these transfers. Colonialism and multinational corporate 'neoimperialism' are considered by neo-Marxists as being phases of such surplus value transfers. According to dependence theory, the exchange price for internationally traded commodities is forced upon the periphery by the core through various (economic, military or other) forms of coercion.

This theoretical construct proposes that it is not the market value of trade but its surplus value which creates imbalances in development. Surplus is the difference between the negotiated (under free will or coercion) exchange price and the true marginal cost (labor-based value, asking price) at the origin or the difference between the exchange (market) price and the real marginal social benefit (the use value, or bid price) at the destination. Although the exchange prices is recordable, it is difficult to estimate surplus value which is an unobservable quantity, the result being that it is not possible to empirically test the central hypothesis of dependence theory.

Although the focus is on inter-national wealth differentials, dependence theory also addresses, along similar lines, intra-national income and population (levels and rates of change) differentials. Further, it views the economic social and political elites of relatively poor nations as the agents acting in the interests of the corresponding elites in richer nations. Finally, the dualism which exists among populations of developing economies is also attributed to intra-nation inter-regional, or inter-social groups surplus value transfers.

Undoubtedly, surplus value transfers play a role in the existence and maintenance of wealth and income differentials among nations, social groups within nations, or regions within national economies. But it can also be argued that a factor in these differentials is successful or unsuccessful speculation. Current economic elites (social groups, regions or nations), according to this argument, acquired their wealth by having successfully competed in a speculative game with prior elites. One may thus hypothesize that another source for these differentials is conflict in interest resolution among the wealthy. Then, contrary to the Marxist thesis, the course of human history is not the outcome of inter-class or inter-regional struggle between the haves and have-nots but instead can be viewed as the outcome of conflict resolution in intra-class interest differences – specifically differences in interest between the outgoing economic elites (social groups, nations or regions) and the emerging ones.

To support this alternative viewpoint, one may cite the fact that the bulk of economic wealth (at least in the form of capital or land assets) is in the hands of the currently wealthy (and thus it is from them that one can extract it), rather than in the hands of the workers (the Marxist source of exploitation). In the current state of economic development in industrialized nations (particularly the US) the source for significant economic profit seems to have shifted in the 1980s from production and ownership of tangible assets to transacting in financial assets. This shift is likely to significantly impact the structure of economies. An important role played in this shift is that of large institutions which hold workers' pension funds. Of course, the workers possess the worth of their labor which is another asset and consequently a potential source for conflict, too. In this case the argument can be construed on the premise that various elites compete for access to that labor resource.

KEYNESIAN MACROECONOMICS, MARXISM, *LAISSEZ-FAIRE* AND ECOLOGICAL DETERMINISM

Here, the topic of private and public interaction is addressed in more detail within the framework of ecological determinism. This framework is broad enough in that it allows for a classification of macroscale public–private sector interaction models. These models have been advocated by a number of influential theorists and are found in the major social science theories of the past century or so. Major ideologies and political movements are founded on the components of a limited number of interactions.

Based on mathematical ecology, these elementary interactions can be conveniently classified and the conditions under which they occur can be clearly demonstrated in fundamental dynamical terms. In analyzing these interactions certain basic components of governance, including rules of choice, might be inferred. They further set the stage for an effort to be undertaken toward a possible restructuring of social (individual and collective, public and private) decision making and action, from a dynamically stable, 'predatory,' standpoint. This effort, however, goes far beyond the scope proper of this monograph, and thus it will not be elaborated in any detail.

Ecological determinism (briefly outlined earlier and developed more extensively in Appendix 1) imposes restrictions on what type of public policies and actions can occur. Specifically, it determines that public intervention necessarily obeys underlying prevailing ecological interdependencies applicable at the particular spatial–temporal–sectoral level and at the specific point in space–time. In summary, it implies that public actions are determined by the broader ecological forces they obey.

Mathematical ecology contains the tools to fully address the interaction between the public and the non-public (private) social sectors. In doing so, it is capable of providing a broader analytical framework in which the major theories currently available regarding this particular interaction can be placed. *Laissez-faire*, Keynesian macroeconomics, and Marxism are three broad macro theories which can be viewed as different types of dynamic ecological interactions between the private and public sectors of a social system at a given point in time. Their dynamic stability properties can be fully analyzed.

In the theory of ecological determinism the term 'determinism' was analytical and substantive. Analytical aspects in the concept were related to the fact that macro dynamic patterns of the particular type examined here, namely by the world's largest urban agglomerations in reference to the world economy, are not found to be random. Substantive aspects of determinism, found in Appendix 1, rely on the economic determinism of market functions. Thus, the notion must be clearly separated from the type of determinism found in Marxism. This is discussed, for example, by M. Harris (see this appendix, pp. 298–302), in the form of 'historic inevitability,' or 'historicism' as it is referred to by K. Popper (1957). Because of this distinction it is not necessary to review the substantive Marxist statements on this issue in any detail here, but instead to focus only on the analytical aspects of private–public sector interactions.

Prominent among macro socio-economic models have been the Keynesian theory (and its versions) in macroeconomics, which has provided the basis for widespread public sector policy making and action in the past half century or so; and the Marxist model (also accompanied by its numerous versions) which certainly has been more interdisciplinary than the Keynesian model and has attempted to provide a theory of capitalism and the socialist State. Next, the Keynesian model will be briefly reviewed since it provides the opportunity for analytical treatment and precise discussion of certain central issues in public–private sector interaction modeling.

In his *General Theory* Keynes set the stage for public intervention, under fiscal and monetary instruments, to regulate national economic growth for the purposes of attaining full employment in isolated national economies. Keynes provided a speculative theory to explain why at times economies do not operate under full employment even when it might be 'rational' to do so from a macroeconomic standpoint. He expressed, through his *General Theory*, utter contempt for speculators. Through public policies on investment, interest, and taxes, the public sector was viewed as an important actor in stimulating economic growth (through various spending, taxing, and investment multipliers), in redistributing income and wealth, and in guiding the economy toward that stage of full employment when it strays away from such a state. Parentheticlly, the objectives of many socialist governments on the globe, were, and are in principle, similar, only differing in the means to attain them.

Keynesian macroeconomics set the stage for a large-scale experiment in the western economies, a direct challenge to the *laissez-faire* and fully decentralized market decision making. It provided the theoretical foundation for large-scale, long-term governmental fiscal and monetary policy making. The Keynesian model was a significant innovation, as was the concomitant Lenin–Stalin experiment in centralized planning and decision making underway in socialist economies. Both were reactions to *laissez-faire*.

Keynesian-type public intervention called for governmental policy making and action parallel with market decision making, so that macroscale and long-term social objectives would be pursued in cooperation with the market. The Marxist experiment (through its Lenin–Stalin version) took a position competitive to the market and proceeded to force its elimination. Thus, both the cooperative and competitive ecologies between the market and political decision making have been tested. Mathematical ecology is insightful at this point. It addresses the type of interaction prevailing between two species and is able to comprehensively describe the effects of these interactions. In effect it offers a systematic means to classify the broad types of linkages between the two aggregate sectors of a nation's economy.

Mathematical ecology demonstrates that among all possible basic ecological associations (six in all: cooperative, competitive, predatory, isolative, commensal, amensal; see R. May (1974) for a full and excellent presentation of these associations), only one association between two interacting species is stable: predation between them may allow both to stably coexist under certain conditions. Transferring this principle, in method but not in substance, to the interaction between the market and political decision-making processes permits one to classify various national economies and outline their intra-national, inter-sectoral stability properties. The transfer also enables one to classify in the broadest possible terms the various modes of economic organization within national economies. Further, it allows identification of possible voids in these organizational structures. A case in point is the predatory association.

A stable predatory association calls for one sector's influence in the economy (say, the private sector) to be positively linked to the other sector's influence, i.e., when the private decision-making impact grows/declines the public sector's impact grows/declines. At the same time, when the public decision-making impact grows/declines, the private sector's impact declines/grows. The roles between the public and private sectors could also be reversed. To detect an existing economic system which might obey this type of association between its private and public sector requires the identification of key variables and the recording of their dynamics.

There are many macroeconomic variables which can be used to look for the presence of this predatory model in a national economy, and gauge the influence of these two sectors. Later in this section, the mathematical exposition of the ecological model is supplied, and two central variables in macroeconomics are used as proxies for the bundle of variables and their multifaceted interactions: interactions are explored between the marginal product of capital and the social rate of discount, two variables which extensively influence the course of a nation's economy.

Their detailed links in the capital sector of a national economy are analyzed in abstract for three broadly defined modes of economic organization: the Keynesian, Marxist and *laissez-faire* models. According to these linkages and their associated dynamics one can identify the ecological type of any economy, and suggest its dynamic properties. It can also search for other modes which may not have been experimented with yet. It is concluded that the Keynesian model corresponds to an ecologically cooperative association between the two sectors, whereas, in mathematical ecology terms, both *laissez-faire* and Marxist models correspond to an isolative association between the two decision-making processes. In the case of Marxism, the dominant factor is the influence of the State with the market sector being extinct. The opposite holds under a *laissez-faire* mode.

These competitive exclusion outcomes, one according to Marxism the other according to Spencer-type social Darwinism, see H. Spencer (1867–8), which is a representative model among the *laissez-faire* types, are the outcome of competitive ecologies between the two sectors leading to two different results.

A similarity in dynamics is found in Keynes, Marxism and *laissez-faire*: all are the outcome of dynamically unstable interactions between the private and public sectors of a mixed economy. Given enough time, that is, the influence of one will grow exponentially to dominate the other. Eventually, one of them will become extinct. Their differences lie in the fact that the Keynsian model is prescriptive of a particular ecology, whereas the Marxist and Spencer models are descriptive, both claiming to represent market economies. In mixed economies, the Keynesian call for long-term, large-scale intervention has gradually given way to short-term, small-scale public action at the margin (i.e., to incremental, disjointed planning). Public sector incrementalism, coupled with the presence of a strong stock market, seems to be the mode of an emerging predatory association between the two decision-making bodies.

Some of the evidence seems to be qualitatively supportive. In mixed economies, during the 1980s, a movement toward 'privatization' has been observed, switching considerable quantities of stock previously managed by the public sector to the private sector. These actions occurred in a variety of nations, from Europe to North America, to the Far East. Parallel to these actions in mixed economies some initial experiments toward a greater role for the private sector has been undertaken in numerous socialist economies, particularly the USSR and the People's Republic of China. The case of PRC, a pioneer in these reforms, is of interest: the opening in the 1980s of a number of special economic provinces, zones, districts within urban areas, and even particular establishments within these urban and rural settings, are cases in point. How far these reforms will go is an open question at present.

Whether experimentation with the roles of market and public decision-making processes may lead to an ecologically stable predatory-type interaction between the public and private sectors of nations is an open question. Keynesian, Marxist, and *laissez-faire* models, each proclaiming a way for societies to develop (collectively in the case of Keynes and Marx, individually in the case of *laissez-faire*) have proven to be temporary.

In the abundance of dynamic instability observed in social events it would be a rare event in the course of social evolution were the dynamically stable predatory model of interdependence to prevail between the private and public sectors of national economies. An evolutionary transition between any of the unstable ecologies (competitive or cooperative) to a stable one (predatory) must be considered unlikely. Were it to be tried, the pervasive presence of instability at present suggests that it will not last long – not because the inherent (fast) dynamics may not be stable, but because the underlying ecological bundle of forces affecting the (slow) dynamics responsible for altering either the form of the interaction or its parameters may dictate so.

According to Keynesian and Marxist macroeconomics stable dynamics do not survive and the predatory (and only ecologically stable) relationship between the private and public sectors is absent. Instead, unstable (cooperative, competitive) dynamics prevail in the economies–ecologies according to Keynes and Marx, respectively. Thus, one is led to a paradox: the stable (long-lasting) pattern may be instability, and stability may not be survivable. Periods of stability may simply be 'intermittent' events.

It is uncommon to argue for an unstable interaction to be selected in ecologies over long-term spatio-temporal horizons. Usually stable associations are expected to survive in these horizons. The question thus remains as to why unstable modes have been so dominant? This topic is very broad and quite unexplored, only lending itself to empirical testing and verification to a comparatively limited extent. It certainly is open to a fair amount of more detailed exposition and numerous suggestions for further development, as well as to many speculative hypotheses.

PUBLIC AND PRIVATE SECTORS: THEIR ECOLOGICAL ASSOCIATIONS

In this section a more technical exposition of the arguments advanced earlier is supplied. Extensive use of mathematical ecology and Keynesian macroeconomics is made.

A mathematical ecology model can be used to simulate all the possible dynamic interactions between the two basic components of decision making in any social system: the political decision-making and public policy-making and action component, and the market decision-making element. In general, the Volterra–Lotka ecology of the form:

$$\dot{x} = (a_{10} + a_{11}x + a_{12}y)\,x$$

$$\dot{y} = (a_{20} + a_{21}x + a_{22}y)\,y$$

where $x, y > 0$ and $-\infty \leq a_{10}, a_{11}, a_{12}, a_{20}, a_{21}, a_{22} \leq +\infty$, presents the forum to classify all possible ecological associations. Of all these interactions, only the *predatory* is dynamically stable. The specifications:

$$\dot{x} = (-a_{10} - a_{11}x + a_{12}y)\,x$$

$$\dot{y} = (a_{20} - a_{21}x - a_{22}y)\,y$$

where x stands for the influence of the public decision-making sector; y stands for the influence of the private decision-making sector; dot stands for time derivative; and a_{10}, a_{11}, a_{12}, a_{20}, a_{21}, a_{22} are interaction parameters, is a dynamically stable predator–prey model, further satisfying the conditions:

$$-\infty \leq a_{10}, a_{20} \leq +\infty$$

$$0 \leq a_{11}, a_{12}, a_{21}, a_{22} \leq +\infty$$

Under these specifications, the model states that the growth/decline in x is associated with growth/decline in y; whereas, growth/decline in y is associated with decline/growth in x. Variables x and y could be any macrovariables depicting the influence of either the public or private sector respectively. There are numerous macrovariables one could use to gauge their cross-sectoral effects and thus qualitatively verify or reject the presence of this association in a social system.

A particular variable designation lending itself to testing, using data from mixed economies, could be a private–public capital-formation-related variable, namely the rate of return on capital. On the private sector side, y could stand for the average yield on private capital investment or for the expected (neoclassical) marginal product of capital. The lower bound of this variable is the market rate of interest (the opportunity cost of private capital). Marginal product of capital could be approximated by a stock market index performance which includes appreciated dividends, so that capital value appreciation and yield are included.

On the public sector side, x could stand for the opportunity cost of public investment, i.e., what is referred to in the US as the (social) rate of discount. Typically, such a macrovariable is controlled by the central bank in mixed economies (the Federal Reserve Board in the US). It represents the rate at which the central bank lends capital to member (private and/or other public) banks.

Were such an ecological–economic model to be found descriptive enough of capital behavior for a particular economy at a particular time-period, then the qualitative dynamics would indicate that the two sectors interact in a dynamically stable ecological framework, under a predator–prey interaction. The public sector, represented by x, is the 'predator,' while the private sector, represented by y, plays the ecological role of the 'prey,' without meaning these terms to imply any substantive equivalences but only methodological similarities. A stable, spiral-type fixed-point dynamic equilibrium (x^*, $y^* > 0$) is the outcome, resulting in a steady state in the long run.

This result is analytically due to the positive cross-sectoral interaction ($+a_{12}$) in the first differential equation, and the negative cross-sectoral effect ($-a_{21}$) in the second differential equation. Under the above assumptions, a decline/growth in the discount rate would result in a growth/decline in the marginal product of capital according to the second differential equation. This is an often observed relationship in mixed economies, confirmed repeatedly at different time-periods. At the same time, an increase/decline in the marginal product of capital causes an increase/decline in the discount rate according to the first differential equation.

This is consistent with widely practiced policies in mixed economies, so that when the rate of return on capital investment declines, the central banks lower their discount rates to stimulate production; when the rate of return on capital increases, the willingness by the central banks to raise their discount rate increases to relieve inflationary pressures. According to this model, there is a different role to be played by both interacting sectors in the four phases of the interaction (see Figure A.2.1). Starting at some point, say in quadrant I, of Figure A.2.1, the private sector's rates of return is declining. As a stimulus to the economy, at this phase the public sector lowers its discount rate. New businesses start with an upturn in the rate of return on private investment: phase II.

Demand for private capital has now picked up and enough capital has been infused into the capital markets to force the central banks to increase the rate of discount to avoid inflationary pressure: phase III. The marginal product of capital starts to decrease while the central bank keeps increasing its discount rate at decreasing rates, pushing the system into phase IV. A decline in marginal productivity in the private sector forces the central bank to lower its rate of discount, pushing the system into phase I again and into a new cycle. The spiralling sink-type dynamic equilibrium brings the economic system toward a point of steady state at some equilibrium E dependent upon values of the composite model parameters.

Note that the above model contains only one macrovariable, the marginal product of capital in the public and private sectors. Another central macrovariable, the total quantity of money supplied by the public sector, could also be incorporated in the model. This addition

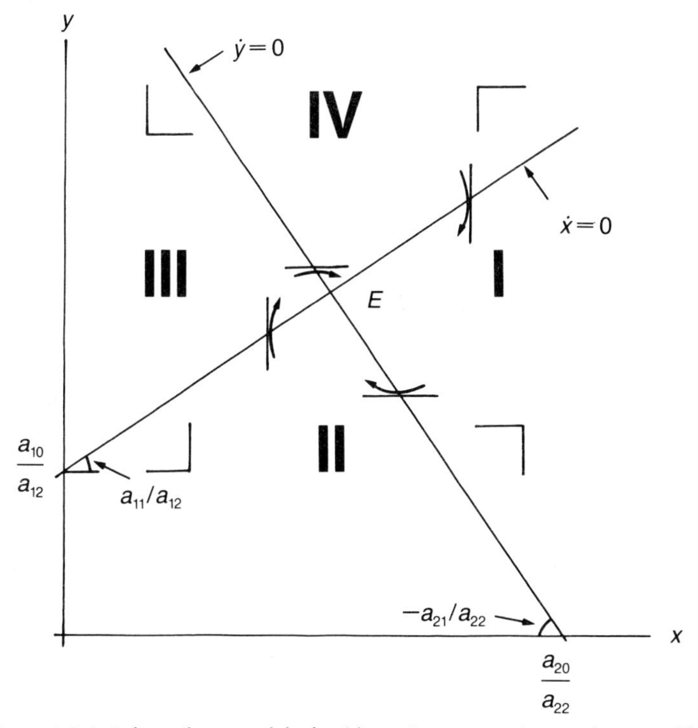

Figure A.2.1 A four-phase model of public–private sector interaction: a stable spiralling sink-type dynamic equilibrium in a predator–prey formulation. Variable x is the public rate of discount and y is the private marginal return on capital

would have enriched the analysis, without affecting the previous results significantly.

One can read different economic theories in the various combinations of cross-sectional effect signs of the above ecological model. The Marxist version of the public–private sector interdependence calls for a *competitive* (dynamically unstable) ecology, where at equilibrium $y^* = 0$. A *laissez-faire* model would require that $x^* = 0$, another competitive exclusion ecology. Ecological specifications of a competitive nature can be drawn up, giving rise to the Marxist and *laissez-faire* theories.

According to the mathematical ecology specifications resulting in the $y^* = 0$ Marxist equilibrium, the following must hold:

$$\dot{x} = (-a_{10} \pm a_{11}x - a_{12}y)\,x$$

$$\dot{y} = (a_{20} - a_{21}x \pm a_{22}y)\,y$$

whereas, the specifications of a competitive ecology producing $x^* = 0$, the *laissez-faire* equilibrium, is of the form:

$$\dot{x} = (a_{10} + a_{11}x - a_{12}y)\,x$$

$$\dot{y} = (-a_{20} - a_{21}x \pm a_{22}y)\,y$$

All parameters in the above two competitive specifications are positive. These models represent economies–ecologies where at some point in time (initial state) the two sectors coexisted; through social political and economic processes, depicted by the phenomenological (bundle-like) parameters of the ecological model, the two sectors moved to competitive exclusion. At the post-competitive exclusion state, the economic system is such that one of the sectors exists in an isolative economy–ecology. Herbert Spencer's (1850) vision of the withering away of the State, and the emergence of a pure free market (Samuelson 1970: 710), traces the dynamics of an unstable competitive economy, as well.

Finally, the Keynesian model, in this author's view, called for a *cooperative* ecology–economy between the two sectors. Analytically, this ecology is unstable. Keynes was apparently unaware of dynamic stability properties in the two-way interdependence between the two rates, and certainly of their mathematical ecology implications. Keynes presented his *General Theory* largely verbally, and thus analytical functions have to be inferred.

In Keynes's model, one finds two statements pertinent to this work, the first being: '*If* the rate of interest were to rise *pari passu* with the marginal efficiency of capital, there would be *no* stimulating effect from the expectation of rising prices. For the stimulus to output depends on the marginal efficiency of a given stock of capital rising *relatively* to the rate of interest' (Keynes 1964: 143). The term 'marginal efficiency of capital' has had as many interpretations as the number of books cited in the section on macroeconomics in this appendix. For the author, it is taken to be equivalent to 'marginal product of capital,' whereas 'rate of interest' is taken to be equivalent to the 'rate of discount' as employed here.

The above statement is seemingly equivalent to the first differential equation of the ecological model. It links a rise of the discount rate (Keynes's rate of interest) to a (sharper) rise in the marginal product of capital. Thus, it leads to a positive cross-sectoral effect ($+a_{12}$).

A second condition, Keynes noted, is that 'an expectation of a future fall in the rate of interest will have the effect of *lowering* the schedule of the marginal efficiency of capital' (Keynes 1964: 143). This statement apparently supplements the first statement applicable to the first differential equation of the ecological model. It links a decline in the discount rate to a decline in the marginal product of capital. Keynes did not explicitly address the second differential equation, although from the two, it is the least objectionable. Thus, through a positive cross-sectional effect, Keynes presented a cooperative ecological–economic association between the two sectors. This is even more clear in the underlying policy statements found throughout the *General Theory*, a base for advocating a large-scale role of the public sector.

It is of interest to briefly attempt a synthesis of all these cooperative, competitive, and predatory models, to the principle of ecological determinism as outlined in Appendix 1. The central feature in such a link is that the evolution of the two sectors' influence within a particular economy at a particular time-period can be attributed to specific points in the parameter space of a deterministic macro ecological model. In this ecological landscape, all these different types of interactions between the two sectors can be depicted (Dendrinos (with Mullally) 1985: 45–53).

In each instance, for a particular set of parameter values, the dynamic equilibria experienced could lead to any of the three types of ecologies (actually four, since there are two different subspecies of competitive exclusion, one leading to the extinction of the private sector, the other to the elimination of the public sector). Each of these dynamics, according to the terms used in this book, involves developmental change. If the parameters of the model change sufficiently, then evolutionary change may occur whereby one type of ecology is converted to another – for example a Spencer-type economy is transformed into a Marxist one.

Through its phenomenological coefficients the model depicts a combination of forces at work: economic, sociological, political, environmental, cultural, etc. It subordinates merely economic decision making to a more composite bundle of forces shaping social evolution through the ecological–economic determinism. The long-term pure economic objectives of the various social agents calling for one particular mode of private–public interaction or another are slaved by the broader political, demographic, and cultural forces at work affecting the social system in space–time.

Finally, one might note that the Keynesian cooperative and the predatory models may be more likely to occur in MDCs. In the case of LDCs the only viable (although dynamically unstable) models may be the various versions of the purely competitive (or isolative) types. This is so because in MDC economies the roles of the private financial institutions and their central banks do affect each other significantly. For LDCs with informal currencies and weak financial institutions it is very unlikely that any of these two interactions can occur.

What socio-spatial rules, institutions, practices, choices, actions, constraints, objectives, and other socio-economic conditions would prevail under a 'predatory' association between the private and public sectors of a spatial setting is not the subject of this book. One can have a fertile ground on which to propose a variety of alternative legal, ethical, moral, or financial systems to accommodate this ecology, and to study its implicit societal benefits and costs. It is offered as a research program suggestion to the interested reader.

TWO THEORIES FROM ANTHROPOLOGY

Anthropology is a field of study not heavily interacting with spatial economics and geography. However, some of the writing by two prominent anthropologists, M. Harris and C. Levi-Strauss, bears directly on the subject at hand. Anthropologists have enjoyed the privilege of dealing with social, economic, political, and psychological factors liberally, without necessarily adhering to any orthodoxy. In this sense, this book might appeal more to them than to core social scientists. Consequently, a few comments will be made on these two anthropologists' theses, with a brief critique.

M. Harris (1979) proposes a broad framework by which to view a social system, a frame of reference largely resting on Marxist ideas. He acknowledges the existence of a 'universal pattern' in social action. He states that 'the universal structure of socio-cultural systems posited by cultural materialism rests on the biological and psychological constraints of human nature, and on the distinction between thought and behavior, and emics and etics' (p. 51).

His structure is briefly the following: expanding on a thesis by Marx, he accepts the three levels of 'etic' behavior given by:

1 infrastructure, a notion which involves a mode of production (environment, ecosystem, work pattern) and a mode of reproduction (demography, mating, fertility, natality, mortality, nurturance of infants, medical control, contraception, abortion, infanticide);
2 structure a concept which includes the domestic economy (family structure, domestic division of labor, domestic socialization, encul-

turation, education, ages and sex roles, domestic disciplines, hierarchies, sanctions), and the political economy (political organizations, factions, clubs, associations, corporations, division of labor, taxation, tribute, political socialization, enculturation, education, class, caste, urban–rural hierarchies, discipline, police or military control, war);

3 superstructure which contains art, music, dance, literature, advertising, rituals, sports, games, hobbies, science (p. 52).

Harris notes that 'the etic behavioral modes of production and reproduction *probabilistically determine* the etic behavioral domestic and political economy, which in turn *probabilistically determine* the behavioral and mental emic superstructures. For brevity's sake, this principle can be referred to as the principle of *infrastructural determinism*' (pp. 55–6) (emphasis mine).

He is close to discovering the principle of discounting, discussed in Appendix 1 in reference to social group deterministic ecological interactions. His persistent adherence to the Marxist classification of 'infrastructure, structure, superstructure' and his failure to recognize the anticipatory, expectations-based, game-theoretic speculative-social actions involving the discounting process prevented him from going further from infrastructural determinism to state the notion of 'ecological determinism.'

Another criticism of the Harris scheme (and Marx's underlying structure) is their failure to recognize, and their indifference to, the rich dynamical aspects of evolution. Further, Harris argues in a non-consistent manner in his attempt to address disaggregate and aggregate behavior. Although he employs a Marxist (aggregate) frame of reference in his analysis, he also states: 'one kind of behavioral option is more likely than another not in terms of abstract pushes, pulls, pressures and other metaphysical "forces," but in terms of concrete bio-psychological principles pertinent to the behavior of the individuals participating in the system' in order to establish a link between individual behavior and aggregate social response (p. 60). It seems that he tries to adopt an 'individual'-based view of a social class and structure. But then he adds: 'But I don't mean to dismiss the possibility that many socio-cultural traits are selected for by the differential survival of whole socio-cultural systems – that is by group selection' (p. 60).

Harris thus seems not to recognize the foundation of microbehavior from macrobehavior, and the dependencies of macrobehavior to microbehavior in the framework of ecological determinism – see a summary discussion on this topic by Dendrinos (with Mullally) (1985), pp. 122–9.

A final point of interest is Harris's critique of Malthus's contention that 'no change in political economy would eliminate poverty.' Harris states that this is a 'reactionary view of history' (p. 70). He further adds: 'Marx cut his followers off from pursuing developments in a theory of human demography and ecology,' as a reaction to Malthus. Maybe, the cost of this isolationist view cannot be underestimated.

More evolutionary, but still primitive under the current stock of knowledge in dynamics, is the work by C. Levi-Strauss (1963) in 'structural anthropology.' Levi-Strauss identified one element out of the two found in modern evolutionary theory. He identified the *structure*, or the psychology's gestalt (i.e., the code as used in the urban evolutionary theory), but failed to remark on the *dynamics* the code contains, i.e., the developmental and evolutionary aspects of the code.

'Levi-Strauss is primarily concerned with *universals*, that is, basic social and mental processes of which cultural institutions are the concrete external projects and manifestations' (p. ix, from the translator's Preface). Further, 'He rejected the atomistic and mechanistic interpretations of evolutionism and diffusionism, as well as the naturalistic and empirical approach of British functionalism' (p. xi).

Levi-Strauss also makes the following remark on urban ecology, referring to the Chicago School:

> ecology and social structure ... both have to do with the spatial distribution phenomena. But social structure deals exclusively with those 'spaces' the properties of which are of purely sociological nature, that is *not affected* by such natural determinants as geology, climatology, physiography, and the like. This is the reason why the so-called urban ecology should hold great interest for the social anthropologist; the urban space is *small enough and homogeneous enough* (from every point of view except the social one) for all the differential qualitative aspects to be ascribed mostly to the action of internal forces accessible to structural sociology ...
>
> (Levi-Strauss 1963: 291; emphasis mine).

This is an example of Levi-Strauss's poor understanding of the sociospatial sciences, although in his defense it must be noted that at the time of his writing they were not well developed. On the other hand, to his credit, he uses terms like 'cycles,' 'discontinuities' and 'dynamics', albeit in simplistic ways, viewed from a contemporary non-linear dynamic theory perspective. He also notes that 'my interest in our society is only a secondary one. Those societies which I seek first to understand are the so-called primitive societies with which anthropologists are concerned' (p. 388). By doing so, he recognized that 'modern' societies and 'primitive' may differ in terms of speed of change and dynamical state. His static structural theory may be the result of his interest in primitive, steady state, cultures that over some relatively prolonged time horizon had reached a state of super stability and ecological isolation.

This brings about a key point in scientific endeavor: the selection of events or data sets or time-periods a social scientist makes, may slave the observer into a particular (partial, local, or erroneous) view of a broader socio-spatial code of reality. One may attach this criticism to this work, as well, in spite of the effort to be global and comprehensive. At the same time, however, one must supply proof that another more comprehensive system and data set exist.

Levi-Strauss and Harris attempted in their analysis to be interdisciplinary. To involve, that is, a very large and exhaustive list of variables in their models. To do so, they attempted to incorporate many variables from different fields, rather than trying to identify central ones in their discipline. This meets with insurmountable problems of empirical verification, and this is the reason why the research program such approaches suggest is too ambitious to carry out with present social science research infrastructure.

The Harris and Levi-Strauss analyses point to an interesting problem in social sciences. What role do specialized disciplinarian forces (economic, sociological, political, pyschological, religious, etc.) play in social action? And, are there in reality 'strictly disciplinarian' forces? M. Harris, using the Marxist model identifies an 'infra-structure' as a bundle that is of economic and social forces as the basic (lower level) ones, moving to the 'superstructure' identifying a bundle of forces associated with 'higher level' cultural ones. C. Levi-Strauss, on the other hand, deals with universals (or bundles) of forces, without any explicit discrimination among them.

The view taken by this book is that if there are strictly disciplinarian forces, they are found strongly linked in a bundle of forces applicable to a specific socio-economic stock at a specific point in space–time, under a particular circumstance, and dependent upon the socio–spatial–temporal scale under consideration. Within this bundle, dominant forces emerge. Dominance shifts as the scale, or space–time-framework, or circumstance (problem) changes.

There is, however, a general element common to all social factors independent of their disciplinary attributes. This element is utilized in economic theory and thus is commonly associated with economic forces. It is the element of demand–supply-market interaction where quantities are exchanged and prices are set for a variety of economic commodities. These economic forces, demand–supply-market price setting, are shared by non-economic stocks, as well. In this sense, economics may provide a common social science conceptual foundation; it does not, however, supply a common substantive one. The substance of social action is composed of the full variety of variales *and their interactions* from which different disciplines make their subjects of investigation in an isolated manner.

3

DATA AND SOURCES

TIME SERIES USED

The following is the set of time series used in the study, cataloged by urban agglomeration and year when both or either of the two counts was available. In these series, x is the metropolitan area's share of the world's current population, and y is the nation's per capita current product or the urban area's per capita income (for the case of US MSAs) over the world's market economics' current prevailing average product.

Table A.3.1 Relative population and per capita product (income) of nineteen urban agglomerations, 1950–80

Buenos Aires (Argentina)

	x^1	y^1
1980	0.002239	(0.936[2])
1979	0.002249	(0.961[2])
1975	0.002111	1.029
1974	–	1.169
1970	0.002445	1.075
1960	0.002330	1.133

Rio de Janeiro (Brazil)

	x	y
1980	0.001132	(0.770[2])
1979	0.001225	0.611
1975	0.001216	0.622
1970	0.001165	0.511
1968	0.001205	0.345
1960	0.001073	0.375
1950	0.000956	–

São Paulo (Brazil)

	x	y
1980	0.001564	(0.770[2])
1979	0.001908	0.611
1975	0.001802	0.622
1970	0.001421	0.511
1968	0.001629	0.345
1960	0.001053	0.375
1950	0.000837	–

Peking (People's Republic of China)

	x	y
1979	–	0.155
1976	0.002100	0.195
1970	0.002074	–
1957	0.001435	–
1953	0.001087	–

Shanghai (People's Republic of China)

	x	y
1979	–	0.155
1976	0.003045	0.195
1970	0.002964	–
1957	0.002469	–
1953	0.002436	–

Cairo (Egypt)

	x	y
1979	–	0.118
1976	0.001255	0.198
1975	0.001482	0.178
1974	0.001469	0.173
1970	0.001359	0.221
1966	0.001259	–
1960	0.001115	0.226

Paris (France)

	x	y
1979	–	3.622
1975	0.002155	3.414
1970	–	2.832
1968	0.002349	2.883
1962	0.002456	–
1954	0.001819	–

Calcutta (India)

	x	y
1979	–	0.061
1976	0.001997	0.076
1971	0.001897	0.099
1968	0.001454	0.095
1961	0.001435	–
1951	0.001878	–

Bombay (India)

	x	y
1979	–	0.061
1971	0.001611	0.099
1961	0.001353	–
1951	0.001165	–

Jakarta (Indonesia)

	x	y
1976	0.001383	0.121
1971	0.001235	0.110
1961	0.000947	–

Teheran (Iran)

	x	y
1976	0.001112	1.010
1973	0.001036	0.577
1971	0.000982	0.458
1966	0.000811	–
1958	0.000582	0.317
1956	0.000552	–

Tokyo (Japan)

	x	y
1979	0.002653	2.915
1977	0.002747	2.694
1975	0.002909	2.378
1973	0.003001	2.521
1971	0.003107	2.052
1969	0.003217	1.728
1965	0.003262	1.201
1960	0.003223	0.833
1955	0.002923	–
1950	0.002605	–

Seoul (South Korea)

	x	y
1980	0.001860	(0.569[2])
1976	0.001802	0.338
1975	0.001722	0.309
1970	0.001517	0.271
1966	0.001132	–
1960	0.000814	0.282
1955	0.000585	–

London (United Kingdom)

	x	y
1979	0.001561	2.430
1977	0.001637	1.991
1976	0.001738	1.535
1973	0.001774	2.059
1971	0.002011	2.328
1961	0.002666	–
1951	0.003424	–

Chicago (USA)[3]

	x	y
1980	0.001579	(5.585[2])
1979	–	3.850
1978	–	3.640
1977	0.002014	3.840
1976	0.001730	3.956
1975	0.001754	3.776
1974	0.001797	3.926
1973	0.001814	4.115
1972	0.001865	4.710
1971	0.001896	–
1970	0.001911	–
1969	–	4.979
1965	0.002008	4.757
1960	0.002053	–

Los Angeles (USA)[3]

	x	y
1980	0.001663	(5.564[2])
1979	–	3.435
1978	–	3.581
1977	0.001648	3.722
1976	0.001732	3.843
1975	0.001739	3.661
1974	0.001782	3.710
1973	0.001797	3.886
1972	0.001843	4.486
1971	0.001907	–
1970	0.001929	–
1969	–	4.984
1965	0.002030	4.599
1960	0.001993	–

New York (USA)[3]

	x	y
1980	0.002028	(5.434[2])
1979	–	3.334
1978	–	3.386
1977	0.002199	3.566
1976	0.002349	3.765
1975	0.002391	3.741
1974	0.002481	3.841
1973	0.002525	4.142
1972	0.002624	4.810
1971	0.002691	–
1970	0.002733	–
1969	–	5.081
1966	0.002832	–
1965	–	4.861
1960	0.003149	–

Mexico City (United States of Mexico)

	x	y
1979	0.003328	0.591
1976	0.002953	0.641
1975	0.002838	0.708

1974	0.002768	0.659
1970	0.002353	0.675
1960	0.000942	0.626
1950	0.000969	–

Moscow (USSR)

	x	y	
1980	0.001824	1.869[4]	(1.814[2])
1977	0.001902	–	(1.712[2])
1976	0.001958	1.393	(1.691[2])
1975	0.001935	1.643	
1974	0.001935	1.799	
1973	0.001920	1.995	
1971	0.001970	2.627	
1959	0.002081	–	
1958	–	3.400	

Notes: [1]Represents calculated counts based on data exclusively from United Nations, *Statistical and Demographic Yearbook* sources; in the case of y they are the ratio of the corresponding nation's current per capita gross *domestic* product to the world's market economies per capita current product in US$. The reason the study period is designated as starting from 1950 is that this is the first year for which the United Nations started supplying per capita product counts for various nations. All undesignated counts represent United Nations sources. In the case of x, they represent the ratio of the population provided for the metropolitan agglomeration (with the exception of the US MSAs) to the world's total population as given by the latest count from the US Bureau of the Census, *Statistical Abstract of the United States*, International Statistics.
[2]Represents calculations based on data solely from the Population Reference Bureau, *World Population Data Sheet* sources; in the case of y they are the ratio of the corresponding nation's current per capita gross *national* product to the world's average per capita current product in US$. Since there is some difference between the UN and PRB counts only UN time series on y have been used in the study. The PRB counts are provided here only for illustrative purposes.
[3]Data for the three US MSAs are from previous studies by the author. They have been obtained from the US Department of Commerce, Bureau of the Census, *Census of Population* for decennial counts on population and per capita income), and from P-25 series for mid-decade estimates for per capita income. Mid-decade estimates for per capita income are also obtained from the *Survey of Current Business*.
[4]Per capita income counts for the USSR have been obtained from the US Congress, Joint Economic Committee, *Soviet Economy in a New Perspective: Third World Block and Soviet Economic Power Growth-Achievements Under Hardships*, US Government Printing Office, 1976: Table 1, p. 246. Counts are in constant 1975 US$. The corresponding counts from the PRB Data Sheets, for selective years, are also shown in parenthesis for illustrative purposes.

DATA ON CITIES AND METROPOLITAN AREAS FROM OTHER CENSUSES

Concern over the size of Mexico City is detected in the Mexican Census's interest in large urban agglomerations population, worldwide. According to the Mexican Census, *Agenda Estadistica 1983*, SPP, Table 5.2.2, a number of urban centers' *city* size was given (in thousands) as shown in Table A.3.2. Their metropolitan size is shown in Table A.3.3. In these data series, when United Nations sources are cited, it is indicated by (UN).

According to the Mexican Census, and given that the list of urban areas in Tables A.3.2 and A.3.3 is not overlapping, one must assume that the reporting by their respective census as either 'city' or 'metropolitan' area was the only qualifier used to include any of these cities in these two categories.

Comparing the above series with the data used in this study, there are significant population size differences for all US MSAs, Mexico City, Tianjin and Cairo. Further, in the list of urban agglomerations with a population greater than 5 million in 1980 the following areas are added: Karachi, Bangkok, Delhi; whereas, in the list of emerging 5 million size urban agglomerations, Bogota and Santiago must be added.

What these counts contain can be inferred by looking, selectively, at

Table A.3.2 City size of fifteen large urban areas of the globe, 1980–82

City	*Year*	*Population ('000)*
Bogota	1981	4,486
Bombay	1981	8,227
Buenos Aires	1981	2,985 (UN)
Karachi	1981	5,103
Lima	1981	3,969 (UN)
Mexico City	1982	10,061
Peking	1982	9,231
Rio de Janeiro	1980	5,093
Santiago	1982	4,039
São Paulo	1980	7,034
Seoul	1980	8,367
Shanghai	1982	11,860
Teheran	1980	6,000
Tianjin	1982	7,764
Jakarta	1980	6,450

Table A.3.3 Urban metropolitan size for thirteen large centers of the globe, 1980–82

City	*Year*	*Population ('000)*
Bangkok	1982	5,397
Cairo	1982	7,258
Calcutta	1981	9,166
Chicago	1980	7,890
Delhi	1981	5,714
Philadelphia	1980	5,548 (UN)
Leningrad	1982	4,719
London	1981	6,703
Los Angeles	1980	11,498 (UN)
Moscow	1982	8,301
New York	1980	16,121
San Francisco	1980	5,180
Tokyo	1981	8,335

a few national censuses. The populations of Rio de Janeiro's and São Paulo's metropolitan regions for 1970 and 1980 are listed as:

	1970	*1980*
Rio de Janeiro	7,082,404	9,018,637
São Paulo	8,137,401	12,588,439

by the *1981 Anuario Estatistico Do Brazil*, SPPR, 1981. Whereas, the case for Karachi seems to be confirmed by the Government of Pakistan, Statistics Division, Federal Bureau of Statistics *10 years of Pakistan in Statistics: 1972–1982*, where the population of Karachi is listed as 5.3 million in 1981 and 3.5 million in 1972.

The point demonstrated by these various counts is that city size, as any other social variable, is fuzzy. Many measurements can be obtained on it, depending on the agent carrying out the measurement and reporting it; the time the measurement was taken and the conditions under which it was taken; and the area included in the measurement. However, if these conditions prevail uniformly then some meaning can be obtained by comparing measurements among cities.

SOME CHINESE METROPOLITAN REGIONS AND THEIR INTERNAL STRUCTURE

Population growth patterns in China's largest metropolitan agglomerations

This section takes a closer look into the largest Chinese urban agglomerations. Table A.3.4 provides a list of the twenty largest metropolitan regions of the People's Republic of China in 1983, ranked in accordance with their *urban population size*. The source of the data is the *Statistical Yearbook of China*, 1984, p. 48, (in Chinese). The names of these regions are provided in this Appendix according to the Yale System of the romanized Chinese characters.

The twenty largest urban regions are shown in Figure A.3.1, ranked according to urban population size as of 1983. According to Huan Yong Hu (1986), a demarcation line linking Heihe of Heilongjiang Province in the north-east and Tengchong of Yunnan Province in the south separates 94.4 percent of the People's Republic of China's population and 19 out of the 20 largest urban regions in the east from the 5.6 percent in the west which contains only one, Lanzhou, of the twenty largest urban regions. This points to two allied conclusions: first, the pronounced dualism in spatial population distribution within a very large nation, and second, the accompanied significant difference in their total population density. The eastern part has an average prevailing population density of 232 persons per square kilometer, whereas the western part's density is about 10 persons per square kilometer.

In 1983 the twenty largest metropolitan agglomerations contained 11.18 percent of the total Chinese population. Shanghai, the largest metropolitan region, contained 1.165 percent of the People's Republic of China population and about 10 percent of all twenty regions. With 80 percent of the Chinese population being rural, there is no discernible urban primacy.

Land distribution among the urban regions is informative, as it reveals the current policy by the central government with regard to these regions. With these metropolitan regions accounting for about 2.2 percent of the total space, when the top fifteen are considered Shanghai's and Nanjing's shares are second and third to last respectively, with only Harbin's being at a lower level. On the other hand, the area of the Chongqing metropolitan region ranks first, with Changchun second and the Beijing and Guangzhou regions ranking a very close third and fourth respectively. These are areas of primary economic and political emphasis by the central government, on issues of urban development. Chongqing (the birthplace of Deng Xiaoping), Changchun and Guangzhou are urban areas in which considerable land annexation took place in the 1977–83 period.

Shanghai, with only 230 square kilometers of urban space enjoys an urban density of 27,780 people per square kilometer, fifteen times higher than the average among the other nineteen regions. In fact, its rural density ranked the highest among all other regions' rural areas; is about nine times higher than the People's Republic of China's average density in 1983; and is higher than the *urban density* of three of the metropolitan regions in the top twenty (Taiyuan, Lanzhou, Kunming). These indicators are testimony to the efforts by the central government toward decelerating the growth of Shanghai, and accelerating the growth of medium-size urban agglomerations. These conclusions are further enhanced by a look into the economic performance of these urban centers.

An index of economic activity is provided in Table A.3.5, for all twenty Chinese metropolitan regions for 1983. Again, Shanghai's prominent role is underscored with a per capita value of output at 8,140 yuan – considerably higher than that of any other region. Quingdao's urban per capita production ranks second at 5,200 yuan per capita, followed by Anshan and Tianjin. The very strong economies of scale found in Shanghai are thus clearly demonstrated. As a result, the task by the central government to de-emphasize the region and tightly control its population growth is formidable. On the other hand, there are ambiguous signals on the future of Shanghai; the above indicators and the decelerating efforts by the central government to the extent that the manufacturing and population base of Shanghai is concerned, are contrasted to its increasing importance as a financial center and its trade position in the South China Region. For example, according to Huan Yong Hu, Shanghai is projected to be the largest port in the Pacific Ocean and a metropolis of 30 million people (Huan Yong Hu 1986: 60).

Plans have been implemented and are currently underway for regional population redistributions. At present, China is in the midst of large-scale selective population migration: from the eastern and south-eastern coastal regions, particularly the Yangtze River Basin (which includes Shanghai and Nanjing), and the Huang (Yellow) River Basin (which includes the urban regions of Beijing and Tianjin, to the northern Jin-Shaan-GanNing and the Inner Mongolia–Xinjiang regions of the northern and north-western plateaux (Huang Yong Hu 1986: 53).

Table A.3.4 The twenty largest metropolitan regions in China, ranked by urban population size, 1983

Rank		Total population ('000)	(%)	Total area (sq. km)	(%)	Aggr. density persons/sq. km	Urban pop. ('000)	Urban area (sq. km)	Urban density ('000/sq. km)	Rural population ('000)	Rural area (sq. km)	Rural density ('000/sq. km)	Comments
1	Shanghai	11,940	1.165	6,186	0.065	1,930	6,390	230	27.78	5,550	5,956	0.93	Growing fast; very high density
2	Beijing	9,330	0.91	16,807	0.175	555	5,670	2,701	2.10	3,660	14,106	0.26	
3	Tianjin	7,850	0.766	11,305	0.118	694	5,220	4,276	1.22	2,630	7,029	0.37	Fast growth
4	Shenyang	5,210	0.508	8,515	0.089	612	4,080	3,495	1.17	1,130	5,020	0.23	
5	Wuhan[1]	5,940	0.58	8,216	0.086	723	3,280	1,557	2.11	2,660	6,659	0.40	
6	Guangzhou	6,840	0.667	16,657	0.174	411	3,160	1,345	2.35	3,680	15,312	0.24	Special region; fast growth
7	Chongqing	13,890	1.355	22,341	0.233	622	2,690	1,521	1.77	11,200	20,820	0.54	Special region; very fast growth
8	Harbin	3,730	0.364	6,769	0.061	551	2,560	1,637	1.56	1,170	5,132	0.23	
9	Chengdu	8,490	0.828	12,614	0.132	673	2,510	1,447	1.73	5,980	11,167	0.54	
10	Xian[2]	5,350	0.522	9,853	0.103	543	2,220	861	2.58	3,130	8,992	0.35	
11	Nanjing	4,560	0.445	6,516	0.068	700	2,170	867	2.50	2,390	5,649	0.42	

12	Taiyuan	2,250	0.22	6,988	0.073	322	1,790	3,044	0.59	460	3,944	0.12	Extremely low densities
13	Changchun	5,800	0.566	18,881	0.197	307	1,770	1,116	1.59	4,030	17,765	0.23	A 5 million urban area
14	Talien[3]	4,770	0.465	12,574	0.131	379	1,520	1,003	1.52	3,250	11,571	0.28	
15	Lanzhou	2,400	0.234	14,414	0.150	167	1,430	2,123	0.67	970	12,292	0.08	Very low densities
16	Jinan[4]	3,390	0.331	4,875	0.051	695	1,360	483	2.82	2,030	4,392	0.46	Potentially a 5 million urban area
17	Kunming	2,020	0.197	6,546	0.068	309	1,320	2,199	0.60	700	4,347	0.16	Very low densities
18	Anshan	2,560	0.25	4,642	0.048	551	1,240	622	1.99	1,320	4,020	0.33	
19	Qingdao[5]	6,200	0.605	10,654	0.111	582	1,210	244	4.96	4,990	10,410	0.48	Becoming a 5 million urban area
20	Fushun	2,080	0.203	10,816	0.113	192	1,200	675	1.78	880	10,141	0.09	
			11.181		2.246								
	China	1,024,950		9,579,000		107							

Source: Administration of National Statistics, 1984, *Statistical Yearbook of China*, Chinese Statistical Publishing Agency, Beijing.

Notes: [1]It includes the three cities of Hangyang, Wuchang and Hankou.

[2]It is pronounced 'ci'an.'

[3]Sometimes spelled 'Dalyan.'

[4]Pronounced 'chinan' or 'tsinan.'

[5]Pronounced 'chingdao.'

Figure A.3.1 Location of the twenty largest metropolitan regions in the People's Republic of China, by urban size rank, 1983

Regional and rural–urban population and per capita income (or product) imbalances are very pronounced in the People's Republic of China. These imbalances are quite obvious even at metropolitan region scale. As shown in Table A.3.5, intra-metropolitan per capita value of product variations are significant between the urban and their rural parts, as seen by the difference between the average region-wide and the urban levels.

The population growth of several of the largest Chinese metropolitan regions is given in Table A.3.6. Shenyang is currently growing at the fastest rate among the regions whose total land area has not been affected by annexation in the 1981–83 period. Growth in Guangzhou is partly fueled by its special economic district status and the associated quantity of land added to its hinterland.

Population growth rates for the top among the largest urban regions in China has slowed down considerably. For example, in the case of the metropolitan regions of Shanghai, Beijing, Tianjin and Shenyang, the average annual population growth rates during the 1950–77 period were: 4.12, 8.33, 6.05 and 3.75 respectively, without involving any land

Table A.3.5 Per capita value of output produced in the twenty largest Chinese urban regions, 1983

Metropolitan region	*Per capita region-wide (1,000 yuan)*	*Per capita urban-wide (1,000 yuan)*
1 Shanghai	5.68	8.14
2 Beijing	2.69	4.16
3 Tianjin	2.92	4.29
4 Shenyang	2.22	2.79
5 Wuhan	2.09	3.58
6 Guangzhou	1.74	3.45
7 Chongqing	0.72	2.57
8 Harbin	1.77	2.33
9 Chengdu	0.87	2.41
10 Xian	1.13	2.40
11 Nanjing	1.87	3.56
12 Taiyuan	2.20	2.73
13 Changchun	0.78	2.33
14 Talien	1.77	4.26
15 Lanzhou	2.05	3.16
16 Jinan	1.43	3.32
17 Kunming	1.90	2.46
18 Anshan	2.45	4.60
19 Qingdao	1.25	5.20
20 Fushun	2.48	4.07

Source: *Statistical Yearbook of China*, 1984, p. 48.

annexation (see L.J.C. Ma (1981), and L.J.C. Ma and E.W. Hansen (1981: 224–6)).

During the 1977–81 period these rates were down to: 1.89, 2.17, 3.58 and 3.29 respectively (Ma (1981), Ma and Hansen (1981: 224–6), and Table A.3.4). Since then, the decelerating pace has continued for the urban regions at the top of China's urban hierarchy.

What can one conclude, in linking these findings to the notions discussed in Chapters 2 and 3 on 'relative parities'? They seem to point out that as the relative parity must remain at a comparable international and national level in these urban regions, and as this relative parity contains components associated with potential for significant increase in economic growth since carrying capacities in population and income (among other variables) have not yet been reached, then these economic and demographic forces must be countered by a rise or acceleration of political forces in applying political, social and cultural pressure to control the population and economic growth potential in these regions.

Table A.3.6 Growth rates in absolute population for seven large Chinese metropolitan regions, 1981–83

Rank[1]	*1980*[2] *('000)*	*Area*[2]	*1983*[3]	*Area*[3]	*% average annual growth rate (pop.)*	*(area)*
1 Shanghai	11,628	6,186	11,940	6,186	1.30	–
2 Beijing	9,019	16,807	9,330	16,807	1.73	–
3 Tianjin	7,628	11,305	7,850	11,305	1.46	–
4 Shenyang	5,029	8,515	5,210	8,515	1.80	–
6 Guangzhou	5,537	11,757	6,840	16,657	11.76	41.68
7 Chongqing	6,434	9,848	13,890	22,341	57.95	126.86
13 Changchun	5,689	18,881	5,800	18,881	0.98	–
China	985,000[4]		1,023,300[4]		1.95	

Notes: [1]Ranking according to Table A.3.4.
[2]Source: *Statistical Yearbook of China*, 1981, EIA, Hong Kong, 1981.
[3]Source: As Table A.3.4.
[4]Source: Population Reference Bureau, *World Population Data Sheets*, 1981, 1983.

The industrial base of a Chinese city

According to the *Statistical Yearbook of China, 1981*, the People's Republic of China's gross output value in 1980 prices is estimated at 7,490 Rmb (yuan) 100 mill. With a population of 996 million in 1981 it corresponds to a per capita 749 Rmb. The gross output value in the Peking province in 1981 is estimated at 235 Rmb 100 mill, whereas Shanghai's level stood at 642 Rmb 100 mill. Given an estimated population level of 9.02 million and 11.63 million respectively, the per capita gross output value stood at 2.6 Rmb 1,000 and 5.5 Rmb 1,000 respectively. Given that only agriculture and industry are included (not government output, one of the most important sectors in Peking) one expects the per capita gross output value to be about the same. Compared with the national totals they amount to about seven times higher than the national average. This seems to being these urban agglomerations to the critical level of the code ($\bar{w} = \ln y = 2$, see Figure 3.3). Is China possibly developing two societies within its border – one urban and one rural? Empirical evidence enforced by the per capita product differentials of Table A.3.5, intra-metropolitan-wise and for urban and rural areas in these regions, in combination with qualitative inferences from popular press reports, may point to such an event.

A pilot industrial employment survey was conducted in the industrial areas of Wuxi City (on the Yangtze river and to the west of Shanghai), in Jiangsu Province, to identify employment diversification of an urban area in contemporary China. The results are aggregated here (see Table A.3.7) from a very fine level of disaggregation (found in the above cited source) so that they conform to the US-used SIC code, with the exception of the finance, insurance and real estate category.

As this is the first time we have obtained a detailed glimpse at the industrial base of a Chinese city it is rather informative. It shows an industrial town at its pre-service period of development, with 44.7 percent of the workforce employed in manufacturing (31.2 percent in durable manufacturing), and only 15.4 percent in services (being the second highest sector none the less).

WORLD POPULATION COUNTS

Time series

For purposes of comparison, Table A.3.8 gives the world population counts for the 1950–87 period from the US *Statistical Abstract* (USSA) and from the Population Reference Bureau (PRB), *World Population Data Sheets*, various years. The latter is used to estimate the exponential growth rate in population, the biological force of Chapter 1, pp. 00–00. It is found to be approximately 2.08 percent per annum, on the average, through a log-linear regression equation: $\ln X(t) = \ln X(o) + at$, $t = 0, \ldots, 27$, $(= 2.08)$.

In Table A.3.9 the current world average per capita product is given for the period 1974–85 and in constant 1972 US dollars. The US GNP deflator has been used, since the US GNP accounts for about one-fourth of the total and it is the only one available with some consistency through the 12-year period.

Sources

Time series of the world's population, over which the urban areas' population has been normalized for various years, have been obtained from the following sources: US Bureau of the Census, *Statistical Abstract of the United States*, Comparative International Statistics; World Summary. Sources employed in these Tables are exclusively from the United Nations. They were selected for consistency purposes, since metropolitan population and per capita product counts used here are

Table A.3.7 The industrial base of Wuxi City at the third digit SIC code, 1980

		Total	*Percent*
All industries		457,022	100.0
Agriculture		39,616	8.7
Farming	34,118		
Forestry	358		
Animal husbandry	2,107		
Fisheries	3,033		
Mining		3,541	0.8
Manufacturing		204,633	44.7
Durable	(142,371)		
Metal-related	3,821		
Machinery	85,586		
Petroleum, natural gas	254		
Chemical and allied	14,084		
Electrical, electronic	19,610		
Glass, pottery, chinaware	2,196		
Building materials	2,997		
Timber, bamboo products	9,280		
Paper products	4,543		
Non-durable	(54,612)		
Beverages	617		
Food	3,454		
Textiles	35,354		
Apparel, silk reeling embroidery	13,713		
Leather products	1,474		
Other	(7,650)		
Construction		11,691	2.6
Transport, communications, public utilities		29,286	6.4
Transportation	(20,040)		
Railway	710		
Road transport	10,606		
Water transport	8,724		
Communications	(1,184)		
Ports, telecommunications, radio	1,184		
Public utilities	(8,062)		
Power plant	5,599		
Public utility	2,463		
Trade		35,085	7.7
Retail, wholesale	12,898		
Storage and warehousing	11,833		
Loading	10,354		
Services		70,230	15.4
Eating establishment	16,300		
Business service	9.791		
Handymen	13,591		

Cleaning	363		
Repair	752		
Professional services	17,339		
Education	9,817		
Cultural	814		
Entertainment	1,463		
Government		52,267	11.4
Management	(33,904)		
Production	21,493		
Statistical work	3,049		
Accountants	8,582		
Financial business	780		
Administration	18,363		
Other		9,560	2.1

Table A.3.8 The world population, 1950–87 (in billions)

	PRB	*USSA*
1987	5.08	
1986	4.94	
1985	4.85	4.86
1984	4.76	4.76
1983	4.68	4.72
1982	4.59	4.59
1981	4.49	4.51
1980	4.41	4.43
1979	4.32	4.36
1978	4.22	4.28
1977	4.08	4.21
1976	4.02	4.04
1975		3.97
1974	3.89	3.89
1973		3.86
1972		3.78
1971	3.71	3.71
1970		3.65
1969		3.56
1968		3.49
1967	3.41	3.42
1966	3.35	3.35
1965		3.30
1964	3.28	3.23
1963		3.17
1962	3.11	
1961		3.07

1960	2.99
1959	2.91
1958	2.85
1957	2.80
1956	2.74
1955	2.69
1954	2.65
1953	2.55
1952	2.47
1951	2.44
1950	2.41

Table A.3.9 The world per capita product, 1974–85

	in current US$	*US GNP deflator 1972 = 100*	*in 1972 US$*
1985	2,880[1]	231.86[2]	1,242
1983	2,760	215.34	1,282
1982	2,800	207.38	1,350
1981	2,754	195.14	1,411
1980	2,620	178.64	1,467
1979	2,340	162.77	1,438
1978	2,040	150.05	1,359
1977	1,800	141.70	1,270
1976	1,650	132.90	1,241
1975	1,530	127.18	1,203
1974	1,360	116.41	1,168

Sources: [1]Population Reference Bureau, *World Population Data Sheets*, 1976–87. [2]US Dept of Commerce, *Survey of Current Business*, November issue, various years 1975–86 (February).

also from United Nations sources. Various yearly counts and the tables they are taken from are found in:

1950 count from 1954, Table 1123, p. 955.
1951 count from 1954, Table 1123, p. 955.
1952 count from 1954, Table 1123, p. 955.
1953 count from 1955, Table 1153, p. 949.
1954 count from 1956, Table 1163, p. 948.
1955 count from 1957, Table 1172, p. 938.
1956 count from 1958, Table 1183, p. 926.

1957 count from 1959, Table 1212, p. 928.
1958 count from 1960, Table 1223, p. 926.
1959 count from 1961, Table 1239, p. 921.
1960 count from 1969, Table 1240, p. 829.
1961 count from 1962, Table 1279, p. 909.
1963 count from 1968, Table 1247, p. 835.
1964 count from 1969, Table 1240, p. 829.
1965 count from 1969, Table 1239, p. 829.
1966 count from 1969, Table 1240, p. 829.
1967 count from 1969, Table 1240, p. 829.
1968 count from 1971, Table 1267, p. 793.
1969 count from 1971, Table 1267, p. 793.
1970 count from 1976, Table 1442, p. 866.
1971 count from 1975, Table 1389, p. 835.
1972 count from 1975, Table 1389, p. 835.
1973 count from 1975, Table 1389, p. 835.
1974 count from 1976, Table 1442, pp. 865–6.
1975 count from 1976, Table 1442, pp. 865–6.
1976 count from 1976, Table 1442, pp. 865–6.
1977 count from 1978, Table 1546, p. 898.
1979 count from 1980, Table 1570, p. 896.
1980 count from 1981, Table 1541, p. 866.
1981 count from 1981, Table 1541, p. 866.

The time series for the major metropolitan areas (except the US MSAs) have been obtained from United Nations, *Statistical Demographic Yearbook*, 1970, ST/STAT-D, pp. 432–72 for the period 1950–70; and United Nations, *Statistical and Demographic Yearbook*, 1981, ST/ESA/STAT/SER. R/11, pp. 265–87 for 1970–81 (superseding any prior count, as well).

Time series on nations' gross domestic product and world market economies gross domestic product average (both in current US$) were obtained from: United Nations, *Statistical and Demographic Yearbook*, ST/ESA/STAT/SER. 5/8, pp. 151–5 'National Accounts,' 1981, 1980, 1979, 1977, 1976, 1974, 1973, 1971, 1969 (later counts always superseding prior count available).

SOURCES OF DATA FOR THE UNITED STATES OF AMERICA

1 US Department of Commerce, Bureau of the Census, *Statistical Abstract of the US*, Washington, DC, various years.

2 US Department of Commerce, Bureau of Economic Analysis, *Survey of Current Business*, Washington, DC, various years.
3 US Department of Commerce, Bureau of the Census, *Commodity Exports and Imports as Related to Output*, Washington, DC, various years.
4 US Department of Commerce, Bureau of the Census, *Census of Population: General Social and Economic Characteristics*, Washington, DC, various years.

SOURCES OF DATA FOR ECONOMIC INDICATORS OF THE UNITED STATES OF MEXICO

1 *X Censo General de Poblacion y Vivienda 1980*, SPP.
2 *Boletin Mensual de Informacion Economia*, SPP, vol. VII, no. 6, August 1983.
3 *Agenda Estadistica*, 1983, SPP.
4 *Sistema de Cuentas Nacionales de Mexico, 1980*; Secretaria de Programacion y Presupuesto (SPP), Coordinacion General de los Servicios Nacionales de Estadistica, Geografia e Informatica.

NOTES

PREFACE

1 Thomas Robert Malthus was the first social scientist who posited a socio-spatial dynamic event in mathematical terms. His simple exponential growth model is a basic feature even in contemporary non-linear dynamics. In this sense all socio-spatial scientists are working in a Malthusian tradition, significantly modified since Malthus's original statement of his exponential growth in population, and the (much less celebrated and uninteresting from a dynamic viewpoint) linear growth in foodstuff model found in his 1798 *An Essay on the Principle of Population.* See also note 24 of Chapter 1.

2 The term 'ecological' is used here in its *mathematical* ecology sense. It is not related to the so-called 'Chicago school of human or urban ecology,' associated with the work (among others) by E.W. Burgess (1925). That earlier and qualitative school of urban ecology bears certain similarities to the much earlier economic work by Von Thünen (1826) and to the later work by W. Alonso (1964) and the 'new urban economics' school, for a very recent compendium of which see E.S. Mills (1987). It also resembles that by B.J.L. Berry and J.D. Kasarda (1977). The work in this book is a modified version of the mathematical ecologists' construct since it contains space and deals with humans. Mathematical ecology as set by R.H. MacArthur (1972) can be construed as a separate branch of the broader ecological framework, from which this line of work emanates; the work by R.M. May (1974) can be thought of as a link between traditional mathematical ecology and socio-spatial mathematical ecology.

3 The subject of primacy is a centerpiece of this work and a phenomenon considered to be the result of unstable spatial dynamics, according to a basic hypothesis suggested in the book and elaborated in Chapters 2 and 3. Primacy can be seen as playing an important role in the forces behind the breakdown of large (and medium size) countries which apparently are not characterized by primacy. An example is that of the Soviet Union. At the time this book went to press, in August 1991, significant events were unfolding inside the Soviet Union and also in Yugoslavia. Both were apparently breaking down into separate independent or quasi-independent republics, each republic characterized by primacy within it. The unification

of Germany and the emergence of Berlin as potentially a primate city there; the ongoing ethnic and religious unrest in India among and within its states; the political, ethnic, religious and cultural strife among and within provinces in China – all of these cases represent events pointing toward a principle (or a requirement) involving interregional conflict and spatial primacy: spatio-temporally stable nations may require primate urban agglomerations within them. If so, the principle may foretell future interregionally unstable dynamic events for all nations without apparent primacy at present. These cases may include the Russian Republic, as well as Canada and even the US, in the long run.

INTRODUCTION

1 J.N. Biraben (1985) lists the human population in 1900 at 1.634 billion; J. Matras (1973) lists the 1900 population of the globe at 1.571 billion. The variance in the count is indicative of the fuzziness in numbers.
2 Population Reference Bureau, 1988 *World Population Data Sheet.*
3 See J. Matras (1977), p. 5 and P.R. Ehrlich and A.H. Ehrlich (1970), p. 6, 7. Also, in Matras (1973) very rough estimates are shown of the human population in the last million years, reproduced here from Tables on pp. 17 and 21, and from the source in note 2:

Year	*Population ('000)*
1,000,000 BC	125
300,000	1,000
25,000	3,340
10,000	5,300
6,000	86,000
2,000	133,000
AD 14	256,000
1000	280,000
1850	1,091,000
1900	1,571,000
1960	3,000,000
1970	3,600,000
1987	5,000,000

4 See T. Chandler and G. Cox (1974) for the sizes of the major cities of the globe over the last 3,000 years up to 1974, and note 12 of Chapter 1 for the population estimates of the world's largest urban agglomerations as of 1985.
5 See the earliest counts on per capita income for 1973 in the Population Reference Bureau, 1975 *World Population Data Sheet.*
6 Population Reference Bureau, 1989 *World Population Data Sheet.*
7 The definition of a 'steady state' draws here from dynamical analysis. It is a state in which either a stationary (non-changing) level of population and

per capita wealth is reached; or a state in which all variables in the system change at a constant rate. An example of the latter is the 'golden age' state of neoclassical economic growth where all indicators (output, savings, investment and the like) grow at the same (exogenous) rate that population (labor) does. Demographers, as will be seen in Chapter 1, have been extensively preoccupied with such steady states (stable and stationary conditions) in population theory; however, the fine differences between these are not the concern of this book. Roughly, in demographic theory steady state and stationary population conditions commence when the birth-rate remains constant for a time-period prolonged enough to become equal to the reciprocal of the life expectancy of the population and equal to the death-rate, see N. Keyfitz and W. Fleiger (1971), p. 133. For our purposes it suffices to assume a steady state in population growth when crude birth-rates equal crude death-rates (globally); or when crude birth plus net in-migration rates equal crude death-rates for a specific region (nation or urban area.) Another issue which will not preoccupy this book is the age distribution of a population stock; in the spirit of the dynamics reported here, and at variance with demographic age cohort models, ecological interactions govern the linkages among cohorts within a social system. See the arguments in Dendrinos and Sonis (1990).

8 Interdependencies are here defined as the collection (bundle) of forces connecting various components of human societies as distributed in space–time. These interdependencies can be captured analytically, when precisely defined, by mathematical models. It is noted here and emphasized at the outset that these mathematical models do *not* constitute 'laws' of human socio-economic spatial dynamics, in the sense that such laws were sought after during the last two centuries by economists and demographers, particularly at the beginning of social science development. The mathematical models of socio-spatial dynamics do not remain unchanged in space–time. Instead, they may be altered and at times quite frequently. These models are mere interpretations of socio-spatial dynamic events; their form, further, affects the perceived behavior of the system modeled in space–time. This is a key argument in this book.

9 See Dendrinos (with Mullally) (1985) for a discussion of relative dynamics and interdependencies.

10 The specific definition and function of relative parity are supplied in Chapters 2 and 3. Relative parity is viewed here as an organizing principle of socio-spatial activity in time. According to an observer, who might use it to model aggregate flows of social stocks in space and time, it drives the individual and collective behavior of agents distributed in space.

11 This finding is due to R.M. May (1973). It was used in the context of urban dynamics by D.S. Dendrinos (with H. Mullally) (1985).

12 Discontinuities of interest in dynamical analysis are of two types: those that involve sudden and sharp changes in a system's state variable(s) when smooth, small and continuous changes in the (exogenous) system's parameters occur. Such discontinuities were first analyzed in the framework of Catastrophe Theory (see note 16 below). Far more interesting are those discontinuities which involve abrupt and sharp change in the nature of the dynamic equilibrium. These discontinuities are the subject of the

broader bifurcation theory (see notes 15 and 17 below).

13 The distinction among periodic, quasi-periodic and non-periodic (chaotic) cycles in simple forms of dynamic models was first made by the theory of chaos in the mathematics of continuous and discrete maps. Periodic motion is defined as that particular type of oscillatory behavior which is exactly repeatable (thus dynamically stable) so that one, without knowing the specific kinetic conditions, could forecast the next time-period form of the motion only by knowing the present and immediately relevant past; for example, knowledge of the state variable(s) at a single time-period is enough in the case of a fixed point; whereas in the case of two (or n) periodic cycles knowledge of two (or n) consecutive time-periods is required, no matter what the exact model specification (form). Non-periodic is defined as that oscillatory motion which does not have a repeatable pattern (thus it is dynamically unstable); one, without knowing the exact form of the kinetic conditions *and* the starting values, cannot predict the future form of the oscillation no matter how extensive the knowledge he might have of the current and past behavior of the system. Quasi-periodic movement is a hybrid of the two. In it the observer may detect a non-regular but identifiable (stable) ring structure involving a non-repeatable but strongly fluctuating motion. Under quasi-periodic conditions the observer may be able to predict approximately, the immediate future state of the system if extensive enough knowledge about its present and past dynamic behavior is available.

14 Phase transitions imply change in the type of dynamic equilibirum process at work. Typically, bifurcations are found when smooth changes in a system's parameter result in a change in the type of dynamic motion recorded in the phase portrait, see also note 17.

15 The notion of dynamic 'slaving' is attributed to the 'synergetics' school and the work by H. Haken (1979), and involves so-called 'adiabatic' conditions, whereby a hierarchy of fast/slow motion in parameters sets in. Its simple and basic spatial and temporal form was examined by D.S. Dendrinos and M. Sonis (1990).

16 Catastrophe Theory was established by René Thom and popularized by his (1972) book. Early contributions were also made by E. Christopher Zeeman (1977), T. Poston and I. Stewart (1978), and R. Gilmore (1981). Catastrophe theory is addressing the problem of identifying elementary forms of discontinuities in state variable space when an optimizing principle (called a gradient) governs the behavior of the system. By itself these elementary catastrophes are static constructs, although dynamic paths can be admitted by exogenously changing the (slow-moving) paramaters. Catastrophe theory first supplied the grounds to distinguish between slow and fast movement (Zeeman 1977).

17 A popular book on chaos and its founders is that by J. Gleick (1987). For a technical presentation of the subject, see J. Guckenheimer and P. Holmes (1983), D.A. Rand and L.S. Young (eds) (1980), and A.V. Holden (ed) (1986) among many others. This theory is an exploding field, currently, in mathematics. The theory of chaos examines bifurcations in the behavior of the dynamic equilibrium in continuous or discrete maps involving among other features period-doubling cycles, non-periodic and quasi-periodic

oscillations, and their transitions. A central component of the theory is the identification of strange attractors and containers in the phase portrait of a dynamical system. These attractors–containers identify states which either attract or restrain the dynamic motion of the system in an asymptotic non-periodic or quasi-periodic fashion. Statistical verification of chaotic attractors or containers in socio-spatial dynamics is not feasible at least at present. Thus, one cannot be certain that chaotic attractors or containers do exist in socio-spatial dynamics, and those who try to identify the presence of a strange attractor in the stock market (presumably to make money by trading appropriately) are likely to be disappointed. Consequently, one must talk about 'model chaos' rather than 'actual chaos' particularly since one cannot be sure that what the theory of model chaos produces is what real chaos contains. For an application of the theory of chaos in the socio-spatial sciences see Dendrinos and Sonis (1990). Chaotic dynamics may be rare (and thus interesting) events, occurring unexpectedly at a very small number of points in space–time. Though unpredictable, not lasting long or covering large areas of space, they have far-reaching effects in space–time. Socio-spatial systems, vast in spatial extent and characterized by relatively prolonged periods of stability, look quite different once interrupted by brief and confined-in-space sparks of chaos (instability). The pre- and post-chaos dynamic stability regimes are rarely identical i.e., the relevant state variables describing such socio-spatial systems have in general quite different (stable) values prior to and after the chaotic event(s).

18 The subject of speculation and the associated issues of futures markets is a vast field in the disciplines of both economics and business, as well as law. From an economic standpoint, of interest here is the notion of 'rational expectations' as originally discussed by J. S. Mill (1848), where in Chapter 9, section 3 of Book II the precursor of the notion of rational expectations is addressed. In contemporary post-Keynesian economic analysis the beginning of the analytical treatment of the subject was founded in the paper by P. A. Samuelson (1957). In the paper, the issue (and myth) of efficient (rational) speculative behavior is defined and the notion of a 'bubble' is identified. Samuelson asked the question whether speculation is stabilizing or destabilizing, profitable or non-profitable, and to whom. He put forward the view that destabilizing speculation is irrational, which is contrasted by the position taken by neo-Keynesian economists that destabilizing speculation could be rational. Variations of both themes can be found in an extensive literature where subjects of 'sunspot equilibria' bubbles and the self-fulfilling prophecies nature of speculation in business cycles are dominant issues – see for example C. Azariadis (1981), and in a more popularized manner the book by C. P. Kindleberger (1978). The discussion on speculative behavior of economic agents in the field of economics has taken amusing directions, requiring at times, among other things, assumptions involving agents which live forever, an infinite number of individuals, perfect foresight, and the like. This is, however, an area in which the theory of chaos stands to make a significant contribution, at least at the theoretical front if not at the empirical.

19 The debate on whether social systems and their dynamics must be looked at in a deterministic or stochastic manner is long-standing and based on the

contention that stochastic models afford modeling individual rather than aggregate behavior which deterministic models usually do. Recently, work from physics has been transferred to stochastic spatial migration dynamics by Weidlich and Haag (1983). In microeconomics, stochastic analysis appeared in the theory of consumer behavior in the 1970s by introducing in static terms the random utility disaggregate choice theory (McFadden 1976) and in the vast literature that followed, particularly within the urban and regional economics field. The claim that stochastic static or dynamic models depict individual (heterogeneous) behavior will be addressed later, in Chapter 1. It is noted that often the qualitative features of identifiable (as opposed to random) patterns emerging from stochastic dynamical analysis are not appreciably different from those of the deterministic dynamics (Dendrinos (with Mullally) 1985). Further, although stochasticity is normally associated with strictly individual actions, the stochastic models mentioned are still aggregate (group) models. Social science modeling of large population size stocks cannot be individually based (purely disaggregative) due to the very large number of individuals involved, vast differences in behavior among individuals, and lack of knowledge regarding all individuals' behavior. Thus, all models claiming that they stimulate individual behavior (i.e., they are microbehavioral) are placing misleading claims.

20 All markets private or public, under any social political cultural environment are imperfect, of course, notwithstanding assumptions to the opposite by neoclassical economists on economic markets, or Marxists on the political markets of socialist states. It is partly so because humans are imperfect beings and choice makers, individually and even more so collectively. Market imperfections are central in identifying the conditions under which individual agents or collectives operate. Examples of factors which result in market imperfections include (without being limited to): externalities, thinness of buyers, suppliers and distributors of the commodities, policies, ideas, actions or whatever is being exchanged resulting either in monopolies or monopsonies, or at least oligopolies or oligopsonies, imperfect knowledge and understanding, limited time horizons, etc.

21 Fast dynamics are those associated with changes in the endogenous state variables of the system, while the (exogenous) parameters remain constant.

22 Slow dynamics are those associated with changes in the parameters (exogenous factors) of the system. There could be a hierarchy of slow-moving (slaving) parameters in a model.

23 Relative parity in attraction is a non-recordable quantity, only open to an outside observer's perception. It is only inferred and measures the attraction of a location in reference to other locations when a bundle of attributes are evaluated. An observer will possibly use this index to explain flows of stocks among locations. The notion is vaguely similar to that of 'utility' in neoclassical economics, and to Adam Smith's 'invisible hand' in regulating markets.

24 This is the subject of a vast literature in the field of social science epistemology. For a review of these issues see, among others, F. Machlup (1980, 1984) and R. W. Miller (1987), among many others. Whereas

Machlup provides an economic foundation (in a non-speculative exchange mode) of human knowledge (from metaphysical to scientific), Miller penetrates into the positivistic and non-positivistic theories of knowledge. In a Popperian deductionist tradition he views natural and social science knowledge as falsifiable hypothesis testing. One may not strongly disagree with a positivistic view of natural sciences, even if one is willing to disregard considerable obstacles like the difficulties found in deriving a logically self-consistent natural science theory, and in the impurities of its linguistics. Such problems plagued logicians and philosophers from R. Carnap to B. Russell to K. Popper. However, one is likely to disagree with a strict positivistic view of knowledge in the social sciences, given among other things the strong feedback existing between the theorist, the theory, and the subject under investigation. A dialectical approach to science (with its accent on natural sciences) is also provided by R. J. Ackermann (1985), an attempt to resolve the conflict between T. Kuhn and the more traditional epistemologies concerning subjectivity in sciences.

25 Humans have been at all points in space–time subject to a variety of constraints, and everywhere they always have objectives on how to define social events and on how to describe, interpret and react to them. At times and in certain places these constraints have been quite binding; often, the collectively as well as individually held objectives are rather uniformly distributed among the population. Periods and locations where either highly oppressive (political, religious or other) regimes ruled, or intellectual stagnation prevailed, or highly homogeneous cultural norms existed are cases in point. In contemporary societies (at least in the West), the recording and description of social events and their constituent elements are considered as 'opinions' or interpretations. Individuals are offered options (rights) to dealing them in interpretation markets and to (legally) react to social events. Academics are offered options in presenting their theories, at least to the extent they can muster an audience.

26 Social events are fuzzy along many dimensions, particularly along the number of variables one chooses in recording and describing them. Thus the definition and use of the term 'fuzzy' here are broader than those found in fuzzy sets theory, see note 27 below.

27 For a recent set of papers on the element of fuzziness in cognitive uncertainty, perception and computing see M.M. Gupta and T. Yamakawa (eds) (1988). However, the problem with this theory is that it makes rather specific boundary statements on what a fuzzy element is and this results in numerous difficulties. For example, in J. J. Buckley (1989), difficulties emerge when bounded, closed intervals for the model's parameters coefficients are assumed which result in the I–O coefficients not equaling, necessarily, one for each industry. In H. J. Zimmermann, L. A. Zadeh and B. R. Gaines (1984), the limited use for only certain types of vagueness in decision making is stressed; in the article by C. Carlsson (1988) 'On the relevance of fuzzy sets in management science methodology', the connections between fuzziness (as assumed in fuzzy set theory) on the one hand and uncertainty, generality, imprecision (which it is argued that fuzziness is not) and vagueness (which it is argued that it is) on the other, are discussed. The existence of 'tolerance limits' in fuzzy sets (however

arbitrary they may be) bring into question the notion of 'fuzziness' as used in formal theory of fuzzy sets. Use of fuzziness in social science theories brings about a potential internal logical inconsistency in trying to precisely (analytically or mathematically) deal with vagueness, a problem which may apply to the work at hand, too. To address this difficulty a 'quantum type' theory of socio-spatial dynamics might be appropriate to develop, although the task of doing so falls outside the scope of this book.

28 See Machlup (1980: vol. I, Pt 3, 155–240).

29 How options and futures markets work and under what conditions such markets exist are topics widely addressed in the business and legal literature. For an introductory text on the fundamentals of options and futures markets see, for example, B. Branch (1976).

30 It is of interest to note that, in spite of the strong resemblance between stock market participant behavior on the one hand and social policy-making behavior and social science theory formation and development on the other, not much has been written on the subject. A systematic exploration of the equivalences between these markets does not exist to the author's knowledge. Advances in the theory of chaos are likely to draw more attention to these similarities because the theory of chaotic systems supplies general principles of interest to both.

31 All major contributors to the nature of scientific knowledge and its development, evolution or revolutions, from T.S. Kuhn (1962) to K. Popper (1963b), W. Salmon (1967), R. Miller (1987) and others, failed to note the speculative nature of social science formation. Such a shortfall is particularly acute when the intrinsic link between the theorist and the system under investigation is so pervasive in social science. Social (present and expected future) events render themselves open to multiple interpretations on which agents take bets and therefore, through betting, affect their perceived outcome.

32 Speculators are here defined as those agents who are expecting an increase in the price of an option or a futures contract; as a result they are willing to enter the options or futures market and purchase call options or sell put options or buy futures contracts. On the other hand, hedgers are agents who expect the price of these contracts to drop and as a result they are willing to sell call (or buy put) options or sell futures contracts they currently possess. There are other strategies speculators and hedgers may adopt besides buying or selling options or futures contracts (for example, 'going short' in a market). For our purpose, however, such differentiation in strategies is not very important.

33 To what extent the past, present or future conditions of a social system are clear (meaning that the number of possible interpretations about it, i.e., options traded on it, is rather small) is of interest in social sciences. One must expect this to be a rare occurrence and to only happen in rather small-scale short-lived events. When the stakes are very high the options must be greater in number and the volume must be significant. The possible paths that the system has or could follow must be multiple and the picture of the system in its phase portrait must be unclear and confusing. The theory of non-linear dynamics has potentially a lot to contribute to the qualitative properties of this picture of confusion, particularly since it might reveal the

presence of order in this apparent picture of disorder. This, however, implies that what is termed 'chaos' in the mathematical theory of chaos represents (at least approximately enough) what one calls chaotic in real systems, and it is not simply 'model chaos.' The resolution of this question still awaits further research (empirical and analytical) in social science epistemology. It is noted that the theory of non-linear dynamics receives only a passing reference in Miller (1987), a footnote to I. Prigogine's work on p. 43. Non-linear dynamics may render obsolete many established concepts about knowledge, learning and understanding and this obsolescence may not be confined only to the social sciences. As with almost all topics in this book, these subjects can be explored in far more detail. An axiomatic theory of social events and theories can be attempted, where the links between interdependencies (forces on stocks) and interactions (flows), as well as the connections between social dynamics and speculative behavior can be further explored.

34 Karl Popper sums up the logical argument against scientific prediction of social history as follows: 'the course of human history is strongly influenced by the growth of human knowledge; we cannot predict, by rational or scientific methods, the future growth of our scientific knowledge; we cannot, therefore, predict the future course of human history' (Popper 1957), p. v. This logical argument is of particular interest to economic forecasting, where the term 'rational expectations' is employed in lieu of predictions in defining the very peculiar notion of *homo economicus*. The term implies, in economics, that an individual is 'rational' if he does not systematically predict erroneously (Lucas 1981 and Sargent 1987). Alternatively it is considered that a group of individuals act rationally if on average they do not consistently predict erroneously. Popper's argument, however, logically proves that the term 'rational expectations' is not internally consistent. At least, it renders the construct useless for the individual and unique in time (rather than a series of identical, repeatedly occurring) forecasts, as is the case with the problem at hand. In social sciences one can only deal simply with 'expectations' without making the claim that they are 'rational.' The argument also puts into its proper framework the ongoing debate in microeconomics on individual behavior, optimization and efficiency. Whether individual behavior can be perceived by an outside observer as an optimization process, whereby the observer can set up objectives and constraints proposed to predict the individual's behavior, is the central issue in the debate. If the individual does not fulfill the observer's expectations then two conclusions are drawn: the individual acted irrationally (i.e., in a sub-optimal manner); the individual acted under another set of objectives and constraints than those assumed by the (neoclassical) economic observer. Neoclassical and other formal theories in economics accept the former and reject the latter as 'unscientific,' since it does not allow them to state 'falsifiable hypotheses.' They claim that accepting the second conclusion reduces individual action and the optimization process to a tautology. See an exposition of the arguments in H. Leibenstein (1980). Popper's argument removes the focal point of the neoclassical and other economic efficiency debates from the question on whether individual action is efficient or not, to the real issue which is that

of the observer's inherent inability to 'rationally' predict. Although the Popperian argument is made for social aggregates, the argument can easily be transferred to individual behavior: the course of an individual's history is strongly influenced by the growth in the individual's knowledge; an outside observer cannot predict, by rational or scientific methods, the future growth of an individual's knowledge; the outside observer, therefore, cannot predict the future course of an individual's actions. In a speculative framework the theory of individual consumer behavior can be stated as follows: individuals allocate their current multifaceted resources on items which carry an expected (and discounted at present) series of future returns. These returns are subject to the risks (odds) currently perceived by the individual, these being not currently apparent to an outside observer. These odds are not subject to outside (scientific) predictions, by either the individual facing them or the outside observer. These allocations are not repetitive in time–space. Thus, the current allocation of an individual's present and/or future resources is not subject to 'rational expectations' by either the individual or the observer, although the past allocation scheme could be by either or both. Further, alternative theories about the individual's resource allocation at a point in space–time can be derived, and it is on this basis that the observer may take bets if a taker of the bet can be found. This betting process may result in transfer of money between the agents involved in it and possibly in a change of the individual's behavior. This is by no means, however, a statement on whether knowledge has been gained on the subject's resource allocation behavior. Further, and within the framework of chaotic dynamics, another concern is raised regarding the 'rationality' of individual (utility maximizing) behavior. For an economic–social agent (individual or collective) to maximize utility, resources must be spent under a continually changing and unpredictable environment. A key among these resources is time, in fact *real* time. If the time it takes to derive an optimal solution is longer than the time it takes for the environmental conditions or an agent's preferences and constraints to change, then is it an efficient and effective strategy for individuals to attempt to maximize utility? Further, if the time it takes for an individual to find an optimum allocation of resources affects the individual's perceived worth of time itself, then the problem may involve a non-linear dynamic interdependency which might result in chaotic behavior. The argument points to one of the major sources for instability cited earlier in the main text. Namely, that an action variable (time allotted for finding an optimum solution) may depend, and at the same time affect, decision variables (perceived value of time and time allotted). These questions could hide important implications on the economic behavior of agents which might include qualitative features of model turbulence. Further, the notion of efficient aggregate behavior, say in the context of a market, is also severely questioned in this book as similar problems raised at the individual level can occur in their collective behavior as well.

1 GLOBAL INTERDEPENDENCIES

1 See Dendrinos (with Mullally) (1985), and Dendrinos and Sonis (1990). These works provide a survey of the relevant literature and a history of how the original Malthusian simple population exponential growth model has evolved into one in which population and per capita income are interdependently evolving. It is noted that V. Pareto considered per capita wealth to be one of the most important determinants in social evolution, according to J.J. Spengler, see R.S. Smith, F. de Vyver and W.R. Allen (1972), p. 75. As a result of Pareto's studies, the Pareto income distribution curve was derived. Of interest are also Spengler's views of Malthus, (*ibid.*, pp. 3–65), and Marchall's views on population (*ibid.*, pp. 114–56). Spengler also studied A. Lotka's population theory, see J.J. Spengler (1976).

2 A short list of major recent works and surveys of demographic theory, chronologically listed, must contain the following works: W.S. Thompson and D.T. Lewis (1965), where they cite major population theories by M.S. Sadler (from his 1829 book on *Ireland: Its Evils and their Remedies*, 2nd edn, J. Murray Ltd., London, on the 'natural law' of population growth in contrast to Malthus's law, p. 38), by T. Doubleday (from his 1847 book on *The True Law of Population Shown to be connected with the Food of the People*, 2nd edn, George Pierce, London, who argues that the man's increase in numbers is inversely related to the food supply, p. 39), by H. Spencer (from his 1867–68 book *The Principles of Biology*, D. Appleton and Company, Inc. New York, who argues that there is a natural decrease in reproductive capacity, or fecundity, as 'personal development' occurs, p. 41), by C. Gini (who argues that the growth of nations is linked to the cyclical rise and fall of population size, p. 43), and by K. Marx (who argued that under communism there is no poverty regardless of the population growth rates, p. 48). Thompson and Lewis use the term 'population pressure' in an absolute and relative context to argue that 'the lower the per capita income the greater the absolute population pressure' and that 'there is a relative feeling of deprivation by some when others in the same country have a much higher per capita income,' p. 502; N. Keyfitz, W. Flieger (1968); A. Sauvy (1969) (translated from the French by C. Campos from the original text published in 1966 by Presses Universitaires de France) who argued that the West enjoys a period of 'near equilibrium' population growth, whereas the Third World is 'alarmingly off balance' demographically, p. ix of Preface; M. Spiegelman (1968) where a statistical-logistic theory of population projections is supplied, Chapter 14, pp. 393–422; G.A. Harrison and A.J. Boyce (1972), where in a paper by R.W. Hiorns, 'Mathematical models in demography' a simple logistic model with a carrying capacity is studied; W. Brass (1971), where in a paper by J.G. Skellam the Volterra ecological model with respect to human population and food supply is analyzed as a producer–consumer system; A.F.J. Coale (1972), where a dynamic demographic (by age distribution) theory of convergence to a steady state under various birth- and death-rate scenarios is presented based largely on Lotka's work; T. Frejka (1973), who believes that 'human populations must reach a

nongrowing, or stationary, state at some point in the future, . . . and utilizes the technique of population projection to identify the demographic trends necessary to reach this stationary state,' p. x, Preface; J. Overbeek (1974), where a brief history of the major theories and their main message (whether easing or evoking fears of overpopulation) is given; E. Boserup (1981), where she argues (more like a demographer than an economist) that technology affects and at the same time is affected by population growth; and J.R. Weeks (1978), which contains a survey of the major demographic theories of a more recent vintage. A hybrid between demography and economics (as is the case with Boserup) is the work by H. Leibenstein (1954).

3 Most of the work by economists on this interdependency is found in the field of economic growth in the area of macroeconomics. It mostly addresses issues of per capita income growth from an economic growth viewpoint where the Malthusian and its derivative models have been severely criticized, see C.L.F. Attfield, D. Demery and N.W. Duck (1985) where they present a two-part Malthusian model (on which they base their critique) consisting of a per capita income function increasing at a decreasing rate with population levels, and two rates (a birth- and death-rate) as functions of per capita income (the birth-rate being independent of per capita income, whereas the death-rate is a declining function of per capita income at a decreasing rate); J.L. Simon (1977) where the emphasis is on the effect of population growth on income, and only the outlining of a dynamic feedback model of income on population growth is supplied in pp. 37–41; K.E. Taeuber, L.L. Bumpass and J.A. Sweet (1978) with a brief review by S.H. Preston on the forming of new trends in demographic analysis (pp. 299–313); D.G. Johnson and R.D. Lee (1987) where in an article by D.A. Ahlbung 'The Impact of Population Growth on Economic Growth in Developing Nations: The Evidence from Macroeconomic–Demographic Models' an evaluation of major demographic models is provided. For a recent review of the state in demographic theory see D. Coleman and R. Schofield (1986). Although originally, in the work of the classicists, macroeconomics was concerned with population issues to a great extent, in recent years the term 'population' does not even appear at times in macroeconomics texts. See also Appendix 2, pp. 274–84.

4 The interdependence between population and per capita income, to the extent covered, appears in the migration, urbanization, and spatial development areas – all of which have a vast literature. For a brief set of references on the migration and regional development front, see T. Tabuchi (1988). In the spatial sciences, population growth and economic development tended to be treated separately and in a static or simplistic inter-spatial equilibrium dynamic framework. The urbanization phenomenon was also examined in a simple dynamic context. See for example the discussions on migration or urbanization by P.J. Schwind (pp. 379–93) and W. Alonso and E. Medrich (pp. 349–78), both in L.S. Bourne and J.W. Simmons (eds) (1978); by J. Freidmann (1973) and B.J.L. Berry (1973), and in the work by B. Thomas (1972) where a more interesting perspective of cyclical motion is presented. For a survey of the standard work on spatial development at various points in time see G. Myrdal

(1957); H.S. Perloff *et al.* (1960); J. Friedmann, W. Alonso (eds) (1964); J.G. Williamson (1965); H.W. Richardson (1973). As a bridge between demography and geography is the work by A. Rogers (1968), K.C. Land and A. Rogers (1982), and A. Rogers, F.J. Willekens (1986). All these employ methods and models from mainstream micro and macroeconomics and demography, and thus they are not theoretically different than their intellectual forebears. Of particular interest is the work by Casetti (1989), where a Malthusian trap model is linked to a fold-type catastrophe in describing the start-up of modern economic growth in Europe; a switch from premodern to modern growth conditions is seen as a tipping event in the neighborhood of a critical region in phase space where two different dynamic events are present.

5 See for example, T.R. Ford and G.F. de Jong (1970), W.B. Miller and R.K. Godwin (1977), J. Matras (1977), C.F. Grindstaff (1981) with a bibliography on this section of the literature, and K.C.W. Kammeyer and H.L. Ginn (1986) where the authors present a 'societalist approach' to population growth particularly to the extent of a collective decision by society being able to lower the crude birth-rate.

6 See, among others, B. Spooner (ed) (1972); W. Peterson (1975); N. Howell (1976); R.G. Sipes (1980), where a demographic transition-type model is tested, and a set of references on the effects cultural variables have on population growth issues is supplied; and Appendix 2, pp. 298–302.

7 The new non-linear dynamic theories allow for a technical definition of the notion of 'development,' see D.S. Dendrinos (with H. Mullally) (1985) p. 7; a continuous or discrete path belonging to a specific dynamic equilibrium type (for example, a spiralling sink or a node, a center, or a limit cycle) is a *developmental* trajectory. It involves 'fast dynamic change.' Social events, either ephemeral (i.e, temporally sharp) or durable (lasting a relatively prolonged time-period), define the dynamic *state* of a socio-spatial system. This state can be recorded as a *phase* containing the uninterrupted dynamics of an event in space–time. Such a phase defines the developmental path of the system. Put differently, it records the event lasting over a particular time-period and spatial extent.

8 *Ibid.*, p. 7; the term 'evolution' implies a change in the qualitative properties of the dynamic equilibrium. For example, if a spiralling sink is switched into a spiralling source, then the change is of an *evolutionary* nature. It is associated with 'structural change,' as the nature of the dynamic equilibrium switches types. In chaos theory, a transition from periodic motion to non-periodic or quasi-periodic is an example of an evolutionary change in a system's dynamics. Instantaneous events, non-durable and temporally sharp (of a relatively short duration) interrupt longer-term events. At times, such shocks or perturbations might cause a *phase transition*, linking two different types of socio-spatial events with different qualitative dynamic properties. This is an evolutionary event. If a shock does not result in qualitative different dynamics, but merely in a shift of a dynamic path to a neighboring trajectory in phase portrait, then the event is not considered to be evolutionary. This analysis lends itself to the formation of a theory of evolution, although such an attempt is not the subject of this volume. Only a few additional, but broad, comments are warranted

here. (a) Multiple *interacting* social events occur at any point in space–time. Because of this interaction social events are never exactly reproduced, because a possibly large set of events need to be exactly repeated in order to at least approximately repeat history; such repetition (requiring isolated effects in controlled experiments) is highly unlikely in socio-spatial systems posting a proposition which questions, among other things, the use of inferential Bayesian statistics. (b) Continuous, longer-term events (like for instance a spatial association of a particular type, say a cooperative interaction) may lead to a sharper shorter-term event, possibly in a location's population explosion. A predatory or competitive association on the other hand, occurring over a prolonged time-period between two regions, might lead to an event like war. (c) The time horizon or duration of an event (a phase in socio-spatial dynamics) might be itself a function of the ongoing interactio among the regions concerned. On the other hand, a sharp event, like a war, might set in motion a different spatial interaction than that existing prior to the war, proving to hold potentially evolutionary effects for the two regions. (d) Both definitions, found in notes 7 and 8, regarding 'development' and 'evolution' are analytical and they differ from the biologist's or physicist's definitions. They are of some interest, however, when demographic transitions are addressed. See also the discussion on the dynamics of the aggregate developmental code (ADC) in Chapter 3. The notions of development and evolution here are used quite differently than their qualitative multiple fuzzy and largely ideological definitions found in the social sciences (for example in Marxism, where transitions in the mode of production, from feudalism to capitalism for instance, are often viewed as evolutionary events).

9 Population data on absolute size of the nineteen urban agglomerations used in Figure 1.1, are supplied in Appendix 3 (pp. 303–9), together with their sources. For their size at a particular point in time, 1985, see the discussion in note 12.

10 See D.R. Vining Jr (1985), where he identifies a threshold of approximately US$3,000 in constant 1975 dollars in per capita income as a lift-off level (p. 47). I became aware of Vining's work in April 1985, after his *Scientific American* article appeared. My own work on this subject was undertaken during the Spring of 1984.

11 The two-variable phase portrait and the discrete time structure of the data in Figure 1.1 indicate that the most appropriate way to model these paths must be a discrete time, two state variables, relative dynamics model. The continuous dynamic formulation provided here is only an approximation to the more precise 'discrete' model; it intends to replicate the behavior of each urban area's expected (moving mean) value trajectories, and thus (being of the two-variable, continuous type) it excludes the possibility for reproducing chaotic motion. See Chapter 3, pp. 180–1.

12 The fuzziness of the count is demonstrated for the forty-eight largest urban areas on the globe as of 1985 in Table N.1, which contains estimates derived from United Nations, World Bank, national census, numerous other sources, and my own calculations. Fuzziness in size is mostly due to varying political jurisdictions considered by the sources, different estimation procedures and other factors. For US metropolitan areas the issue of

fuzziness enters into (among other aspects) the designation (as of 1990) of what constitutes a Primary Metropolitan Statistical Area and/or a Consolidated Metropolitan Statistical Area. The Office of Management and Budget allows local agencies to decide whether they wish to be included in such intermetropolitan agglomerations. Uncertainty increases in the ranking as one moves down the list, which contains two major divisions and seven minor subdivisions (groups). Division I includes the urban super-giants, containing a population level likely to exceed ten million. The division includes two groups; group A is comprised of the four biggest urban agglomerations with a population level greater than fifteen million; group B contains nine urban agglomerations with a population most likely to be between ten and fifteen million. In Division II, five groups are included: group C contains four urban areas with a likely population between eight and ten million; group D has six metropolitan areas with a population in the range of 6.5 to 8.5 million; group E includes four urban areas in the 5.5 to 7.0 million range; and group F has fourteen agglomerations likely to range between 4.5 and 6.0 million. Uncertainty regarding the count, coupled with the relatively high number of urban areas in this group, renders the within-group ranking almost meaningless. A group G is provided which contains seven urban agglomerations from the People's Republic of China, India, Indonesia and Zaire. These are the metropolitan agglomerations of Asia and Africa which are rapidly emerging as population centers of five million. Finally, the table contains urban agglomerations in nations as they were defined prior to the considerable socio-political events in Eastern Europe (including Germany) and the Soviet Union of the late 1980s and early 1990s.

Table N.1 The fuzzy population count of the world's largest urban areas, 1985

I	Super-giant urban agglomerations:			
	A	Likely population larger than 15 million		
		1	Mexico City	(18.0–21.0)
		2	Tokyo	(12.0–18.0)
		3	New York	(13.0–17.5)
		4	São Paulo	(12.5–16.5)
	B	Likely population between 10 and 15 million		
		5	Shanghai	(11.5–13.5)
		6	Calcutta	(9.5–13.0)
		7	Seoul	(8.5–13.0)
		8	Bombay	(8.5–12.5)
		9	Buenos Aires	(9.5–11.0)
		10	Beijing	(9.3–12.0)
		11	Jakarta	(7.5–12.0)
		12	Rio de Janeiro	(8.8–11.0)
		13	Cairo	(5.5–12.0)
II	Giant urban agglomerations:			
	C	Likely population between 8 and 10 million		
		14	Los Angeles	(7.5–12.5)
		15	London	(6.7–10.0)
		16	Paris	(8.5– 9.0)
		17	Moscow	(8.0– 8.7)

D	Likely population between 6.5 and 8.5 million		
	18	Chicago	(6.5– 8.5)
	19	Delhi	(6.0– 8.5)
	20	Tianjin	(6.0– 8.0)
	21	Teheran	(6.0– 8.0)
	22	Karachi	(5.5– 8.0)
	23	Baghdad	(5.0– 7.0)
E	Likely population between 5.5 and 7.0 million		
	24	Manila	(2.0– 7.5)
	25	Guangzhou	(5.7– 6.9)
	26	Lima	(4.9– 6.5)
	27	Istanbul	(3.5– 6.0)
F	Likely population between 4.5 and 6.0 million		
	–	Leningrad	(4.8– 5.5)
	–	Hong Kong	(5.1– 5.5)
	–	Taipei	(4.5– 6.5)
	–	Wuhan	(4.5– 6.0)
	–	Bangkok	(5.0– 6.5)
	–	Shenyang	(4.5– 6.0)
	–	Philadelphia	(4.5– 6.0)
	–	Lagos	(4.5– 6.0)
	–	Bogota	(4.5– 5.5)
	–	Dacca	(3.7– 6.0)
	–	San Francisco	(3.5– 5.7)
	–	Madras	(3.5– 6.0)
	–	Caracas	(3.0– 6.0)
	41	Madrid	(3.2– 5.1)
G	Rapidly emerging five million in population urban agglomerations		
	–	Chongqing	(2.7– 6.5)
	–	Changchun	(1.7– 5.7)
	–	Bangalor	(1.5– 4.5)
	–	Kinshasa	(2.5– 4.5)
	–	Surabaya	(2.0– 4.0)
	–	Hyderabad	(3.0– 4.3)
	–	Bandung	(1.5– 4.3)

See also T. Chandler (1987), for a very rough population estimate of major world cities from 3,100 BC to 1975; and T. Chandler and G. Cox (1974). The fuzziness of these counts is noted.

13 It should not be inferred that only large MDCs are the sources of such spatio-temporally propagating social activity – potentially, any nation could be such a source. However, it is reasonable to expect that the number of such events, their speed of motion, their potential spatial and temporal extent of spreading, and their magnitude of impact must be related to the size of their source. Large and relatively wealthy nations might be prime sources for such social experimentation in initiating and affording such events, particularly capital intensive technological innovation.

14 For example, besides economic there are geographical, sociological, political, environmental, legal, moral, ethnographic, psychological, genetic, anthropological, cultural, religious, linguistic, architectural, art, urban planning and historical aspects, theories, models and methods. Issues of development and evolution also fall under the domain of biology, genetics,

paleontology, geology and the various natural sciences. Collectively, the social and natural sciences tell the story of human and natural development and evolution as currently understood. The presence of composite comprehensive bundle forces referred to as 'ecological' (meaning non-specific multifaceted non-disciplinarian) is an amalgamation of phenomenological abstract forces containing causes and effects. The term 'ecological' differs from the way it is used in the field of ecology; it remotely resembles K. Boulding's use of it in his ecological dynamics (1978). This urban-based ecological framework was stated by the author in, among other sources, Dendrinos (with Mullally) (1985), where the theoretical framework together with a list of references is supplied, and in Dendrinos (1984a) and (1984b) in the context of regional dynamics. Both present a theory of continuous spatial dynamics. A similar theory has also been suggested in discrete dynamical terms in, among other sources, Dendrinos and Sonis (1990).

15 It is apparent that demand for research on development-related subjects in all fields of knowledge far outweighs the supply of researchers. It is also apparent that it would simply be futile for anyone to attempt a survey and a synthesis of this voluminous and heterogeneous literature. Only strategically chosen observations can be made.

16 Population and per capita wealth (income or product) series have been collected by various census agencies at different time-periods, and by nations at different phases of national development at the early stages of the census process mainly for taxing and security related purposes. In the US, for example, they occupy prominent roles as evidenced by the US Department of Commerce Bureau of the Census and Labor Statistics publications; and so they do for the United States of Mexico and the People's Republic of China Reports, these being three sources among others used in this study.

17 To realize their perceived importance by government and policy makers and analysts one need only read a 1984 A.W. Clausen (former President of the World Bank) address on population and development, *Address to the National Leaders' Seminar on Population and Development*, The World Bank. The second paragraph of the address reads:

> As all of us here agree, [population growth] is a subject of vast importance and undeniable urgency. While the effects of fast population growth may vary widely, depending on the institutional, economic, cultural, and demographic setting, all the evidence points overwhelmingly to the conclusion that it slows developing countries. And the poor of these countries are the principal victims of the slowdown.

His last paragraph also points to the importance attributed by the World Bank to these variables; it reads:

> World population has grown faster, and to higher numbers than Malthus would ever have imagined. But so have world population and income. If we can correct the current mismatch between population and income producing ability, a mismatch that leaves many of the worlds' people in

> a vicious circle of poverty and high fertility, we may yet evade the doom which Malthus saw as inevitable.

For further evidence of the importance attributed by the World Bank to these variables see also its annual *World Development Reports*.

18 See for example, S. G. Morley (1956); more on the growth and decline cycles of Chichen Itza and other Maya cities is given in Chapter 2 (see pp. 144–5). More recently a collection of papers addressed this issue, some in a non-linear mode, see J. Sablov (ed.) (1980).

19 Their complex socio-cultural factors were studied by, among others, Veblen (1967) who analyzed elements ranging from attributes found to prevail in a culture of leisure and decadence to those found in societies with a dominant work ethic.

20 In mathematical terms the equations of motion for the population and per capita income interactions could be captured by the following two sets of ordinary simultaneous differential equations:

$$\dot{x} = f(x, y)$$
$$\dot{y} = g(x, y)$$

when rates of change depend only on current levels; or,

$$\ddot{x} = f(\dot{x}, \dot{y}, x, y)$$
$$\ddot{y} = g(\dot{x}, \dot{y}, x, y)$$

when acceleration or deceleration in growth rates depend on their first order change and their current levels.

21 The subject of absolute and relative socio-spatial dynamics is extensively addressed in both sets of references by the author, found in note 14 above.

22 See Chapter 3 (pp. 176–7) and Appendix 3 (pp. 321–2).

23 Population Reference Bureau, *World Population Data Sheet*, 1976 and 1988.

24 The original Malthusian model has many interpreters and many analytical formulations. What is currently referred to in the mathematical literature as the Malthusian exponential growth model is simply an exponential function on a stock's size:

$$\dot{x} = A\,x, \text{ or } x(t) = x(0) \exp At,\ 0 < t < T$$

where x(0) is the initial condition, and parameter A is fixed and exogenous and it represents a constant growth rate in the horizon T. Malthus's other model component, that of a linear growth in the food supplied (or produced) in time, has become atrophic as other more interesting formulations have substituted it. The next significant development in the Malthusian growth model was the P. Verhulst (1838), logistic growth:

$$\dot{x} = A\,x\,(1 - x),\ 0 < x < 1$$

which was followed by the R. Pearl (1924) formulation where A is a function of time itself, the so-called Pearl–Reed equation. These equations were further extended by Lotka (1925) (1956) and Volterra's two species formulation, which was elaborated further and transferred to multiple species by Volterra; see his classical papers in the book by Scudo and Ziegler (1978).

But the original exponential formulation was a most distinct contribution to social sciences and it is this basic model which is at the beginning of all non-linear modern mathematical analysis. In this sense, all of us are modified or neo-Malthusians, particularly those who use the Volterra–Lotka formalism and its multiple extensions. These extensions basically take the constant coefficient A and make it a function of stock size. More stocks are also introduced under the same conditions. Recently, discrete dynamic formulations of the above equations have been introduced, in both absolute and relative dynamics, see Dendrinos and Sonis (1990).

25 See Appendix 3, Table A.8.2.

26 Population Reference Bureau, *World Population Data Sheet*, 1980, 1982.

27 There are numerous textbooks on financial instruments, including options. An explanation of European (exercised only at their expiration date) and American (can be exercised any time up till their expiration date) options is given there together with the definitions of calls and puts. Calls are defined as the right but not the obligation to buy from the issuer or writer of options contracts; puts as the right but not the obligation to sell to the aforementioned issuer or writer. One can pick out the definitions of basic option market terms, like 'expiration date,' 'contract price,' and 'striking price,' among others from the above sources. Briefly, the purchaser of an option contract must pay a premium (which is the value or market price of the contract) for the privilege of holding the option to the issuer (or seller) of the option contract. See also B. Branch (1976) p. 23.

28 For example, see the work by M. Harris (1979). Another cultural determinism view is that of the 'structural anthropology' school of C. Levi-Strauss (1963). Both of these authors' work is reviewed briefly in Appendix 2 (see pp. 298–302).

29 In the spatial sciences disaggregation of models often takes place at will. Spatial input–output, econometric, export base, land use and transportation and other types of such models are not thoughtfully disaggregated by either sector or location with the only limitation in such inflation being data availability. From what we now know about the behavior of dynamic models, it is clear that their qualitative dynamical patterns change sharply when one moves from one level of disaggregation to another. Hence, these efforts to arbitrarily disaggregate models must be thought of as at least naïve. Models may be strongly tied to particular levels of disaggregation, their form changing significantly when finer or more coarse levels are considered. An additional and most severe problem for social sciences is the lack of appropriate statistical method; those which have been developed based on Bayesian inferential statistics require homogeneity and repetitive experimental conditions not met in social systems. Indeed, the book on social statistics is still to be written. In searching for a common method to link and possibly unify social science (and physical science) fields one might resort to non-linear dynamics rather than statistics as they now stand.

30 The subject of speculation and expectations is very broad and has been extensively discussed in the fields of macroeconomics and game theory. The difficulty in all analytical approaches to speculation is basically the (insurmountable) difficulty of modeling and estimating aggregate and

individual preference, consumption and demand functions. Demand functions are and will remain unknown to both observers of and participants in social games and dynamics. Further, these functions are highly non-linear and discontinuous, in contrast to the manner in which demand functions have been treated in both micro and macroeconomics so far. As a result, dynamic market conditions (and thus prices fluctuations) cannot be systematically approached. To bypass these difficulties, analysts routinely make strong assumptions regarding the behavior of economic aggregates, like interest rates, or the determination of perceived odds by market participants in taking actions.

31 The slaving principle is described by H. Haken (1983) in its temporal dimension; a simplified version of it has been used by Dendrinos and Sonis (1990) in a spatio-temporal context. More on this is supplied in Chapter 2 (see pp. 133–41).

32 Among other sources, the issue of aggregate behavior affecting and being simultaneously affected by disaggregate behavior is explored in a non-linear mode by Weidlich and Haag (1983) in a one-variable (population) model, and by Haag and Dendrinos (1983) and Dendrinos and Haag (1984) in a two-variable (population density and land rent) model. It is shown that, under relatively broadly defined conditions, stochastic microbehavior could result in deterministic macrodynamics.

33 See for example the arguments by Hicks (1979) and those by many authors found in Hausman (1984).

34 The basic theory is given by McFadden (1973: 105–42).

35 See Dendrinos and Sonis (1990) for a fuller exposition on this issue. The subject of 'fractal' behavior in socio-spatial dynamics is very recent (M. Batty 1989). Roughly, fractal properties occur when a system's qualitative behavior and form does not change as magnification in the scale of observation (not to be confused with disaggregation) occurs. The subject has been popularized (but not invented) by Mandelbrot (1977). In our context, fractal points might be attributed to individuals although the behavior giving rise to the fractals is of a deterministic aggregate type. An issue with fractals that needs further exploration is the limitless detail in which the overall pattern is repeated: in social and physical systems there are obviously limits as to how much detail can be attained. For example, a social system does not behave at a scale lower than that of a single individual, and a hierarchical system of cities cannot exceed (as a floor) the level of the smallest village or farm. But fractal (and those chaotic regimes in non-linear systems which are subjected to fractal dimensions) require no limit to the application of the fractal property. Consequently, one might have to reject some properties of model chaos or the domains of behavior of non-linear systems which reproduce such inadmissible social and physical states, and certainly on these grounds one can question the use of fractal geometry in spatial analysis. This is a fundamental issue in the application of non-linear dynamics in the socio-spatial sciences and needs further study.

36 For an early and introductory work on chaos see R. May (1976), and for a late but still accessible review see Schaffer *et al.* (1988). A popular book on this subject is that by Gleick (1987).

37 In the literature on urban economics there is a plethora of papers on the subject of optimal (normative) and equilibrium (positive) models of congestion. For an introduction see the work by E.S. Mills (1972) where an individual and social marginal cost was introduced in the transportation sector of an urban economy, pp. 96–136. For a recent review see Y. Sheffi (1985) where various efforts of incorporating congestion externality into the user equilibrium structure of transportation networks are discussed.

38 It should be noted that recently the subjects of chaos have started making inroads into particular models in macroeconomic (and microeconomic) theories. For a review see Rosser (1991).

39 See for example Todaro (1982) and Weeks (1978) for comprehensive surveys.

40 Malthus (1798), was the first to argue that high population growth rates cause poverty and thus affect per capita wealth. Since then, neo-Malthusians have, at times, argued along these lines even more strongly than Malthus. Their key concern has been the problem of 'overpopulation.' On this issue see, among others, the original book by P.R. Ehrlich (1968) which presents a classical treatise on overpopulation; P.R. Ehrlich and A.H. Ehrlich (1970); and D.L. Meadows and D. Meadows (1972) which advocates profound changes in socio-economic factors to attain an 'optimum' population size. On the other side is the issue (or problem), as it is perceived by some, of 'underpopulation.' For a representative exposition see: M.S. Teitelbaum and J.M. Winter (1985) where strategies on how to address fears stemming from population decline are presented. They provide some evidence to indicate that fertility in the US declined during the depression of the 1930s and in the recessions of 1973 to 1984 (p. 139). For a summary of the theories regarding over or underpopulation see J. Overbeek (1974, 1977).

41 For example, see C. Clark (1967) and A. Hirshman (1958).

42 The proponents of this debate have been K. Marx (1906–09) originally, and then possibly all neo-Marxists.

43 Arguments also cover along similar lines and delineations the connectance between absolute population levels and per capita wealth; also arguments can be found in the demographic literature covering the effects of the age structure composition on economic growth (see Weeks 1978). The subject of population growth and economic development is a very helpful one to test the validity of the propositions advanced in the Preface about social science data, theories, their markets and their effects on social events.

44 See for example the work by Leibenstein (1963) and note 51 below.

45 This is mainly the Marxist belief, see also note 56 below, as well as the view of neoclassical economists.

46 Todaro (1982), p. 186.

47. *Ibid.*, p. 188.

48 For example, see Clausen's (1984) statement, p. 25.

49 Todaro (1982), p. 187.

50 Data are taken from the Population Reference Bureau, *World Population Data Sheets*, 1980, 1982, 1984, 1986. That *increase* in per capita income might 'solve the population growth problem' is cited by K. Boulding (1964) as a prospect of 'a generation ago' as 'it was observed that the

richer countries had lower birth-rates than the poorer countries, and that the rich classes within each country had a lower birth-rate than the poorer classes,' p. 125; Boulding does not provide citations to document the argument. He uses, however this argument by drawing from the Irish experience and the potato famine of the mid-1800s to recite two 'theorems of misery' which have given economics the title of 'dismal science'; they are based on the assumption that if the per capita income of the poor increases their growth rate will also increase (an argument also supported by Leibenstein, as cited in note 51 below) so that the end result will be that eventually their per capita lot will decrease. The first is the simply dismal theorem: 'if the only thing which can check the growth of population is starvation and misery then the population will grow until it is sufficiently miserable and starving to check its growth,' p. 126. The second theorem, the utterly dismal one is: 'if only starvation and misery can check population growth, then the ultimate result of any technological improvement is to enable a larger number of people to live in misery than before (the innovation) and hence to increase the total sum of human misery,' p. 127. This logic has led to the implementation of a hands-off strategy toward the vast masses of urban poor in many developed and developing countries, the reasoning being that any improvement of slum conditions might lead to eventual deterioration of these very conditions the improvement intended to address. It is also noted that although starvation may be a check on population growth, there is no conclusive evidence to suggest that undernourishment and/or malnutrition are effective population control conditions at least not yet.

51 The Malthusian trap is an interpretation of the original model by Malthus which posits that poor societies are locked into slow development rates as a result of stable internal dynamics. Further theoretical speculation regarding this trap can be found in the work by H. Leibenstein (1963) who argues that 'an increase in per capita product/income in the less developed economies might result in an increase of the growth rate in population'; and in R.R. Nelson (1956) and in W.R. Hosek (1975), p. 260. Thus, the poor could not only be blamed for poverty (because they have high population growth rates which presumably contribute to the economy being trapped), but also, according to the Leibenstein hypothesis, any policy that might be implemented to improve their lot might actually backfire. The conclusion might then be drawn not to do anything for the poor, but to make their lot worse in the hope that per capita income might improve! These are some of the foundations for policy that economic growth theory offers.

52 The exposition is based partly on Todaro (1982).

53 See W.S. Thompson (1935). The work addressed in a comprehensive manner demographic, economic, social, political, and technological factors (at a rough level) affecting population growth. He posed questions on 'optimum population growth' (p. 435), on the 'advantages' and 'disadvantages' of a stationary population level (pp. 444–7), and for a need to adopt a 'population policy' arguing that because we control breeding of cattle we should control human growth rates and fertility (pp. 447–8). Thompson was the first to argue, in 1929, that 'high birth-rates were a

reaction to high death-rates,' Weeks (1978), p. 24.

54 See, among other sources, F.W. Notestein (1944, 1952). See also the discussion of the Notestein model by McNamara (1982) and L. Brown *et al.* (1987). Contributions have also been made in the theory of demographic transition by K. Davis (1948, 1963), and K. Davis and J. Blake (1956). K. Davis takes a disaggregate view of demographic events, arguing that a theory must rely more on the individual decision making and everyday occurrences affecting it.

55 Weeks (1978), p. 22.

56 Marxists argue that during the ultimate state of communism the population growth rate would be irrelevant, since the 'exploitive' nature of population policies by the state (occurring under the capitalist mode, to keep wages low by encouraging an abundant labor supply) will disappear. Communism would be able to accommodate any population growth rates. According to Marxists and neo-Marxists economic development is not at all related to population growth but instead to the broader social, political, economic, and institutional structure of a society. For a concise summary of these positions see Todaro (1982) and Weeks (1978).

57 Many demographic models are geared toward obtaining and studying steady states; see for example those by economist-demographer J.J. Spengler (1978) where he addresses the likelihood that most industrialized countries will be in a stationary state or declining population phase, and argues that this state will dampen oscillatory behavior in a number of economic indicators (p. 187).

2 NATIONS, CITIES, INDUSTRIES, AND THEIR CONNECTANCE

1 US Dept of Commerce, Bureau of the Census, *Statistical Abstract of the US*, 1986, Table 129.

2 All of the above data are from the Population Reference Bureau, *World Population Data Sheets*, various years.

3 See Dendrinos (1984c) and Figure 2.4, in Chapter 2.

4 US Dept of Commerce, *Survey of Current Business*, November 1986.

5 US Dept of Commerce, *Survey of Current Business*, S-16, 17, 18, March 1984.

6 US Dept of Commerce, *Survey of Current Business*, S-16, 17, February 1987; GNP count from *Survey of Current Business*, Table 1.7, February 1987.

7 Economic Information Agency, *Statistical Yearbook of China*, 1981, Hong Kong.

8 Data from the Population Reference Bureau, *World Population Data Sheets*, 1985, 1987.

9 *Agenda Estadistica 1983*, SPP, Tables 4.3, 4, 5.

10 This issue is explored in a paper by Dendrinos (1986). See also note 25 below.

11 The theoretical proposition is attributed to R. May (1974), although earlier evidence through computer simulation was supplied by Gardner and Ashby (1970).

12 See L. Brown *et al.* (1987), Chapter 2.

13 In the analysis, longitudinal and latitudinal coordinates were converted to curved surface airline distances among New York City, Los Angeles, Mexico City and the largest metropolitan areas belonging to all nations on Tables 7 and 1.2. For the case of the US the shortest distance (from either New York or Los Angeles) was chosen for every flow. It should be kept in mind that exports are in f.o.b. prices whereas imports are in c.i.f. prices. Running the stepwise option from SPSS-X as available in 1984 using the default values for PIN (0.05) and POUT (0.1) with the value of exports/imports/their sum as the dependent variable and population, per capita GDP, distance from the country exporting to/importing from as the independent variables one obtains the results indicated (shown in note 14).

14 (i) The US imports/exports: For the year 1958 only population and distance were regressed as independent variables since many counts for per capita GDP for other nations were not available. From the three regressions, only when exports were regressed was any independent variable able to enter the equation. Indeed the only variable entered was 'distance,' a rather surprising result given that the exports are in f.o.b. prices. The R-square statistic was 0.326 whereas F was equal to 5.8 (significant at the 0.033 level.) Exports as the dependent variable was the only regression with results for 1970 as well, with per capita GDP being the only variable entering the equation. The R-square was lower (0.310) and so was the F-statistic (5.4) (significant at the 0.039 level). Distance, not in the equation, was less in its significance than was population size. In 1981 per capita GDP was the only variable entering the equation with exports as the dependent variable. The R-square was higher (0.350) and so was the F-statistic (6.02) (significant at the 0.032 level). Distance effects were not figured in the equation. (ii) Mexico's case: Only one year, 1982, was examined for Mexico but the results were not very different than those obtained for the US imports/exports. Only when exports were regressed were there any results obtained. Per capita GDP was the only variable entering the equation, with R-square (0.382) and F (6.8) significant at the 0.024 level. The effects of distance were not significant enough to enter the picture. The caveats of note 8 of Chapter 1 regarding statistics apply.

15 See for example, Myrdal (1957), Williamson (1965), Alonso (1968), Gilbert (1970), Richardson (1973), Friedmann (1973) and many others.

16 Harrod (1948), Domar (1957), and Solow (1956, 1962) are examples of researchers who at the early stages of the economic, aggregate national (and linear) growth literature paid little attention to this interdependence.

17 As evidenced by the work of Schumpeter (1934).

18 Theoretical shortcomings in the spatial sciences are much more pronounced than those in the non-spatial sciences like macroeconomics. The main cause may be that not much funding ever supported (and thus little has been bet on) their outcomes.

19 In Dendrinos (1984b) an example of this linkage is modeled and tested with data from Mexico City and the United States of Mexico.

20 This has been discussed at length in Dendrinos (with Mullally) (1985), pp. 85–94.

21 From a geographical view see the collection of papers by Bourne and

Simmons (1978). From an economic approach a recent collection of papers is found in Tolley and Thomas (1986).

22 Becker, Mills and Williamson (1983), Kelley and Williamson (1984).

23 Kelley and Williamson (1984), p. 122.

24 *Ibid.*, p. 110.

25 Dendrinos (1986); in this paper the analytical aspects on a general theory of counter-flow spatio-temporal invasion or diffusion of stocks is provided. It states the dynamics of motion in spatially liquid stocks, which could produce chaotic events, as in the case of atmospheric turbulence. This paper is structured around the spatial movement of labor (population) on the basis of interspatial wage differentials; it depicts the fundamental spatial flow of labor incongruity in which labor supply always flows into regions with high real wage rates and labor demand flows into regions with low real wage rates. As a result of these flows, wage rates adjust and in turn affect jobs supplied or demanded. This spatial flow may produce irregular non-periodic motions.

26 Dendrinos (1984b).

27 J. Jacobs (1984), as the title of the book – *Cities and the Wealth of Nations* – markedly suggests.

28 See the references cited in note 15.

29 F. von Hayek (1976).

30 Taiwan is not usually considered a city-state. Geographically, however, its western coast is a continuous urban setting. With a population of 20 million in 1985 residing in an area of approximately 4,000 square kilometers of non-mountainous terrain it had an average population density in that region of about 12,800 persons per square kilometer. For references, see note 48. Its urban agglomeration stretched from the northern port of Keelong, through the Taipei metropolitan region to the southern port of Kaohsiung in an almost uninterrupted fashion with only pockets of farm land. Its aggregate density, including its mountainous region to the east, was about 627 persons per square kilometer – a very close count to that of the Shenyang metropolitan region in the People's Republic of China with a density of 611 in 1985 (see note 48).

31 D. Vining (1985) among others.

32 This issue is well studied in urban economics; see for example Wheaton and Shishido (1981).

33 See Dendrinos (with Mullally) (1985), pp. 67–70, for the notion of 'relative carrying capacity.'

34 These models are further discussed in Chapter 3, pp. 148–57, where the constituent elements of this relative parity notion are defined, and the resulting dynamics identified.

35 Dualism is here considered in its geographic and non-geographic contexts. Geographically, dualism is defined as the sharp differentials in density (in reference to population of built capital stock of any kind – housing, roads and factories being examples of capital stock) in space at any point in time; see the paper by D.S. Dendrinos and J. Zhang (1988) on the spatial population dualism of the People's Republic of China. Dualism in its non-geographic context refers to the sharply uneven distribution of wealth (economic or political – the latter in the form of power) among various

socio-economic groups. In this work the thesis is advanced that the two are closely connected.

36 Henderson and Quandt (1971), pp. 164–202.

37 *Ibid.*, p. 170. The difficulty in dealing with capital is the very definition of 'capital': whether it is an asset which is exclusively used to produce commodities and, not being consumed, thus contributes to wealth but not directly to utility, or it is simply a form of commodity as Sraffa (1960) argues. See the discussion by Hicks (1965) among many others on this issue. Due to these difficulties this work avoided incorporating capital into the macrodynamics studied, viewing it as a lower level factor than population and per capita income, although in the latter the contribution of capital as well as other factors (like knowledge or education) may be quite important. For the discussion which follows it does not matter whether we think of capital (and currency) as a commodity or an asset; what matters is that most central banks think of it as a means to produce or purchase commodities in exchanges, as well as a component of the economic system over which consumers have preferences (for liquidity.)

38 Henderson and Quandt (1971), p. 177.

39 Some of these recent efforts are found in W.A. Barnett, E.R. Berndt and H. White (eds) (1988), and W.A. Barnett, J. Geweke and K. Shell (eds) (1989). See also Day and Huang (1990).

40 Indications about these stocks' quantities in reserve could be the open interest on these currencies (excluding the US dollar) during an average trading day in early 1987 at the International Monetary Market of the Chicago Mercantile Exchange where futures contracts are traded on these currencies. An international hierarchy of currencies exists with these formal currencies at the top.

41 Currently, the Japanese yen is making a run for the top against the other four currencies. As an indication one might cite that, whereas the US placed three among the top ten ranked banks (including Citibank at the top) in terms of assets in 1975, while in 1987 it placed only one (Citibank, ranked 17th) in the top twenty. Dai Ichi Kangio Bank of Japan ranked first with fourteen Japanese banks winding up at the top twenty.

42 Four of these currencies belong to nations forming the Group of 5 (G-5), the fifth being France. As of early 1987 they controlled, according to the IMF, approximately 40 percent of the currencies as foreign reserves. This group seems to have the formidable task of stabilizing (the inherently unstable) currency exchange rates; and to control the national accumulation or decumulation reserves in foreign currencies.

43 A number of nations, like Singapore for example, were seemingly run on the objective of maximizing foreign reserves. Internally, economic production and consumption were in the 1970–85 period appropriately structured by the central government to that end.

44 Even to the casual observer it is surprising to find in some of these nations exquisite quality of labor, capital, and products coexisting with very low-quality domestic-oriented merchandise. Next to slum housing one might see high quality golf courses, mansions, or nuclear reactors. It is not rare to find highly advanced electronics-related production taking place in rundown buildings in shanty towns or in remote fields in the countryside.

This type of production might be an additional reason why the average cost of production is so much lower in these LDCs than in certain MDCs, setting aside the labor cost differential factors which have already been widely addressed in the popular and scientific literature.

45 One finds first rate universities and highly advanced scientific research institutes coexisting with outlets producing information (mostly for the mass media) judged to be of low quality even by their own consumers. All these dualisms are interconnected. Restricting access to high quality education to a vast majority of an LDC's population base is an obvious example where these dualisms feed on one another.

46 *China Post*, 23 October, 1985; the Taipei newspaper quoted the Central Statistics Bureau of Indonesia.

47 See note 12 of Chapter 1.

48 Following is a series of monthly and annual counts of the City of Taipei's population for the period 1974 to September 1985. These counts represent the registered residents within the administrative boundaries of the city of Taipei in that period. The city constitutes about half of the Taipei regional population, which in combination with the Taipei suburban municipalities (Taipei Chen) numbered about 5.5 million in 1985. Data supplied here are from the *Express Report of Important Statistics of the Taipei Municipality*, Bureau of the Budget, Accounting and Statistics, Taipei Municipal Government, ROC September 1985, No. 57, Table 3. One must keep in mind that the non-registered population of Taipei is estimated by local experts to be about one-third of the registered total in the mid 1980s. All registered residents of Taipei carry an identity card allowing them to find housing and employment. These restrictions are not, however, strictly observed. Of theoretical interest in these counts is the variation in the magnitude of population growth within one-year and one-month periods, in absolute numbers. The count cannot be normalized with respect to the national total population because there are no nationwide population counts available with similar frequency. It is demonstrated, in absolute counts, that one-month population growth variations are more pronounced than in those found in one-year counts, smoothing out the monthly variations. Monthly, and thus to a greater extent, weekly and/or daily, population growth rate cycles must not be calm. A much more violent discrete dynamic pattern must characterize urban population growth, than longer periods and lower frequency (annual) population growth rate cycles indicate. This may be an indication of Taipei's vitality, see this chapter, pp. 136–41. In reading this table one must keep in mind that the exactness of these counts may be an illusion, given the comments regarding the fuzziness of urban population data.

Table N.2 Taipei's monthly and yearly population counts, 1974–85

Year	*Month*	*Population*	*% change from preceding period*	
1974		2,003,604	–	
1975		2,043,318	1.982	
1976		2,089,288	2.250	
1977		2,127,625	1.835	
1978		2,163,605	1.691	
1979		2,196,237	1.508	
1980		2,220,427	1.101	
1981		2,270,983	2.227	
1982		2,327,641	2.495	
1983		2,388,374	2.609	
1984	1	2,449,702	2.568	
	9	2,433,996		−0.641
	10	2,438,770		0.196
	11	2,442,884		0.169
	12	2,449,702		0.279
1985	1	2,452,729		0.124
	2	2,456,491		0.153
	3	2,464,248		0.316
	4	2,469,741		0.223
	5	2,475,037		0.214
	6	2,477,704		0.108
	7	2,483,893		0.250
	8	2,489,450		0.224
	9	2,494,642		0.209

49 The issuance of bonds or stock on railroads are examples of the early development stages during the mid-1800s in the US in an effort to develop the West. For more on this issue see also note 11 of Chapter 4. An informative analogy can be established between movement into or out of cities or regions and movement into or out of stocks in stock exchanges. Also, long-term commitment to a region/city by an incomer is depicted by the behavior of a migrant; short-term, quasi-periodic movement is depicted by those referred to as 'circulators' into and out of cities/regions; finally, very short-term and highly periodic motion is experienced by 'commuters' into and out of cities. These three types of movements have been documented for the case of Medan, in north-western Sumatra in Indonesia, by T.R. Leinback and B. Suwarno (1985).

50 A north to south movement of economic activity might also be detected in the People's Republic of China: the northern three provinces of Jilin, Heilongjiang and Liaoning may be experiencing a loss of relative economic activity to the southern provinces, particularly Fujian and Guangdong (including the new province, Heinan).

51 Corporate takeovers and acquisitions of or mergers with existing firms as well as extensive restructuring of firms, industries, and even regions' economies are the events underlying, in the past decade or so, the widespread economic change in the US. Speculative behavior in stock exchanges, increasingly fast turnover in the ownership of stocks associated

with very short time horizons in investment decisions, fast execution of very large blocks traded daily, are a few of the events currently impacting the course of urban and regional economies in time-spans never before thought possible. The study of the new instrument in trading, like computerized program trading on the New York Stock Exchange, present opportunities to analyze the processes of restructuring not only currently underway in the US (and possibly global) financial markets, but even in urban and regional economies. This might be a very fruitful ground to explore further.

52 The cities' vast heterogeneity in socio-economic activities (land uses), their complex ecological interactions and their different growth rates, have been described mathematically by D.S. Dendrinos and H. Mullally (1981), and Dendrinos (with Mullally) (1985), where intra-urban dynamics are addressed. At times, these instabilities convey the impression that cities (or at least the old city of one central business district and of a homogeneous ring of residents surrounding it) break up.

53 US Dept of Commerce, Bureau of Census, *Current Population Reports*, P-26 Series, Local Population Estimates, No. 83-43-C, April 1985.

54 *Ibid.*, No. 85-52-C, August 1986.

55 US Dept of Commerce, Bureau of the Census, *Statistical Abstract of the U.S.*, 1986, Appendix II. This reference cites the Current Population Report, P-25 series, No. 976.

56 *Ibid.*, P-26 Series, Report No. 85-52-C.

57 *Ibid.*, P-25 Series, Report No. 976.

58 *Ibid.*, P-26 Series, Report No. 85-52-C, pp. 1–2.

59 US Dept of Commerce, Bureau of the Census, *Statistical Abstract of the U.S.*, 1986, Table 735.

60 US Dept of Commerce, *Survey of Current Business*, April 1986, Table 2.

61 *Ibid.*, August, Table 2.

62 See for example the Symposium issue of the *Review of Economic Studies*, 1974. Various papers by a number of authors on the subject of optimum and equilibrium depletion of natural resources.

63 I. Scott (1982).

64 *Ibid.*, p. 152.

65 *Ibid.*, p. 124.

66 P. Ward (1981), p. 36.

67 Table 2, *X Censo General de Poblacion y Vivenda 1980*; Resumen General Abreviado.

68 Palen (1987) estimates that Mexico City will reach 31.6 million by the year 2000, if the current 4.4 percent growth rates continue, p. 338. He cites Hauser and Gardiner (1982), p. 6–8, for that estimate. The United Nations projects that by the year 2000 Mexico City will have 26.3 million (with São Paulo a close second with 24 million), Palen, p. 334. Naturally, what will happen to Mexico City depends on what happens to the population of Mexico and the world's population among many other things.

69 *IX Censo General de Poblacion*, 1970, Estimaciones del Consejo National de Poblacion (estimates based on Alternative I), p. 27.

70 Nacional Financiera, s.a. 1977 *Statistics of the Mexican Economy*, Mexico, D.F., Table 1.6, p. 11.

71 *Proyecciones de la Poblacion de Mexico y de las Entidades Federativas: 1980–2010*, Instituto Nacional de Estadistica, Geografia e Informatica, Mexico, D.F., 1985, p. 4.
72 See for example: R.G.D. Allen (1967) *Macro-Economic Theory*, Macmillan, St Martin's Press; particularly the multiplier-accelerator models (pp. 321–40), the trade-cycle theory (pp. 364–82), and the Phillips' model of product cycles (pp. 385–408). As always, the emphasis is on steady states.
73 Nurke (1961), Kahn (1979), Olson (1982), and others, provide a review of the various stages of development in this field.
74 See for example the work by Schumpeter (1939), Kondratieff (1926), Rostow (1978), and Lucas (1981) among others.
75 It is based on pioneering analysis by Mitchell (1913) and by Keynes, and central in it is the work by Lucas (1981).
76 It is estimated that to statistically validate the simply logistic prototype model: $x(t + 1) = a\,x(t)[1 - x(t)]$, requires about 60,000 points in the regime of chaotic behavior, where $3.8 < a < 4.0$, and $0 < x(t) < 1$. It must be noted that chaotic cycles arise in some models with rational expectations, see for example J.M. Grandmont (1985).
77 P. Samuelson (1947).
78. Plotting a number of variables for the US the author found no periodic motion in any of them, with the exception of a possible six-year cycle in population growth rate. Data were used from the earliest date available for each count. Time series included: employment and unemployment rates, GNP and GNP growth rate, interest rate, the price of capital as obtained from the Dow-Jones Industrial monthly average (including compounded dividends), and population growth rates.
79 Urban ecological dynamics are found in Dendrinos (with Mullally) (1985). They include the scant empirical evidence available to date.
80 In Dendrinos (1984d) the case of Madrid obeying similar cycles in reference to Spain's economy was also hinted at, for very limited time series, on population and per capita income.
81 This is supplied in Dendrinos and Sonis (1990).
82 See Dendrinos (with Mullally) (1985).
83 E. Mandel (1980) wonders if current attempts (presumably by non-Marxist social scientists) are to marginalize the two strictly economic forces (rate of profit and capital accumulation) in favor of 'monetary, psychological, or inventive forces' (p. viii, Preface).
84 See notes 79 and 80 and their references. See also the discussion in Dendrinos and Sonis (1990).
85 Dendrinos (1984a), in a study of US regional dynamics, found nodal dynamic stability to be present for each and every one of the nine individual subregions of the US. Similar results are reported by T. Tabuchi (1988) for Japan's regions, and by Dendrinos (1989) regarding the regional structure of the People's Republic of China.
86 Von Hagen (1960) estimates that during the course of the Maya civilization, which lasted between approximately 2000 BC to AD 987 (p. 12), the inhabitants of the Yucatan numbered more than 3 million

people at its low point (p. 29). Since then, and every two hundred years, cities in the Peninsula rose and fell in population; abandoned in 987 they were repopulated and reabandoned in 1194. Then, the cycle repeated itself till 1441 when they were abandoned permanently (pp. 35–6). The causes for these cycles are still debated in the literature. S.G. Morley (1956) pinpoints the flourishing of the Maya civilization as the period between the fourth and sixth centuries AD (p. 3). He estimates the Mayan-speaking population of the Yucatan, in 1950, to be about 2 million (p. 20). It must be noted that the Maya practiced human sacrifice extensively as a matter of public policy. The reasons for this practice, largely associated with religious and sporting activities, is not clear, although population levels could have been affected appreciably. Cycles are also found in the Aztec culture as well, where a 52-year astronomical cycle has been prominent in archeological work regarding their calendars. The precise manner in which these cycles influenced the socio-spatial dynamics of the Aztecs is not precisely known, however.

87 The exact population of Athens during the classical era is largely unknown. J. Matras (1973) puts the population of Athens in the fifth century BC at around 200,000 (p. 18). In T. Chandler and G. Cox (1974) there is a very rough estimate of 155,000 (p. 368) for 430 BC. However, one can make some estimates based on an incomplete record. J.L. Myres (1953) estimates the population of the Greek Peninsula around 330 BC to have been about 4 million (p. 200). According to R.E. Wycherley (1962), p. 14, Attica had in Pericles' time 40,000 citizens and a total free population of about 150,000 of whom less than half lived in the city and its port. To these, one may probably add over 20,000 aliens resident at Athens or the Piraeus for purposes of trade, and over 100,000 slaves. See also A.W. Gomme (1933), and G. Glotz (1929). One can obtain a better feeling of the approximate population of Athens by looking at the estimated population of its rival Sparta. A. Toynbee (1971) estimates Sparta's population during the Peloponnesian War to have contained 8,000 Spartan citizens, plus an army of 45,000 allied with Spartan residents of other city-states in the Peloponnese, plus slaves (p. 10). Thus, one can estimate the total population of the decentralized Sparta metropolitan region (with a core of four small villages and without a dominant center) in the range of 100,000–200,000. These are, parenthetically, good examples of fuzzy measurements involving definition of residents and the spatial extent of urban settings. Considering the total population estimate of the peninsula, urban hierarchy, the size of the second largest urban settlement in the hierarchy challenging the dominance of Athens, and the relative dominance of Athens in culture, commerce, in its *de facto* and *de jure* geographical limits, it can be estimated that Athens in its golden age must have had a population of Athenian, alien residents, and slaves, ranging between 250,000 and 500,000, with one-third of a million being a good bet. It is noted too, that during this period infanticide was widely practiced in the Greek city-states as a matter of public policy, clearly affecting the population levels there as it did in the Mayan cities.

88 A.E. Vacalopoulos (1976), p. 227. He estimates, on the basis of the

Turkish census taken in the sixteenth century, that the Athenian population grew from 12,600 between 1520–30 to 17,600 in the period 1571–80.

89 National Statistical Service of Greece, *Census of Population*, Statistical Annal, 1981.

90 See note 18, Chapter 1.

91 See May (1974) and earlier basic work by Lotka, Volterra, and others found in Scudo and Ziegler (1978), for absolute growth models. More recent mathematical ecology work has been by MacArthur (1972), Pielou (1969), Smith (1974) among others. This work has been applied to urban relative dynamics recently, Dendrinos (with Mullally) (1985), Dendrinos and Sonis (1990).

92 M. Sonis, D.S. Dendrinos (1990).

93 The first paper was by Dendrinos and Mullally (1981), and Dendrinos (with Mullally) (1985).

94 Dendrinos (with Mullally) (1985), p. 67.

95 Dendrinos (1984c).

96 See the cyclical behavior and the role of the major cities of dominant nations in the global urban hierarchy over the last four millennia in the tables found in T. Chandler and G. Cox (1974). The path of the metropolis (core) of the dominant nations–empires were closely associated with (being leading indicators of) the eventual fate of their hinterlands (national economies).

97 Population Reference Bureau, *World Population Data Sheets*, 1987.

98 A closely related event demonstrates the continuous erosion in the US economic position with reference to other market economies in the latter part of the 1980s. At the beginning of the second quarter of 1987 and for the first time ever, the total valuation of stocks on the Tokyo Stock Exchange exceeded in US dollar value that of the New York Stock Exchange. It dropped below it again in the early 1990s.

99 Hutchinson (1978).

100 Dendrinos (1985) examined the urban hierarchy of the Mountain Region of the US and found no significant links between the distance and magnitude of the ecological coefficients. The subject is also addressed by Wei-Bin Zhang (1988) in a paper on intra-urban location ecological dynamics under the effect of spatial diffusion processes. Their presence seems to dampen the cyclical motion of the Volterra–Lotka model, according to Zhang.

3 THE DYNAMIC CODE OF A GLOBAL URBAN HIERARCHY

1 These urban areas were selected among those supplied by the 1981 *UN Demographic Yearbook*, which contains data on population size for cities, and their 'urban agglomerations,' of at least 100,000 inhabitants for all nations in the UN.

2 The estimate on Lima's population is further confirmed by the 1981 census

count on the Lima–Callao metropolitan region. See Sanders (1984).

3 For example, see W. Isard (1975) and E.S. Mills (1980) for definition of the term in spatial economics. According to its economic definition, it contains scale and auxiliary industry clustering effects. These scale effects are mostly tied to the presence of transport costs over space.

4 This notion is qualitatively similar, although not identical, to Alonso's (1964) concept of 'bid-price/rent.' Here, a more composite view to this willingness or unwillingness to locate at a specific point in space–time is taken.

5 Casetti (1984a).

6 In the US, the Bureau of the Census utilizes population (passenger) and commodity (freight) movements when considering MSA designation in certain instances.

7 Webber (1979), Kelley (1982), Kelley and Williamson (1984), Vining (1985).

8 Williamson (1965), Alonso (1968), Richardson (1973), Friedmann (1973), Mera (1973), Gilbert (1976), Dendrinos (1984b).

9 Johnson (1970), Robinson (1971).

10 The case is discussed by R. Fox (1984).

11 Of the six satellite towns planned to be built around Cairo, the Sixth of October and Fifteenth of May are now under construction; whereas in the Tenth of Ramadan, El-Obour, El-Amal and Badr, construction still has not started. See A.Z. Rageh (1984).

12 Pittsburgh, Detroit, Philadelphia, Baltimore and New York MSAs during the 1965–80 period, are cases in point.

13 Population Reference Bureau, *World Population Data Sheets*, 1971 and 1987.

14 See Dendrinos (1991). Quantum-type dynamics in socio-spatial analysis is a very young but promising field of work. It suggests that multiple recordings (states) of a past, present, or future socio-spatial system might coexist in a stochastic environment. These multiple states are due to the different observers, the effort they put into recordings, the area considered, and other factors. It is a much more elaborate version of a fuzzy sets theory of socio-spatial systems; in it real and imaginary dimensions in complex conjugate variables have specific insightful meanings and interpretations. They imply a quantum interference among them, as well as the impossibility of simultaneously obtaining different measurements of them.

15 See for example, Alonso (1964), Mills (1972) among others.

16 See Dendrinos (with Mullally) (1985).

17 Dendrinos (1985) examines the urban hierarchy of the Mountain Region of the US during the period 1890–1980. Unstable dynamic patterns were detected in it.

18 See Dendrinos (with Mullally) (1985), pp. 81–91; and Dendrinos (1985).

19 Population Reference Bureau, *World Population Data Sheets*, 1979–81.

20 UN *Statistical and Demographic Yearbook*, 1981, Table 8 on metropolitan areas.

21 Population Reference Bureau, *World Population Data Sheets*, 1979–81.

22 For an example of such evidence see the Chinese data on cities with populations of more than 5 million in Appendix 3, pp. 314–15.

23 US Dept of Commerce, Bureau of the Census, *Statistical Abstract of the US*, 1982–83, Table 5.
24 See for example Weidlich and Hagg (1983).

4 EPILOGUE

1 See the E. Lorenz (1963) pioneering paper on chaotic dynamics.
2 Quantum-type chaos, in the natural sciences, presents something of a problem: nothing in nature can be smaller than Plank's constant. In relative socio-spatial dynamics, chaotic and quantum-type chaotic quantities along the population dimension also present a problem: relative population cannot be smaller than the inverse of the current worldwide absolute population level. This number, decreasing in time as population grows, sets a limit to the quantum-type socio-spatial world. On the per capita income size, the limit may be even higher. One can conclude thus: that socio-spatial dynamics cannot admit chaotic regimes with fractal properties (i.e., strange attractors) which replicate themselves at infinitely small sizes. One must need a mechanism to set bounds to these non-periodic dynamics, and the manner in which such bounds are set is still to be explored: See Dendrinos (1991).
3 On the doomsday side one is directed to the Club of Rome work by D.L. Meadows and D. Meadows (1972), and P.R. Ehrlich and A.H. Ehrlich (1978). On the optimistic side, see the work by H. Kahn and A.J. Wiener (1967), and H. Kahn (1979, 1982).
4 J.L. Myres (1953), p. 186.
5 Population Reference Bureau, *World Population Data Sheet*, 1987.
6 The paper, 'The Calculus of Variation and the Logistic Curve,' is found in Scudo and Ziegler (1978) pp. 11–17, together with the classical 'Variations and Fluctuations in the Numbers of Coexisting Animal Species,' pp. 65–236. How these classical papers have influenced the theory of spatial (urban and regional) relative population dynamics is addressed in Dendrinos (with Mullally) (1985) and Dendrinos and Sonis (1986).
7 There are no forecasts currently available regarding the population size of these urban agglomerations over such an extended time horizon. However, some indications can be obtained by looking at the Mexican census forecasts about the Mexico City Metropolitan Region (MCMR) as of 1970: it was then projected that in 1980 the MCMR would have 14,445,000 inhabitants; in 1985 17,322,000; and in the year 2000, 23.4 million (source: *IX Censo General de Poblacion*, 1970, Estimaciones del Consejo National de Poblacion, p. 27. Estimates are based on Alternative I projections.) The spatial composition of the MCMR is: the Federal District, and the Atizapan de Zaragoza, Coadcalco, Cuautitlan, Chimalhuacan, Ecatepec, Naucalpan, Netzahualcoyotl, La Paz, Tlalnepantle, and Tultitlan municipalities in the State of Mexico.
8 The phase portraits were derived using an Adams integrator for ordinary differential equations with the aid of the BDP shellscript in the *Dynamical Software II* (version 2.2) package developed by W.M. Schaffer, G.L. Truty, and S. Fulmer (1988), Tucson.

9 US Dept of Commerce, Bureau of the Census, *Statistical Abstract of the U.S.* 1985, Table 33, p. 30.

10 *Ibid.*. 1982–83, Table 30, p. 26; and Table 36, p. 32.

11 Two examples directly bearing on the subject of this book may be cited. One is the current policy to issue options contracts on foreign debt by holders of such positions; the other is the issuance of debentures by the IMF and the World Bank to further finance such debt to Third World nations. More broadly, the availability of put or call options on a nation's debt allows for speculators to profit from national crises. In the US, options on foreign debt were under consideration by the Chicago Mercantile Exchange for US banks with a large exposure to Third World debt, a practice already in place in Japan and western Europe. Hedging or speculating on economic downturns or upturns affords the opportunity to successfully confront fluctuations in the future performance of a socio-economic-spatial system.

12 I owe this to a passing remark made by a colleague, Professor Borje Johansson from the University of Umea, Sweden, at a meeting at Zandvoort, the Netherlands, on 30 March 1985 in one of our discussions. Here I expand on this comment.

13 For example, in US-style football, the recognition that after a forward pass the ball is dead if it touches the ground before securely caught in bounds by an eligible receiver is a widely known rule. Others are not as important, as well known, or obvious. The loss of the designated hitter advantage rule in US American League baseball games is rarely witnessed, for instance. Some rules change yearly, as for example the various penalties on interference calls in US-type football.

BIBLIOGRAPHY

Ackermann, R.J. (1985) *Data, Instruments, and Theory*, Princeton University Press, Princeton, N.J.

Ackley, G. (1961) *Macroeconomic Theory*, Macmillan, New York.

Adelman, I. (1975) 'Development economics: a reassessment of goals,' *American Economic Review*, Vol. 65, No. 2: 302–9.

—— and C.M. Taft (1973) *Economic Growth and Social Equity in Developing Countries*, Stanford University Press, Stanford, Calif.

Ahlbung, D.A. (1987) 'The impact of population growth on economic growth in developing nations: the evidence from macroeconomic–demographic models,' in D.G. Johnson and R.D. Lee (eds), *Population Growth and Economic Development: Issues and Evidence*, The University of Wisconsin Press, Madison, Wis.

Allen, R.G.D. (1967) *Macro-Economic Theory*, St Martin's Press, New York.

Alonso, W. (1964) *Location and Land Use*, Harvard University Press, Cambridge, Mass.

—— (1968) 'Urban and regional imbalances in national development,' *Economic Development and Cultural Change*, Vol. 2: 1–14.

—— and E. Medrich (1978) 'Spontaneous growth centers in twentieth century American urbanization,' in L. Bourne and J.W. Simmons (eds), *Systems of Cities*, Oxford University Press, New York.

Andersson, A.E. and B. Johansson (1984) 'Industrial dynamics, product cycles, and employment structure,' *International Institute for Applied Systems Analysis*, Working Paper 84–89.

Arrow, K.J. (1968) 'Application of control theory to economic growth,' in *Lectures in Applied Mathematics*, Vol. 12, American Mathematical Society, Providence, R.I.

Attfield, C.L.F., D. Demery and N.W. Duck (1985) *Rational Expectations in Macroeconomics*, Basil Blackwell, London.

Azariadis, C. (1981) 'Self-fulfilling prophecies,' *Journal of Economic Theory*, Vol. 25: 380–96.

Banfield, E. (1970) *The Unheavenly City*, Little, Brown & Co., Boston, Mass.

Baran, P. (1962) *The Political Economy of Growth*, Monthly Review Press, New York.

Barnett, W.A., E.R. Berndt and H. White (eds) (1988) *Dynamic Econometric Modeling*, Cambridge University Press, New York.

—— J. Geweke and K. Shell (eds) (1989) *Economic Complexity*, Cambridge University Press, New York.

Bass, F.M. (1969) 'A new product growth model for consumer durables,' *Management Science*, Vol. 15: 215–27.

Batty, M. (1989) 'Geography and the new geometry,' *Geography Review*, Vol. 2, No. 4: 7–10.

Becker, C.M., E.S. Mills and J.G. Williamson (1983) 'Public policy, urbanization and development: a computable general equilibrium model of the Indian economy,' Vanderbilt U. Working Paper No. 83-W14.

——, ——, —— (1984) 'The impact of unbalanced productivity advance of Indian urbanization: some preliminary findings,' *Proceedings of the Fifteenth Annual Pittsburgh Conference*, Instrument Society of America, Research Triangle Park, N.C.

Beckmann, M. (1968) *Location Theory*, Random House, New York.

Bernstein, J. (1991) *Quantum Profiles*, Princeton University Press, Princeton, N.J.

Berry, B.J.L. (1973) 'Contemporary urbanization process,' *Geographical Perspectives on Urban Problems*, Academy of Sciences: 94–107.

—— and J.D. Kasarda (1977) *Contemporary Urban Ecology*, Macmillan, New York.

Bhagwats, J. (1964) 'The pure theory of international trade: a survey,' *Economic Journal*, March 1964: 1–84.

Bienen, H. (1984) 'Urbanization and Third World stability,' *World Development*, Vol. 12, No. 7: 661–91.

Biraben, J.N. (1985) 'Essai sur l'evolution du nombre des hommes,' in H. Le Bras (ed.), *Population*, Hachette, Paris.

Black, C.E. (1966) *The Dynamics of Modernization: A Study in Comparative History*, Harper & Row, New York.

Boserup, E. (1981) *Population and Technological Change*, University of Chicago Press, Chicago, Ill.

Boulding, K.E. (1964) *The Meaning of the 20th Century: The Great Transition*, Harper & Row, (Colophon Books), New York.

—— (1978) *Ecodynamics*, Sage Publications, Beverly Hills.

Bourne, L.S. and J.W. Simmons (1978) *Systems of Cities: Reading on Structure, Growth and Policy*, Oxford University Press, New York.

Branch, B. (1976) *Fundamentals of Investing*, John Wiley, New York.

Brass, W. (ed.) (1971) *Biological Aspects of Demography*, Taylor and Francis Ltd., London.

Brown, L.A. (1981) *Innovation Diffusion: A New Perspective*, Methuen, New York.

Brown, L.R., E.C. Wolf and L. Starke (1987) *The State of the World*, Norton & Co., New York.

Buckley, J.J. (1989) 'Fuzzy input–output analysis,' *European Journal of Operations Research*, Vol. 39: 54–60.

Burgess, E.W. (1925) 'The growth of the city,' in R.E. Park, E.W. Burgess and R.D. McKenzie (eds), *The City*, University of Chicago Press, Chicago, Ill.

Carlsson, C. (1988) 'On the relevance of fuzzy sets in management science methodology,' in M.M. Gupta and T. Yamakawa (eds), *Fuzzy Logic in Knowledge-Based Systems, Decision and Control*, Amsterdam, North Holland: 11–28.

Casetti, E. (1984a) 'Peripheral growth in mature economies,' *Economic Geography*, Vol. 60, No. 23: 122–31.

—— (1984b) 'Manufacturing productivity and Snowbelt–Sunbelt shifts,' *Economic Geography*, Vol. 60, No. 4: 313–24.

—— (1989) 'The onset and spread of modern economic growth in Europe: an empirical test of a catastrophe model,' *Environment and Planning A*, Vol. 21: 1473–89.

Cass, D. (1965) 'Optimum growth in an aggregate model of capital accumulation,' *Review of Economic Studies*, Vol. 32: 223–40.

IX Censo General de Population (1970) Estimaciones del Consejo National de Poblacion, p. 27.

X Censo General de Poblacion y Vivenda 1980, resumen General Abreviado, Table 2.

Chamberlin, E.H. (1938) *Theory of Monopolistic Competition*, Harvard University Press, Cambridge, Mass.

Chandler, T. (1987) *Four Thousand Years of Urban Growth: An Historical Census*, St David's University Press, Lewiston, N.Y.

—— and G. Cox (1974) *3000 Years of Urban Growth*, Academic Press, New York.

Chang, Chang Ann, *The History of Urban Development*, Wuhan University Publishing Company, Wuhan (in Chinese).

Chilcote, R.H. (1982) *Dependency and Marxism*, Westview Press, Boulder, Colo.

China Post, 23 October 1985.

Choi, K. (1983) *Theories of Comparative Economic Growth*, Iowa State University Press, Ames, Ia.

Christaller, W. (1966) *Central Places in Southern Germany* (C.W. Baskin, translator), Prentice-Hall, New York.

Clark, C. (1967) *Population Growth and Land Use*, St Martin's Press, New York.

Clausen, A.W. (1984) *Population Growth and Economic and Social Development*, The World Bank, Washington, D.C.

—— (1984) *Address to the National Leaders' Seminar on Population and Development*, The World Bank, Washington, D.C.

Coale, A.F.J. (1972) *The Growth and Structure of Human Populations*, Princeton University Press, Princeton, N.J.

Cohen, B.J. (1973) *The Question of Imperialism: The Political Economy of Dominance and Dependence*, Basic Books, New York.

Coleman, D. and R. Schofield (eds) (1986) *The State of Population Theory*, Basil Blackwell, Oxford.

Davis, K. (1948) *Human Society*, Macmillan, New York.

—— (1963) 'The theory of change and response in modern demographic history,' *Population Index*, Vol. 29, No. 4: 345–66.

—— and J. Blake (1956) 'Social structure and fertility: an analytical framework,' *Economic Development and Cultural Change*, Vol. 4: 211–35.

Davis, S. (1979) *The Diffusion of Process Innovation*, Cambridge University Press, Cambridge.

Day, R.H. and W. Huang (1990) 'Bulls, bears, and market sheep,' *Journal of Economic Behavior and Organization*, Vol. 14: 299–329.

Dendrinos, D.S. (1984a) 'Regions, antiregions and their dynamic stability; the US case (1929–1979),' *Journal of Regional Sciences*, Vol. 24, No. 1: 65–84.

—— (1984b) 'Absolute and relative growth in spatial dynamics: the phenomenon of primacy,' *Proceedings of the Fifteenth Annual Pittsburgh Conference*, Instrument Society of America, Research Triangle Park, N.C.

—— (1984c) 'The decline of the US economy: a perspective from the theory of mathematical ecology,' *Environmental and Planning A*, Vol. 16: 651–62.

—— (1984d) 'Madrid's aggregate growth pattern: a note on the evidence regarding the urban Volterra–Lotka model,' *Sistemi Urbani*, Vol. XI, No. 2: 237–46.

—— (1984e) 'The structural stability of the US regions: empirical evidence and theoretical underpinnings,' *Environment and Planning A*, Vol. 16, No. 11: 1433–44.

—— (1985) 'Dynamics of unstable urban hierarchies: the case of the US Mountain Region (1890–1980),' Mimeo, University of Kansas.

—— (1986) 'On the incongruous spatial employment dynamics,' in P. Nijkamp (ed.), *Technological Change, Employment and Spatial Dynamics*, Springer-Verlag, Berlin (Lecture Notes in Economics and Mathematical Systems, Vol. 270: 321–39).

—— (1989) 'Growth patterns of the eight regions in the People's Republic of China (1980–1985),' *Annals of Regional Science*, Vol. 23: 213–22.

—— (1991) 'Methods in quantum mechanics and the socio-spatial World,' *Socio-Spatial Dynamics*, Vol. 2, No. 2: 81–109.

—— and G. Haag (1984) 'Toward a stochastic dynamical theory of location: empirical evidence,' *Geographical Analysis*, Vol. 16, No. 4: 287–300.

—— and H. Mullally (1981) 'Evolutionary patterns of urban populations,' *Geographical Analysis*, Vol. 13, 328–44.

—— with H. Mullally (1985) *Urban Evolution: Studies in the Mathematical Ecology of Cities*, Oxford University Press, Oxford.

—— and M. Sonis (1986) 'Variational principles in Volterra's ecology and in urban relative dynamics,' *Journal of Regional Science*, Vol. 26, No. 2: 359–77.

——, —— (1990) *Chaos and Socio-Spatial Dynamics*, Springer-Verlag, New York.

—— and J. Zhang (1988) 'The effects of topography upon population dualism; the case of the People's Republic of China (1936–1985), Mimeo, University of Kansas.

Domar, E.D. (1946) 'Capital expansion, rate of growth and employment,' *Econometrica*, Vol. 14: 137–47.

—— (1957) *Essays in the Theory of Economic Growth*, Oxford University Press, New York.

Doubleday, T. (1847) *The True Law of Population Shown to be Connected with the Food of the People* (2nd Ed) George Pierce, London.

Economic Information Agency, *Statistical Yearbook of China*, 1981, Hong Kong.

Ehrlich, P.R. (1968) *The Population Bomb*, Valentine Books, New York.

—— and A.H. Ehrlich (1970) *Population Resources Environment*, W.H. Freeman & Co., San Francisco.

——, —— (1978) *The Population Bomb*, Valentine Books, New York (revised ed, 12th printing).

Elliott, J.W. (1979) *Macroeconomic Analysis* (2nd ed), Winthrop Publishers, Framingham, Mass.

Findlay, R. (1979) 'Economic development and the Theory of International Trade,' *American Economic Review*, Vol. 69, No. 2: 186–90.

Fisher, J.C. and R.H. Pry (1970) 'A simple substitution model of technological change,' *Technological Forecasting and Social Change*, Vol. 3: 75–88.

Ford, T.R. and G.F. de Jong (1970) *Social Demography*, Prentice-Hall, New York.

Fox, R. (1984) 'The world's urban explosion,' *National Geographic*, Vol. 166, No. 2: 179–85.

Freeman, C. (ed.) (1983) *Long Waves in the World Economy*, Butterworths, London.

Frejka, T. (1973) *The Future of Population Growth*, John Wiley Interscience, New York.

Friedmann, J. (1973) *Urbanization, Planning and National Development*, Sage Publications, Beverly Hills.

—— and W. Alonso (eds) (1964) *Regional Development and Planning: a Reader*, MIT Press, Cambridge, Mass.

Gardner, M.R. and W.R. Ashby (1970) 'Connections of large dynamical (cybernetic) systems: critical values for stability,' *Nature*, Vol. 228: 784–90.

Gilbert, A. (1970) *Development Planning and Spatial Structure*, John Wiley, London.

Gilmore, R. (1981) *Catastrophe Theory for Social Scientists and Engineers*, J. Wiley, New York.

Gleick, J. (1987) *Chaos: Making of a New Science*, Viking, New York.

Glotz, G. (1929) *The Greek City* (translation from the French of *La cite grecque*, 1928).

Gomme, A.W. (1933) *Population of Athens*, Basil Blackwell, Cambridge, Mass.

Gordon, R.J. (1978) *Macroeconomics*, Little, Brown & Co., Boston, Mass.

Grandmont, J.M. (1985) 'On endogenous competitive business cycles,' *Econometrica*, Vol. 53: 995–1045.

Grindstaff, C.F. (1981) *Population and Society: A Sociological Perspective*, The Christopher Publishing House, West Hannover.

Guckenheimer, J. and P. Holmes (1983) *Nonlinear Oscillations, Dynamical Systems, and Bifurcations of Vector Fields*, Springer-Verlag, New York.

Gupta, M.M. and T. Yamakawa (eds) (1988) *Fuzzy Logic in Knowledge-Based Systems, Decision and Control*, Amsterdam, North Holland.

Haag, G. and D.S. Dendrinos (1983) 'Toward a stochastic dynamical theory of location: a nonlinear migration process,' *Geographical Analysis*, Vol. 15, No. 4: 269–86.

Hagen, V.W. von (1960) *World of Maya*, Mentor Press, New York.

Hagerstrand, T. (1969) *Innovation Diffusion as a Spatial Process*, University of Chicago Press, Chicago, Ill.

Hahn, F.H. and R.C.O. Matthews (1964) 'The theory of economic growth: a survey,' *Economic Journal*, Vol. 74: 779–902.

Haken, H. (1979) *Synergetics*, Springer-Verlag, Berlin.

—— (1983) *Synergetics, An Introduction*, Springer-Verlag, Berlin.

Harris, D.J. (1975) 'The theory of economic growth: a critique and reformulation,' *American Economic Review*, Vol. 65: 329–37.

Harris, M. (1979) *Cultural Materialism*, Random House, New York.

Harrison, G.A. and A.J. Boyce (eds) (1972) *The Structure of Human Populations*, Clarendon Press, Oxford.

Harrod, R.F. (1939) 'An essay in dynamic theory,' *Economic Journal*, Vol. 49: 14–33.

—— (1948) *Toward a Dynamic Economics*, Macmillan, London.

Hauser, P.M. and R.W. Gardiner (1982) 'Urban future: trends and prospects,' in P.M. Hauser (ed.) *Population and the Urban Future*, State of New York Press, New York.

Hausman, D.M. (ed.) (1984) *The Philosophy of Economics: An Anthology*, Cambridge University Press, Cambridge.

Hayek, F.A. von (1975) 'The pretence of knowledge,' *Swedish Journal of Economics*, Vol. 77: 433–42.

—— (1976) *Denationalization of Money*, The Institute for Economic Affairs, Hobart Papers, No. 70.

Henderson, J.M. and R.E. Quandt (1971) *Microeconomic Theory: A Mathematical Approach*, McGraw-Hill, New York.

Hicks, J. (1965) *Capital and Growth*, Clarendon Press, Oxford.

—— (1973) 'Mr. Keynes and the classics: a suggested interpretation,' *Econometrica*, Vol. 5: 147–59.

—— (1979) *Causality in Economics*, Basic Books, New York.

Hiorns, R.W. (1972) 'Mathematical models in demography,' in G.A. Harrison and A.J. Boyce (eds), *The Structure of Human Populations*, Clarendon Press, Oxford.

Hirshman, A. (1958) *The Strategy of Economic Development*, Yale University Press, New Haven, Conn.

Holden, A.V. (ed.) (1986) *Chaos*, Princeton University Press, Princeton, N.J.

Hosek, W.R. (1975) *Macroeconomic Theory*, R.D. Irwin Inc., Homewood, Ill.

Hotelling, H. (1929) 'Stability in competition,' *Economic Journal*, Vol. 39: 41–57.

Howell, N. (1976) 'Toward a uniformitarian theory of human paleodemography,' in R.H. Ward and K.M. Weirs (eds) *The Demographic Evolution of Human Populations*, Academic Press, New York.

Hu, Huan Yong (1986) 'The past and future of population growth, economic development of China's eight regions,' Hwa Tong Teachers School Publishing Company, Shanghai.

Hutchinson, G.E. (1978) *An Introduction to Population Ecology*, Yale University Press, New Haven, Conn.

Instituto Nacional de Estadistica (1985) *Proyecciones de la Poblacion de Mexico y de las Entidades Federativas: 1980–2010*, Geografia e Informatica, Mexico D.F., p. 4.

Isard, W. (1965) *Location and Space Economy*, MIT Press, Cambridge, Mass.

—— (1975) *Introduction to Regional Science*, Prentice-Hall, New York.

Jacobs, J. (1984) *Cities and the Wealth of Nations*, Random House, New York.

Jevons, W.S. (1871) *The Theory of Political Economy*, Harmondsworth.

Johnson, D.G. and R.D. Lee (eds) (1987) *Population Growth and Economic Development: Issues and Evidence*, The University of Wisconsin Press, Madison, Wis.

Johnson, E.A.J. (1970) *The Organization of Space in Developing Countries*, Harvard University Press, Cambridge, Mass.

Jones, H.G. (1976) *An Introduction to Modern Theories of Economic Growth*, McGraw-Hill, New York.

Kahn, H. (1979) *World Economic Development*, Westview Press, Boulder, Colo.

—— (1982) *The Coming Boom: Economic, Political and Social*, Simon & Schuster, New York.

—— and A.J. Wiener (1967) *The Year 2000*, Macmillan, New York.

Kaldor, N. (1967) *Strategic Factors in Economic Development*, Cornell University Press, Ithaca, N.Y.

Kammeyer, K.C.S. and H.L. Ginn (1986) *An Introduction to Population*, The Dorsey Press, Chicago.

Kelley, A.C. (1982) 'The limits to urban growth: suggestions for macromodelling Third World economies,' *Economic Development and Cultural Change*, Vol. 20, No. 3: 595–623.

—— and J.G. Williamson (1984) *What Drives Third World City Growth? A Dynamic General Equilibrium Approach*, Princeton University Press, Princeton, N.J.

Keyfitz, N. and W. Flieger (1968) *An Introduction to Population*, Cornell University Press, Ithaca, N.Y.

——, —— (1971) *Population: Facts and Methods in Demography*, W.H. Freeman & Co., San Francisco.

Keynes, J.M. (1964) *The General Theory of Employment, Interest and Money*, Harcourt, Brace and World, Inc., San Diego, Calif.

Kindleberger, C.P. (1978) *Manias, Panics and Crashes*, Basic Books, New York.

Kondratieff, N.D. (1926) 'The long waves in economic life,' *Review of Economic Statistics*, Vol. 17, 1935 (English version).

Kuhn, T.S. (1962) *Structure of Scientific Revolutions*, University of Chicago Press, Chicago, Ill.

Kuznets, S. (1966) *Modern Economic Growth: Rate, Structure and Spread*, Yale University Press, New Haven, Conn.

—— (1971) *Economic Growth of Nations: Total Output and Production Structure*, Harvard University Press, Cambridge, Mass.

Land, K.C. and A. Rogers (1982) *Multidimensional Mathematical Demography*, Academic Press, New York.

Leibenstein, H. (1954) *A Theory of Economic–Demographic Development*, Princeton University Press, Princeton, N.J.

—— (1963) *Economic Backwardness and Economic Growth*, John Wiley, New York.

—— (1980) 'Microeconomics and x-efficiency theory: if there is no crisis, there ought to be,' *The Public Interest*, Special ed: 97–110.

Leinbach, T.R. and B. Suwarno (1985) 'Commuting and circulation characteristics in the intermediate sized city: the example of Medan, Indonesia,' *Singapore Journal of Tropical Geography*, Vol. 6, No. 1: 36–47.

Leontief, V. (1953) *Studies in the Structure of the American Economy*, Oxford University Press, Oxford.

—— (1977) *The Future of the World Economy: a United Nations Study*, Oxford University Press, Oxford.

Levi-Strauss, C. (1963) *Structural Anthropology* (translated by C. Jacobson and B. Grundfest Schoepf) Basic Books, New York (originally appeared in 1958).

Lewis, O. (1966) *La Vida*, Random House, New York.

Lewis, W.A. (1955) *Theory of Economic Growth*, Unwin, Winchester, Mass.

Linn, J.F. (1983) *Cities in the Developing World*, Oxford University Press, Oxford.

Lorenz, E. (1963) 'Deterministic non-periodic flows,' *Journal of Atmospheric Sciences*, Vol. 20: 130–141.

Lösch, A. (1954) *The Economics of Location* (translation of *Die raumliche Ordnung der Wirtschaft*, 2nd Edn, 1944), Yale University Press, New Haven, Conn.

Lotka, A. (1925) *Elements of Physical Biology*, Williams & Wilkins, Baltimore, Md.

—— (1956) *Elements of Mathematical Biology*, Dover Publications, New York.

Lucas, R.E. (1981) *Studies in Business-Cycle Theory*, MIT, Cambridge, Mass.

Ma, L.J.C. (1981) 'Urban housing supply in the people's R.O.C.' in L.J.C. Ma and E.W. Hansen (eds), *Urban Development in Modern China*, Westview Press, Boulder, Colo.

—— and E.W. Hansen (eds) (1981) *Urban Development in Modern China*, Westview Press, Boulder, Colo.

MacArthur, R.H. (1972) *Geographical Ecology*, Harper & Row, New York.

McFadden, D. (1976) 'Conditional logit analysis of qualitative choice behavior,' in P. Zarembka (ed.), *Frontiers in Econometrics*, Academic Press, New York.

Machlup. R. (1980) *Knowlege: Its Creation, Distribution and Economic Significance* (Volume I: *Knowledge and Knowledge Production*), Princeton University Press, Princeton, N.J.

—— (1984) *Knowledge: Its Creation, Distribution and Economic Significance* (Volume III: *The Economics of Information and Human Capital*), Princeton University Press, Princeton, N.J.

McNamara, R. (1982) 'Demographic transition theory,' *International Encyclopedia of Population*, Vol. 1.

Mahajan, V. and R.A. Peterson (1985) *Models for Innovation Difussion*, Sage Publications, Beverly Hills, Calif.

Maitra, P. (1986) *Population Technology and Development*, Gower, Brookfield, Vt.

Malthus, T. (1798) *An Essay on the Principle of Population* (1st edn), Reeves & Turner, London.

Mandel, E. (1980) *Long Waves of Capitalist Development: The Marxist Interpretation*, Cambridge University Press, Cambridge.

—— and A. Freeman (1984) *Ricardo, Marx, Sraffa*, Langston Foundation, New York.

Mandelbrot, B. (1977) *The Fractal Geometry of Nature*, Freeman, New York.

Mansfield, E. (1968) *The Economics of Technological Change*, W.W. Norton & Co., New York.

Marchall, A. (1922) *Principles of Economics*, Macmillan, London.

Marchetti, C. (1975) 'Society as a learning system: discovery, invention and innovation cycles revisited,' *Technological Forecasting and Social Change*, Vol. 18: 267–82.

Marsden, J.E. and M. McCracken (1976) *The Hopf Bifurcation and Its Applications*, Springer-Verlag, New York.

Marx, K. (1906–9) *Das Kapital* (Vol. 1, S. Moore and E. Areling translators), C.H. Kerr, Chicago, Ill.

Matras, J. (1973) *Populations and Societies*, Prentice-Hall, New York.

—— (1977) *Introduction to Population: A Sociological Approach*, Prentice-Hall, New York.

May, R.M. (1974) *Stability and Complexity in Model Ecosystems*, Princeton University Press, Princeton, N.J.

—— (1976) 'Simple mathematical models with very complicated dynamics,' *Nature*, Vol. 261: 459–67.

Meade, J.E. (1961) *A Neoclassical Theory of Economic Growth*, Oxford University Press, New York.

Meadows, D.L. and D. Meadows (1972) *The Limits to Growth*, Universe, New York.

Mera, K. (1973) 'On urban agglomeration and economic efficiency,' *Economic Development and Cultural Change*, Vol. 21: 309–24.

Mill, J.S. (1848) *Principles of Political Economy* (Book II) J.W. Parker, London.

Miller, M.H. and C.W. Upton (1974) *Macroeconomics*, University of Chicago Press, Chicago, Ill.

Miller, R.W. (1987) *Fact and Method; Explanation, Confirmation and Reality in the Natural and Social Sciences*, Princeton University Press, Princeton, N.J.

Miller, W.B. and R.K. Godwin (1977) *Psyche and Demons*, Oxford University Press, New York.

Mills, E.S. (1972) *Studies in the Structure of the Urban Economy*, Johns Hopkins University Press, Baltimore, Md.

—— (1980) *Urban Economics* (2nd edn), Scott, Foresman & Co., Glenview, Ill.

—— (1987) *Handbook of Regional and Urban Economics* (Vol. II, *Urban Economics*), Elsevier, Amsterdam, North Holland.

Mirrlees, J.A. and N.H. Stern (eds) (1973) *Models of Economic Growth*, Macmillan, London.

Mitchell, W.C. (1913) *Business Cycles*, University of California Press, Berkeley.

Morley, S.G. (1956) *The Ancient Maya*, Stanford University Press, Stanford, Calif.

Morishimo, M. (1964) *Equilibrium, Stability and Growth*, Oxford University Press, Cambridge, Mass.

Mueller, D.C. (ed.) (1982) *The Political Economy of Growth*, Yale University Press, New Haven, Conn.

Myrdal, G. (1957) *Economic Theory and Underdeveloped Regions*, Duckworth, London.

Myres, J.L. (1953) *Geographical History in Greek Lands*, Clarendon Press, Oxford.

Nacional Financiera, S.A. (1977) *Statistics of the Mexican Economy*, Mexico, D.F., Mexico City, Table 1.6, p. 11.
National Statistical Service of Greece (1981) *Census of Population*, Statistical Annals, Athens.
Nelson, R.R. (1956) 'A theory of the low-level equilibrium trap in underdeveloped economies,' *American Economic Review*, Vol. 46: 894–908.
—— and S.G. Winter (1974) 'Neoclassical vs. evolutionary theories of economic growth: critique and prospects,' *Economic Journal*, Vol. 84: 887–905.
——, —— (1982) *An Evolutionary Theory of Economic Change*, Harvard University Press, Cambridge, Mass.
Notestein, F.W. (1944) 'Problems of policy in relation to areas of heavy population pressure,' *Demographic Studies of Selected Areas of Rapid Growth*, Milbank Memorial Fund, New York.
—— (1952) 'Population theory,' in B.F. Haley (ed.) *A Survey of Contemporary Economics* (Vol. 2), Irvin, Homewood, Ill.
——, D. Kirk and S. Segal (1963) 'The problem of population control,' in P.M. Hauser (ed.), *The Population Dilemma*, Prentice-Hall, New York.
Nurke, R. (1961) *Equilibrium and Growth in the World Economy: Economic Essays*, Harvard University Press, Cambridge, Mass.
Olin, B. (1933) *Interregional and International Trade*, Harvard University Press, Cambridge, Mass.
Olson, M.L. (1982) *The Rise and Decline of Nations*, Yale University Press, New Haven, Conn.
Overbeek, J. (1974) *History of Population Theories*, Rotterdam University Press, Rotterdam.
—— (ed.) (1977) *The Evolution of Population Theory*, Greenwood Press, Westport, Conn.
Palen, J.J. (1987) *The Urban World*, McGraw-Hill, New York.
Palma, G. (1978) 'Dependency: a formal theory of underdevelopment, or a methodology for the analysis of concrete situations of underdevelopment,' *World Development*, Vol. 6: 881–924.
Papageorgiou, Y.Y. and T.R. Smith (1983) 'Agglomeration as local instability of spatially uniform steady states,' *Econometrica*, Vol. 51: 1109–20.
Pearl, R. (1924) *Studies in Human Biology*, Williams & Wilkins, Baltimore.
Perlman, J.E. (1976) *The Myth of Marginality*, University of California Press, Berkeley.
Perloff, H.S., E.S. Dunn, E.E. Lampard and R. Muth (1960) *Regions, Resources and Economic Growth*, Johns Hopkins University Press, Baltimore.
Peterson, W. (1975) *Population*, Macmillan, New York.
Peterson, W.L. (1980) *Principles of Economics: Macro* (4th edn), R.D. Irwin Inc., Homewood, Ill.
Phelps, E.S. (1966) *Golden Rules of Economic Growth*, W.W. Norton & Co., New York.
Pielou, E.C. (1969) *An Introduction to Mathematical Ecology*, John Wiley, New York.
Popper, K. (1957) *The Poverty of Historicism*, Routledge & Kegan Paul, London.

—— (1963a) *The Logic of Scientific Discovery*, Harper & Row, New York.
—— (1963b) *The Open Society*, Routledge & Kegan Paul, London.
—— (1974) *The Poverty of Historicism*, Routledge & Kegan Paul, London (reprint of the original 1957 book).
Population Reference Bureau, 'Composite of World Population Data Sheets,' Washington, D.C., various years.
Poston, T. and I. Stewart (1978) *Catastrophe Theory and its Applications*, Pitman, London.
Rageh, A.Z. (1984) 'Changing pattern of housing in Cairo,' Mimeo, General Organization of Housing, Building and Planning Research, Cairo, September.
Rand, D.A. and L.S. Young (eds) (1980) *Dynamical Systems and Turbulence*, Springer-Verlag, New York.
Resnick, S.A. (1975) 'State of development economics,' *American Economic Review*, Vol. 65, No. 2: 317–22.
Rhodes, R.I. (ed.) (1970) *Imperialism and Underdevelopment: A Reader*, Monthly Review Press, New York.
Ricardo, D. (1891) *Principles of Political Economy and Taxation*, G. Bell, London.
Richardson, H.W. (1973) *Regional Growth Theory*, John Wiley, New York.
Robinson, R. (ed.) (1971) *Developing the Third World*, Cambridge University Press, Cambridge.
Rogers, A. (1968) *Matrix Analysis of Interregional Population Growth and Distribution*, University of California Press, Berkeley.
—— and F.J. Willekens (1986) *Migration and Settlement*, D. Reidel Publishing Co., Holland.
Rogers, E.M. (1983) *Diffusion of Innovations* (3rd edn), Free Press, New York.
Rosser, J.B. (1991) *From Catastrophe to Chaos: A General Theory of Economic Discontinuities*, Kluwer Academic Publishers, Boston, Mass.
Rostow, W.W. (1960) *The Stages of Economic Growth*, Cambridge University Press, New York.
—— (1978) *The World Economy: History and Prospects*, University of Texas Press, Austin.
Rusinow, D. (1986) 'Mega-cities today and tomorrow: is the cup half full or half empty?,' *Universities Field Staff International Report*, No. 12, UFSI, Indianapolis.
Sablov, J. (ed.) (1980) *Simulations in Archeology*, University of New Mexico Press, Albuquerque.
Sadler, M.S. (1829) *Ireland: Its Evils and their Remedies* (2nd edn), J. Murray Ltd., London.
Salmon, W. (1967) *The Foundations of Scientific Inference*, University of Pittsburgh Press, Pittsburgh, Pa.
Samuelson, P.A. (1947) *Foundations of Economic Analysis*, Harvard University Press, Cambridge, Mass.
—— (1957) 'Intertemporal price equilibrium: a prologue to the theory of speculation,' *Weltwirtschaftlichnes Archiv*, Vol. 79: 181–219.
—— (1959) 'Efficient paths of capital accumulation in terms of the calculus of variations,' in K.J. Arrow, S. Karlin and P. Suppes (eds), *Mathematical Methods in the Social Sciences*, Stanford University Press, Stanford, Calif.

—— (1970) *Economics* (8th edn), McGraw-Hill, New York.

Sanders, T.G. (1984) 'Peru's population in the 1980s,' *Universities Field Staff International Report*, No. 27, UFSI, Indianapolis.

Sargent, T.J. (1978) 'Estimation of dynamic labor demand schedules under rational expectations,' *Journal of Political Economy*, Vol. 86: 1009–44.

Sauvy, A. (1969) *General Theory of Population*, Basic Books, New York.

Schaffer, W.M., G.L. Truty and S.L. Fulmer (1988) *Dynamical Software: User's Manual and Introduction to Chaotic Systems*, Dynamical Systems Inc., Tucson, Ariz.

Schumpeter, J.A. (1934) *The Theory of Economic Development*, Harvard University Press, Cambridge, Mass.

—— (1939) *Business Cycles: A Theoretical, Historical, and Statistical Analysis of the Capitalist Process*, McGraw-Hill, New York.

Schwind, P.J. (1978) 'The spatial structure of migration behavior,' in L. Bourne and J.W. Simmons (eds), *System of Cities*, Oxford University Press, New York.

Scott, I. (1982) *Urban and Spatial Development in Mexico*, Johns Hopkins University Press, Baltimore, Md.

Scudo, F. and J. Ziegler (eds) (1978) *The Golden Age of Theoretical Ecology* (Lecture Notes in Biomathematics, Vol. 22), Springer-Verlag, New York.

Sen, A.K. (ed.) (1970) *Growth Economics*, Penguin, Baltimore.

Seers, D. (1981) *Dependence Theory: A Critical Reassessment*, Frances Pinter, London.

Sheffi, Y. (1985) *Urban Transporation Networks: Equilibrium Analysis with Mathematical Programming Methods*, Prentice-Hall, New York.

Shell, K. (1967) *Essays on the Theory of Optimal Economic Growth*, MIT Press, Cambridge, Mass.

Siepel, B.N. (1974) *Aggregate Economics and Public Policy*, R.D. Irwin, Inc., Homewood, Ill.

Simon, H. (1955) 'On a class of skew distribution functions,' *Biometrika*, Vol. 42: 425–40.

Simon, J.L. (1977) *The Economics of Population Growth*, Princeton University Press, Princeton, N.J.

Sipes, R.G. (1980) *Population Growth and Society*, HRAF Press, New Haven.

Skellam, J.G. (1971) 'Human population dynamics considered from an ecological standpoint,' in W. Brass (ed.), *Biological Aspects of Demography*, Taylor & Francis Ltd, London.

Smale, S. (1966) 'Structurally stable systems are not dense,' *American Journal of Mathematics*, Vol. 87: 491–6.

Smith, A. (1776) *The Wealth of Nations*, Valladolid, Santander.

Smith, J.M. (1974) *Models in Ecology*, Cambridge University Press, Cambridge.

Smith, R.S., F. de Vyver and W.R. Allen (1972) *Population Economics: Selected Essays of Joseph J. Spengler*, Duke University Press, Durham, N.C.

Solow, R.M. (1956) 'A contribution to the theory of economic growth,' *Quarterly Journal of Economics*, Vol. 70: 65–94.

—— (1962) 'Technical progress, capital formation, and economic growth,' *American Economic Review*, Vol. 52: 76–86.

—— (1970) *Growth Theory: An Exposition*, Oxford University Press, Oxford.

Sonis, M. (1983) 'Competition and environment – a theory of temporal innovation diffusion,' in D. Griffith and T. Lea (eds), *Evolving Geographic Structures*, Martinus Nijhoff, Amsterdam.

—— (1989) 'Innovation diffusion, Schumpeterian economic competition and dynamic choices: a new synthesis,' Paper presented at the IISA meeting on Diffusion of Technologies and Social Behavior, Laxenburg, Austria, June.

—— and D.S. Dendrinos (1990) 'Multi-stock, multi-location relative Volterra–Lotka dynamics are degenerate,' *Sistemi Urbani*, Vol. XII, No. 1: 7–15.

Spencer, H. (1867–8) *The Principles of Biology*, D. Appleton & Co., New York.

—— (1965) 'Poverty purifies society,' in R.E. Mills and H.G. Vatter (eds), *Poverty in Affluence: The Social, Political, and Economic Dimensions of Poverty in the U.S.* (1st edn), Harcourt, Brace and World, Inc. (The article was originally written in 1850.)

Spengler, J.J. (1976) 'Alfred James Lotka's vision of the population problem,' in H. Richards (ed.), *Population, Factor Movements and Economic Development*, University of Wales Press, Cardiff.

—— (1978) *Facing Zero Population Growth*, Duke University Press, Durham, N.C.

Spiegelman, M. (1968) *Introduction to Demography*, Harvard University Press, Cambridge, Mass., Chapter 14, pp. 393–422.

Spooner, B. (ed.) (1972) *Population Growth: Anthropological Implications*, MIT Press, Cambridge, Mass.

Sraffa, P. (1960) *The Production of Commodities by Means of Commodities*, Cambridge University Press, Cambridge, Mass.

Steedman, I. (1977) *Marx after Sraffa*, Unwin Brothers, The Greshman Press, Surrey.

Stewart, H.B. (1982) *Technology Innovation and Business Growth*, Nuterco, San Diego.

Stohr, W.B. and F. Taylor (1981) *Development from Above or Below?*, John Wiley, New York.

Tabuchi, T. (1988) 'Interregional income differential and migration: their interrelationships,' *Regional Studies*, Vol. 22, No. 1: 1–10.

Taeuber, K.E., L.L. Bumpass and J.A. Sweet (1978) *Social Demography*, Academic Press, New York.

Teitelbaum, M.S. and J.M. Winter (1985) *The Fear of Population Decline*, Academic Press, New York.

Thom, R. (1972) *Stabilisé Structurelle et Morphogénèse*, Benjamin, New York.

Thomas, B. (1972) *Migration and Urban Development: A Reappraisal of British and American Long Cycle*, Methuen, London.

Thompson, W.S. (1935) *Population Problems* (2nd edn), McGraw-Hill, New York.

—— and D.T. Lewis (1965) *Population Problems* (5th Edn), McGraw-Hill, New York.

Thünen, J.H. von (1826) *Der isolirte Staat in Beziehung auf Nationalekonomie und Landwistschaft*, reprinted by G. Fischer, New York, 1966.

Tiebout, C.M. (1956) 'Exports and regional economic growth,' *Journal of Political Economy*, Vol. 64: 160–4.

Todaro, M.P. (1982) *Economic Development in the Third World* (3rd edn), Longman Inc., New York.

Tolley, G.S. and V. Thomas (eds) (1986) *The Economics of Urbanization and Urban Policies in Developing Countries*, The World Bank, Washington, D.C.

Toynbee, A. (1970) *Cities on the Move*, Oxford University Press, Oxford.

—— (1971) *An Ekistical Study of the Hellenic City-State*, Athens Center of Ekistics, Athens.

United Nations (1981) *Statistical and Demographic Yearbook*, New York.

United States Department of Commerce, *Survey of Current Business*, various issues and years.

—— Bureau of the Census, *Statistical Abstract of the US*, 1986, Table 129.

Vacalopoulos, A.E. (1976) *The Greek Nation: 1453–1669*, Rutgers University Press, New Brunswick.

Vasko, T. (ed.) (1987) *The Long Wave Debate*, Springer-Verlag, Berlin.

Veblen, T. (1967) *The Theory of the Leisure Class* (originally published in 1899), Viking Press, New York.

Verdoorn, P.J. (1949), 'Fattori che regolano lo sviluppo economico della produttiva del lavoro,' *L'Industria*, Vol. 23: 45–53.

Verhulst, P. (1838) 'Notice sur la loi que la population suit dans son accroissement,' *Correspondence Mathematique et Physique*, Vol. 10: 113–21.

Vining, D.R. Jr (1985) 'The growth of core regions in the Third World,' *Scientific American*, Vol. 252, No. 4: 42–9.

Walras, L. (1874) *Elements of Pure Economics* (translated by W. Jaffe, 1954, Irwin, Homewood, Ill.

Ward, P. (1981) 'Mexico City,' in M. Pacione (ed.), *Problems and Planning in Third World Cities*, St Martin's Press, New York.

Weber, A. (1929) *On the Location of Industries* (translation of *Uber den Standort der Industrie*, 1909), University of Chicago Press, Chicago, Ill.

Webber, M. (1979) 'Social policy for the exploding metropolis: what roles for the social sciences,' in M. Pimeri le Moraes and J.M. Pimerta (eds), *Urban Networks, Development of Metropolitan Areas*, EDUCAM, Rio de Janeiro.

Weeks, J.R. (1978) *Population: An Introduction to Concepts and Issues*, Wadsworth Publishing Co., Belmont, Calif.

Weidlich, W. and G. Haag (1983) *Concepts and Models of a Quantitative Sociology: The Dynamics of Interacting Populations*, Springer-Verlag, Berlin.

——, —— (1987) 'A dynamic phase transition model for spatial agglomeration processes,' *Journal of Regional Science*, Vol. 24, No. 4: 529–69.

Wheaton, W.C. and H. Shishido (1981) 'Urban concentration, agglomeration economies, and the level of economic development,' *Economic Development and Cultural Change*, Vol. 30, No. 1: 17–30.

Williamson, J.G. (1965) 'Regional inequality and the process of national development – a description of the pattern,' *Economic Development and Cultural Change*, Vol. 13, No. 4: 8–84.

World Bank, *World Development Reports*, Washington, DC, various years.

Wycherley, R.E. (1962) *How the Greeks Built Cities*, W.W. Norton & Co., New York.

Zeeman, E.C. (1977) *Catastrophe Theory: Selected Papers*, Addison-Wesley, Reading, Mass.

Zhang, Wei-Bin (1988) 'Pattern formation of an urban system,' *Geographical Analysis*, Vol. 20: 75–84.

Zimmerman, H.J., L.A. Zadeh and B.R. Gaines (1984) *Fuzzy Sets and Decision Analysis*, Elsevier, Amsterdam, North Holland.

Zipf, G.K. (1949) *Human Behavior and the Principle of Least Effort*, Addison-Wesley, Reading, Mass.

INDEX

Abuja 108
accessibility 114, 157
Ackerman, R.J. 331
Ackley, G. 271
Adelman, I. 285
adjustment 132, 200
agents 194, 198, 236–7; *see also* economic agents; social agents
agglomerátion 88, 91, 161
agglomeration effects 169, 183, 268
agglomeration forces 163–75
aggregate development code (ADC) 180–92, 199, 200, 202, 207, 218, 225–6
aggregate models 274
aggregate and disaggregate analysis 36, 37–48
aggregate and disaggregate behaviour 299–300, 344
Ahlbung, D.A. 336
Alaska 123
Albuquerque 128
Alonso, W. 282, 325, 336, 357
Amsterdam 106
Anshan 313, 315, 317
anthropology 298–302
arbitrage, spatio-temporal 57, 129, 130, 158, 166, 205; flows of 92–5
archeology 16, 138
Argentina 19, 72, 77, 108, 114
Arizona 128
Arrow, K.J. 277
Ashby, W.R. 347
Asia, Southern 114
Athens 144, 355–6
Atizapan de Zaragoza 358
Atlanta 108
Atlantic Seaboard 113
Attfield, C.L.F. 336
attraction 92, 93, 203; *see also* net attraction
attraction index 196–7, 200, 202, 226
attractors 185, 186; strange 23, 329
austerity 229, 230
Austin 122, 124–5, 128
Australia 281
Austrian School 275
Aztecs 355

Badr (Cairo) 357
Baghdad 340
balance of trade 284
Baltimore 111, 357
Bandung 340
bandwagon effect 28, 91, 167, 168
Banfield, E. 255, 256
Bangalor 340
Bangkok 106, 310, 311, 340
Bank of America 123, 124
Baran, P. 285
Barnett, W.A. 350
barriers 245; to exchange in preferences 255; to international trade 6, 104–5
barter system 102, 244
baseball 235, 359
Bass, F.M. 278
Bayesian statistics 338, 343
Becker, C.M. 86

Beckmann, M. 282
behavior 34, 42, 46–7, 248; aggregate and disaggregate 299–300, 344; individual 333–4; in space–time 2, 192–201; *see also* collective behavior; consumer behavior; macrobehavior; microbehavior
Beijing 20, 21, 106, 186, 313, 317; growth rate in 318; population of 304, 310, 312, 314, 316, 339
Belgium 281
Berndt, E.R. 350
Berry, B.J.L. 325, 336
betting 39, 237, 254, 332, 334
Bhagwats, J. 281
bifurcation theory 5, 18, 137, 328
biological force 30, 31, 37, 40, 54–5, 57, 92, 177, 249–50
Biraben, J.N. 326
birth rates 52, 269, 327, 336, 346
Bogota 310, 340
bolivar (Venezuelan) 129
Bombay 21, 22, 106, 161, 186, 305, 310, 339
border towns 129, 130
Borneo 266
Boserup, E. 278, 336
Boston 106, 112
Boulding, K.E. 341, 345, 346
boundaries 59, 61, 88; and externalities 264; national 56, 158
Bourne, L.S. 336
Boyce, A.J. 335
Brass, W. 335
Brazil 70, 72, 77, 108, 114
Brazilia 108
Bretton Woods accord 98
Brown, L.A. 278
Brown, L.R. 253, 255
Buckley, J.J. 331
budget deficit 272
Buenos Aires 20, 21, 310, 339
Bumpass, L.L. 336
bundle of forces 32, 39–40, 149, 244; dominance among 41; and ecological determinism 26–7, 30–1, 297, 341; in relative parity 93, 94, 193; and spatial trade 284–5
Burgess, E.W. 325
business cycles 131–3, 136
butterfly effect 221

Cairo 20, 21, 174, 186, 305, 357; future of 226; pollution in 93, 94; population of 311, 339
Calcutta 20, 21, 186, 305, 311, 339
California 128
Cananda 70, 71, 77, 281
capital 11, 169, 224, 245, 277, 350; accumulation of 42–3, 86, 123, 139, 274; in ecological model 293, 294; marginal efficiency of *see* marginal product of capital; and restructuring of urban industries 89, 112; *see also* capital flow
capital flow 3, 25, 56, 86, 100; as a factor in production 96–7; and international exchanges 59, 60, 61, 78; and urban growth 91, 93
capital stock conversion 134
Caracas 340
Carlsson, C. 331
carrying capacity 94, 95, 184, 198–9, 203–4, 205, 206, 224; in mathematical ecology framework 153
cartels 158
Casetti, E. 260, 337
Cass, D. 277
catastrophe theory 5, 327, 328
catch-up process 278
census information 14, 135
Chamberlin, E.H. 282
Chandler, T. 326, 340, 355
Changchun 312, 313, 315, 317, 318, 340
change, rate of 61–2
chaos 333; ordered 2, 22–3, 37, quantum-type 358; theory of 196, 221–2, 328–9, 337; *see also* dynamic chaos
chaotic behavior 10, 31
chaotic motion 181
Chengdu 314, 317
Chicago 20, 21, 106, 107, 111, 113; population of 209, 307, 311, 340

Chicago Mercantile Exchange 350, 359
Chichen Itza 144
Chilcote, R.H. 285
Chimalhuacan 358
China, People's Republic of 67, 72, 77, 339; economic activity in 318–19, 352; metropolitan regions of 108, 114, 148, 226, 312–18; migration policies in 262; population of 210, 211, 260; private sector in 291
Choi, K. 272, 279, 280
Chongqing 312, 313, 314, 317, 318, 340
Christaller, W. 208, 282
Cincinnati 111
Citibank 350
cities 38, 158, 160; economic activity in 84, 87–8, 106–8, 146; populations of 135–6; *see also* urban agglomerations
city states 90
civilizations 38
class 266, 286; *see also* elites; underclass
Clausen, A.W. 123, 341
Club of Rome 358
coaches (sport) 238
Coadcalco 358
Coale, A.F.J. 335
Cobb–Douglas form 85, 275
cocoa beans 97
codes 6, 22, 190, 231; mutation of 231; *see also* aggregate development code
Cohen, B.J. 285
Coleman, D. 336
collective action 194, 196, 250–1
collective behavior 34, 160, 161
Colombia 114
colonialism 286
Colorado Springs 112
commodities 36, 68, 169, 224; dualism in 104; flow of 56, 59, 78; trade in 3, 91, 94
communism 347
comparative advantages 91, 156, 163–4, 204–5, 206, 282
competition 236, 243, 282
competitive ecology 295–6, 297
competitive exclusion 290–1, 296
composite price 35, 244, 252, 258
composite relative parity 3–4, 197
composite willingness 170
conflicts 22, 105, 246, 251, 255; in development 110; economic 147; resolution of 258; social 13
congestion 91, 170, 183, 345
constraints 34, 35, 42, 195, 199, 333; of human nature 298, 331; and market determinism 245, 246, 247–8, 251, 256; and public policy 259
consumer behavior 45, 46, 49, 245–6, 334
consumers 248; money and 97, 102
convergence to uniform austerity scenario 229, 230
cooperation 236, 243, 246, 296
cooperative association 290
cooperative ecology 129, 296, 297
core region 167, 168, 170, 172–3
costs 198; access 255; maintenance 254; *see also* opportunity costs
Cox, G. 326, 355
Cuautitlan 358
cultural anthropology 55
cultural norms 31, 139
cumulative human life 224
currencies 57, 88, 157, 158, 350; dualisms and 95–105; and exchange rates 98–9, 100; formal and informal 99–100, 102; and *numèraires* 97
cycles 22, 58, 113, 131, 300; form of 133–41; spatio-temporal 231; urban 141–5; *see also* business cycles; economic cycles; periodic cycles; slaving principle; spatial cycles

Dacca 340
Dai Ichi Kangio Bank 350
Dallas 108, 112, 128
data 6–7, 8–11, 12, 175
Davis, K. 347
death rates 52, 269, 327, 336

debt 359
decentralization 127
decision-making: in ecological model 292–3; and the market 289–90; rules of 244, 245, 248, 256; social 35
Delhi 310, 311, 340
Demery, D. 336
demographic models 48–53
demographic transitions 31, 52, 112
demography 4, 25, 160, 224–5
Dendrinos, D.S. 31, 155, 207, 260, 300, 327, 328, 329, 341, 344, 353, 354, 356, 357
Denver 112
dependence theory 55, 285–7
depreciation 102
determinism 5, 30, 222, 249; *see also* ecological determinism; market determinism
Detroit 106, 120, 146, 175, 209, 357
development 82, 108–10, 337, 340–1; code of 153, 161; in LDCs 50; policies of 84; *see also* urban development
development theory 52, 53–4, 271–85
developmental threshold 19, 161, 176, 186, 188, 218
disaggregation 38–43, 189, 268, 343; *see also* aggregate and disaggregate analysis; aggregate and disaggregate behavior
disciplinarian forces 301
discontinuities 4, 5, 172, 300, 327
discounting process 299
disequilibrium 11, 130, 131, 150, 229
dispersal 165
distance 82, 157, 167, 170
diversification 105–6
division of labor 282
dollar (US) 89, 100, 123, 129, 154
Domar, E.D. 274, 348
domestic consumption 31, 43, 60, 85, 90
dominance 27, 57, 157, 208; of ecology 80–1; political 3; in production 79
dominant forces 39, 40, 41, 145, 301
Doubleday, T. 335
Dow Jones industrial average index 354
dualisms 17, 33, 157, 169, 225, 260–7, 349–50; and currencies 95–105; regional 223; socio-spatial 57, 205, 262–8; in wealth 232
Duck, N.W. 336
durability 134
dynamic chaos 57, 133, 136, 137, 196, 207, 329
dynamic equilibrium 85, 130, 131, 184, 186, 202–3, 297, 337; in national economy model 85; spatial arbitrage and 130
dynamic instability 2–3, 4, 135, 145, 233, 291–2; in American cities 110–11; in foreign trade 79, 80–1; of national economies 56, 57, 60, 61, 101; and spatial dualisms 157–8, 160
dynamic stability 79, 80, 83, 84, 135, 269
dynamics: absolute 123, 130; continuous 30, 150–2, 154, 280; discrete 31, 137, 152, 207, 280; fast and slow 6–7, 192, 223, 226, 229–30, 232–3, 330; of stocks 24–6, 28–9, 133–5, 261; *see also* macrodynamics; relative dynamics; urban dynamics

Ecatepec 358
ecological associations 151–2, 289–90, 292–8
ecological distance 167
ecological determinism 5, 54–5, 158, 243–70, 287–92, 297; interdependencies and 24–55; speculative behavior and 34–7
ecological factors 35, 153
ecological framework 189, 279
ecological interdependencies 81, 146, 151
ecological niches 156
ecology 4, 80, 297; *see also* urban ecology

econometrics 132
economic agents 103, 132, 265, 273
economic cycles 83, 136
economic development 61, 101
economic factors 205, 245
economic growth 61; and business cycles 131–3; public sector and 289; theory of 53–4, 61, 82–3, 124, 274–7
economic variables 59–60, 133, 137
economic–demographic (E–D) force 32–3, 37, 40, 52, 54–5, 57, 92, 177, 184
Egypt 72, 77
Ehrlich, A.H. 326
Ehrlich, P.R. 326
eigenvalues 79, 181
El-Amal (Cairo) 357
elasticities 28, 85, 147, 156
elites 38, 104, 266, 286; currency and 99, 101, 102
Elliott, J.W. 272
El-Obour (Cairo) 357
emics 298, 299
employment 127, 133
endogenous forces 163
endogenous variables 86
environment 28, 231
environmental catastrophes 3, 140, 203
epistemology, in social sciences 330–1
equilibrium 51–2, 166; bound 52; individual 194–5; observer 201–9; stable 51, 132; Walras's 97, 98; *see also* dynamic equilibrium
errors, in dynamic adjustment 132
ethnic minorities 232–3
etics 298–9
European Community 158
evolution 299, 337, 340–1; and ecological determinism 47; of urban areas 6, 105, 189, 191–2
exchange rates 57, 58, 78, 88, 89, 98, 100, 101
exogenous checks 13, 199
exogenous forces 163
exogenous variables 86, 131
expectations 14, 35, 95, 233, 251, 258, 343; rational 132, 333–4
exploitation 109
explosion phenomenon 171–2
exponential growth 30, 31, 32, 83, 177, 223, 325, 342–3
exports 283, 284; of Mexico 75, 77; of USA 67, 68, 69, 70
externalities 91–2, 127, 263–5; *see also* congestion; pollution
extinction 83, 134, 270
ex-urban communities 109

family planning 49, 231
family size 169
famine 346
Far East 70, 71, 82, 114, 291
Federal Reserve Bond (US) 293
feedback 14, 34, 208
fertility rate 52, 225
Fifteenth of May (Cairo) 357
financial markets 233
Findlay, R. 281
Fisher, J.C. 278
Flieger, W. 335
fluctuations 94, 156, 164–5, 196, 283; of price 132, 133
fluid factors 93, 158, 166, 199
food supply 32
football (American' 235, 359
forces 28–9, 29–33, 135; *see also* biological force; economic–demographic force
Ford, T.R. 337
Ford Motor Company 129
forecasts 14, 333
foreign trade *see* international trade
fractals 344
franc (French) 100
franc (Swiss) 100
France 72, 77, 281, 350
Freeman, A. 271
Freeman, C. 278
Frejka, T. 335
Friedmann, J. 336
Fushun 315, 317
future 14, 222–39
futures markets 5, 329, 332
fuzziness 9–10, 23, 135, 162, 193, 311, 321–2, 338–9

gain, composite net 197, 198, 200, 202, 231–2, 251
game theory 147
Gardner, M.R. 347
geography 234
geology 82
Germany 77, 80, 171, 281
Geweke, J. 350
Gini, C. 335
Ginn, H.L. 337
Gleick, J. 328
global misery scenario 229
Glotz, G. 355
Godwin, R.K. 337
gold 58, 98
golden age 276, 327
Gomme, A.W. 355
Gordon, R.J. 255, 272
governments 12, 87; control of information flow by 103; intervention by 55, 87, 88, 267
Grandmont, J.M. 354
Grindstaff, C.F. 337
gross national product 33, 281
growth rates 101, 102–3, 147; *see also* exponential growth
Guangzhon 130, 312, 313, 314, 316, 317, 318, 340
Guckenheimer, J. 328
Gulf Coast 114, 127
Gupta, M.M. 331

Haag, G. 165, 330, 344
Hagen, V.W. von 354
Hagerstrand, T. 278
Haken, H. 328, 344
Hansen, E.W. 317
Harbin 312, 314, 317
Harris, D.J. 172
Harris, M. 288, 298, 299, 300, 301
Harrison, G.A. 335
Harrod, R.F. 274, 348
Harrod–Domar model 274, 275
Hausman, D.M. 277
Hayek, F.A. von 89, 275
hedgers 13, 36, 251, 332
heterogeneity 45, 59, 225, 232; locational 163–4; socio-spatial 261
Hicks, J. 246, 271, 275, 350
hierarchies 60, 81, 208; *see also* urban hierarchy
high yield bonds 124
hinterland 84, 87, 88, 90–1, 168
Hiorns, R.W. 335
history 16, 26, 333
Holden, A.V. 328
Holmes, P. 328
homo economicus 333
Hong Kong 90, 99, 146, 340
Hotelling, H. 282
housing 86, 258
Houston 113, 120, 128
Hu, H.Y. 312, 313
Huang (Yellow) River 313
human settlement 4, 18, 135, 152, 164–5
Hyderabad 340
hyper-concentrations 91, 104, 160, 204, 208; and agglomeration gradients 163–75
hypo-concentrations 160, 204, 208; and agglomeration gradients 171

impedance *see* distance
imports 283, 284; of Mexico 75–6, 77; of USA 65–6, 67–9, 70–1
India 72, 77, 114, 148, 210, 211, 226, 339
Indonesia 72, 77, 114, 339, 352
industrial bases 61, 127
industrial revolution 127
industrialization 112
industries 38, 84; dynamics of 89–90; in USA 64, 65
inflation 101, 102, 124
informal sector 262
information 224, 225; and currency flows 103; flows of 3, 12, 25, 56, 59, 61, 96
infrastructural determinism 299
infrastructure 127, 134, 298
Inner Mongolia 313
input factors 10–11, 88, 156; in production 98
input–output analysis 41, 64
instability 2, 7, 17–18, 84, 100, 223; *see also* dynamic instability

interactions: between interpretations 13; ecological 244, 246, 268–70, 284; and interdependencies 3, 24–9; international 146–8; inter-urban 56, 84, 146; of populations 79–80; private and public 287–98; spatial 3–4, 25, 57, 59–62, 78; and urban hierarchies 207–8; *see also* capital flow; commodities, flow of; information, flow of; labor flow
interdependencies 2, 3, 327; between wealth and population 17, 26–8, 37, 43, 48–53; between urban areas 56, 60, 83–4; and ecological determinism 24–55; global 160
interest rates 98, 124, 132, 291, 293, 294; Keynes and 296, 297
International Bank of Reconstruction and Development 26
international markets 58–62
International Monetary Fund 100, 359
international trade 4, 55, 57, 60–2, 90; barriers to 6, 104–5; currencies and 101; distance and 82; and economic stability 78–82; and internal growth 62–3, 66–7; shifts in 113–14; theory of 280–5
interpretation markets 11–13
inter-regional flows 28, 57
intra-national trade 90, 104–5
investment 103, 108, 274, 285, 293
Iran 71, 72, 77
Isard, W. 282, 357
isolation 243, 289, 290
Israel 77
Istanbul 106, 340

Jacobs, J. 87, 89
Jakarta 20, 21, 104, 186, 226, 262, 266, 306; population of 310, 339
Japan 70, 72, 77, 82, 114, 279; foreign workers in 233; GDP of 281; and oil prices 123
Jevons, W.S. 272
Jinan 315, 317
Johansson, B. 359
Jones, H.G. 277
Jong, G.F. de 337
junk bonds 124
jurisdiction 87, 88

Kaldor, N. 260, 275
Kammeyer, K.C.S. 337
Kansas City 111
Kaohsiung 349
Karachi 310, 311, 340
Kasarda, J.D. 325
Keelong 349
Kelley, A.C. 85
Keyfitz, N. 335
Keynes, J.M. 272, 277, 288–9, 296, 297
Keynesian macroeconomics 53, 272
Kinshasa 340
Kondratieff, N.D. 131, 137
Korea, Republic of 72, 77, 99, 233
Kunming 313, 315, 317
Kuznets, S. 131, 277, 278

labor 11, 123, 245
labor flow 25, 56, 86, 96, 260, 349; and international exchanges 59, 78, 96; and urban economies 106–7, 129
labor market 85–6, 91, 112
lagging 113, 260
Lagos 226, 340
laissez-faire 53, 288, 289, 290–1, 295
Land, K.C. 337
land prices 170, 200
land use patterns 41, 56
landscapes 164
Lanzhou 312, 313, 315, 317
La Paz 358
Latin America 70, 71, 114
Leibenstein, H. 33, 336, 346
Leinbach, T.R. 352
Leningrad 162, 209, 311, 340
Lenin–Stalin experiment 289
Leontief, V. 131
less developed countries (LDCs) 19, 33, 50, 52, 200, 276, 350–1; currency in 98, 100, 101, 297–8; future of 227, 228–9; populations

of 49, 148, 209–12; urban agglomerations in 176, 180, 186; urban interactions in 83, 215
Levi-Strauss, C. 298, 300, 301
Lewis, D.T. 335
Lewis, D. 255, 256
lifespan 133, 134
Lima 162, 209, 310, 341, 356
linkages 7, 13, 29; between macro and micro socio-spatial worlds 192–209; economic 25; *see also* interactions; interdependencies
Linn, J.F. 262
Lisbon 106
location theory 156–7, 174
locational equilibrium 194–5
London 20, 21, 106, 186, 307, 311, 339
Lorenz, E, 221, 358
Los Angeles 20, 21, 107, 113, 146, 175, 220; pollution in 93; population of 308, 311, 339; service sector in 118; trade and 113–14; transformation in 115, 118–20
Losch, A. 208, 282
Louisiana 123
Lucas, R.E. 131

Ma, J.C. 317
MacArthur, R.H. 325
Machlup, F. 331
macrobehavior 45, 46, 218
macrodynamics 5–6, 14, 37–48, 218; and business cycles 130–41; and intranational urban cycles 141–5; relative dynamics and 145–8; urban 14–15, 158–9, 161, 180–1, 225, 226–7
macroeconomics 45, 97, 271–4; Keynesian 288–92
macrovariables 44, 140, 180, 189, 198, 218; in ecological model 290, 293, 294
Madagascar 266
Madras 340
Madrid 106, 340, 354
Mahajan, V. 278
Maitra, P. 278
Malthus, T.R. 30, 31, 32, 271, 272, 278, 300, 325, 335, 345
Malthusian population growth 30–1, 32
Malthusian trap model 50–2, 176, 189, 342, 346
Manchester 106
Mandel, E. 271, 354
Mandelbrot, B. 344
Manila 340
Mansfield, E. 278
manufacturing goods: in Mexico 74–5; in USA 64–5
manufacturing industries to services transition 107, 116, 117, 118
Marchall, A. 272, 335
Marchetti, C. 278
marginal product of capital 293, 294, 296, 297
marginality 262–3
mark (German) 100
market determinism 234, 243–4, 247
market imperfections 5, 330
market theory 45
markets 34, 236, 330; and decision making 289, 291; domestic 60
Marx, K. 246, 278, 298, 299, 300, 335, 345
Marxism 43, 53, 139, 338; and public–private sector interdependence 288, 292, 295
Marxist anthropology 37, 298, 299, 301
mathematical ecology 4, 58, 79, 83, 154, 268, 287–8, 325; model of 292–8; *see also* Volterra–Lotka formalism
Matras, J. 326, 337, 355
Matthews, R.C.D. 272
May, R.M. 154, 269, 289, 325, 327, 347
Mayan cities 27, 97, 144, 354–5
Meade, J.E. 275
Meadows, D. 345
Medrich, E. 336
mergers 352
metropolitan areas 209, 211; in China 312–18; identification of 163–4

Mexican miracle 73
Mexican Census Bureau 128
Mexico, United States of 72; economy of 73; employment composition of 74, 76; foreign trade of 73, 75–8, 82; GNP of 73; population of 73, 74, 78
Mexico City 20, 21, 106, 125–30, 186, 266, 308; employment in 125–6; future of 226; pollution in 93, 94; population of 128, 146, 188, 310, 339, 353, 358
Miami 108
microbehavior 44, 45, 46, 218
microdynamics 38
microeconomics 41, 42, 45, 46–7, 49, 97
microvariables 44
Middle East 70
migration of labor 85, 86, 115
Milan–Turin agglomeration 171
Mill, J.S. 271, 272, 282, 329
Miller, R.W. 331
Miller, W.B. 337
Mills, E.S. 282, 345, 357
Minneapolis 111
Mirrlees, J.A. 275, 277
misery 229, 346
mode of production 244, 298, 299
models: aggregate 274; behavioral 46–7; demographic 48–53; disaggregation of 343; dynamical 17; economic growth 124; Malthusian trap 50–2, 176, 189, 342, 346; mathematical ecology 292–8; observer type 5, 195–209; predator–prey 151–2, 270, 293–4; Walrasian 97, 98
modernization 112
money 97, 98, 133; denationalization of 89
monopolies 169, 170
monopsonies 11, 169
Monterrey 127
more developed countries (MDCs) 19, 33, 50, 200, 297–8, 340; currencies in 99; future of 227, 228; populations of 209–12, 232; urban agglomerations in 176, 180, 186, 215
Morishimo, M. 277
Morley, S.G. 342, 355
Moscow 20, 21, 161, 186, 309, 311, 339
Mullally, H. 300, 327, 335, 353
multinational corporations 286
Myrdal, G. 336
Myres, J.L. 355

Nanjing 312, 313, 314, 317
national economies 84–90
national growth rate 33
nations 38, 146–8; differences between 22; dominant 154, 156, 157; interdependence with urban centers 83, 154; primate 90–2
natural gas 74
natural resources 3, 82, 88, 101, 156, 224, 245
Naucalpan 358
Nelson, R.R. 272, 278
neoclassical economics 43, 46, 275
net attraction 4, 6 ,92–3, 158, 165–6, 200, 201–6; aggregate 25, 94, 95, 198
Netzahualcoyotl 128, 174, 358
New Mexico 128
New York 20, 21, 106, 107, 113, 146, 175, 357; population of 209, 308, 311, 339; service sector in 117; transformation in 115–17
niche dynamics 88, 110, 156
Nigeria 108
non-linear dynamics 4, 83, 164, 196, 332–3, 337
non-linear interdependency 4
non-linear theory 17
non-periodic cycle 4, 328
non-periodic motion 328
norms 34, 245; cultural 31, 139; relative 92, 93, 94, 199; social 3
North America 109, 111, 291; trading bloc 78
Notestein, F.W. 52, 53, 347
numèraire 97, 98, 99, 103

observer equilibrium 201–9

observer type model 5, 47, 95, 195–209, 234–9, 283; Harrod–Domar 274–6
odds 334
Ohlin, B. 282
oil: price of 58, 76, 112, 123, 124–5; reserves of 74, 123
Oklahoma 123
Oklahoma City 111
oligopolies 170
oligopsonies 11, 170
opportunity cost 172–3, 174, 293
optimization process 248, 256
optimum city size 263, 266, 345
options 35, 36, 251, 252, 258, 343; markets 11, 34–5, 36; social 34–5
oscillations 22; *see also* cycles
oscillatory behavior 166
outputs 26, 133, 245, 276; social 35, 38
Overbeek, J. 336
overfills 94
overpopulation 345
overreaction 111–12

Pacific Ocean 114
Palen, J.J. 353
Palma, G. 285
Pareto, V. 335
Paris 20, 21, 106, 186, 305, 339
participant agent 95
participant type model 95, 192–4, 234–9
Pearl, R. 342
Pearl–Reed type 30, 279, 342
Peking *see* Beijing
per capita gross product 18, 32, 37, 48, 93, 152, 175–6, 281
  of Mexico 73
  of USA 63, 154
per capita income 1–2, 4, 29, 33
  in China 316, 317
  data of 309
  distribution of 214–15
  in the future 226–7, 228–9
  in Mexico 129
  and population 29–31, 32, 33, 48–50, 51, 80, 142, 175–6
  in Texas 122, 124
  of urban agglomerations 142, 169, 183
periodic cycles 135, 136, 137, 138–9, 140, 328
periodic movement 22, 23, 134, 135, 141–3, 328
periphery 170, 286
Perlman, J.E. 262
Perloff, H.S. 337
Peru 72
peso (Mexican) 58, 78, 129
Peterson, R.A. 278
Peterson, W.L. 273
petroleum *see* oil
petty production 262
phase transitions 4, 60, 261, 328, 337
Phelps, E.S. 277
Philadelphia 106, 111, 120, 175, 209, 311, 340, 357
Philippines 233
Phoenix 108, 112, 128
Pittsburgh 106, 107, 111, 120, 357
planning 234
pollution 91, 93–4, 170
Popper, K. 288, 333
population 14, 26, 27, 93; and ADC 181, 183, 184; changes in 174; control of 263, 346; cycles of 144–5; density of 107, 225; distribution of 165; and dynamic instability 2–3; explosion of 128; flows of 59, 61, 78, 96, 104; interaction of 79–80; and per capita income 29–31, 32, 33, 48–50, 51, 80, 142, 175–6; of urban cycles 209–12; and wealth 17, 32, 48–53, 180–9; *see also* population dynamics; population growth; population migration
population dynamics 17, 80, 134–5, 224–5
population growth 1, 4, 18, 30–1, 115, 146–7, 341–2; in China 316–17; and communism 347; in Texas 124; and theorem of misery 346; of urban agglomerations 142, 176–7, 180, 181, 183

population migration 3, 28, 105, 128, 137, 204–5, 231; in China 313; and the informal sector 262
Population Reference Bureau 176, 177, 209, 210, 225, 309
"Population and the Urban Future" (Conference in Barcelona, May 1986) 253
pound sterling 100
poverty 3, 50, 91, 223, 253, 300, 335, 346
power 3, 104, 245
predations 236, 243, 246, 289–90
predator–prey models 151–2, 270, 293–4
predatory association 290, 291, 297, 338
predictions 333–4
preference functions 248, 249, 250, 251, 252, 254–5
preference markets 249, 250, 251, 252, 253
preferences 41–2, 245, 248–57; accumulation of 254; bundle of 249–50; and public policy markets 257–60
prices 86, 89, 102, 233, 244, 245, 246; composite 35, 244, 252, 258
primacy 83, 168, 325–6
primate cities 83, 84, 90–2, 168
private sector 5, 290, 291, 297; in ecological model 293–8
privatization 291
problem definition 267
production: interference in 88; mode of 244, 298, 299; processes 61, 106, 275; theory of 45
profit 11, 139, 167, 245, 246; composite 197, 198, 200, 202, 231–2, 251; functions 248
Pry, R.H. 278
psychological forces 31, 41, 133, 138
psychology 25, 34
public action 257–60, 291
public intervention 55, 234, 267, 291; in national economies 87, 88, 288–9
public policy 55, 234, 244, 257–60, 289
public sector 5, 288, 289, 290, 291, 297; in ecological model 293–8
public utilities 258

Qingdao 313, 315, 317
quantum theory 222
quasi-periodic cycle 23, 328
quotas 88

Rageh, A.Z. 357
rainforest ecology 266
Rand, D.A. 328
random utility 330
rank-size distribution 212–13
rate of discount *see* interest rate
rate of return 294
rational expectations 132, 333, 334
rationality 249, 334
redundancy 103
regional development 108–9
relative development threshold 6
relative dynamics 3, 28–9, 58, 80, 145–8
relative macrodynamics 189; mathematical formulation of 148–57
relative parity 3–4, 6, 8, 57, 160, 220, 327, 330; in China 317; global 110; and spatial arbitrage 92–3, 94, 95, 105, 130, 158; *see also* carrying capacity; net attraction
relative per capita income/product distribution 6, 19
relative population distribution 6, 212–14
religious forces 31, 244
rents 170
reproduction 31, 298
repulsion 25, 165, 169
resonance 148
revolutions 13
reward 11
Rhein–Ruhr region 171
Ricardo, D. 271, 272, 282
Richardson, H.W. 337
Rio de Janeiro 20, 21, 161, 186, 304; population of 310, 311, 339
risk management 251
road construction 42–3

road use 46–7
Rogers, A. 337
Rogers, E.M. 278
Rome 106, 144
Rostow, W.W. 278
rules 235, 238, 244, 264
rural populations 210, 211, 255
Rusinow, D. 253

Sablov, J. 342
Sadler, M.S. 335
St Louis 111, 120
St Petersburg (*formerly* Leningrad) 162, 209, 311, 340
Samuelson, P.A. 132, 275, 282, 284, 296, 329
San Antonio 128
San Francisco 107, 114, 120, 146, 175, 209; population of 311, 340
Santiago 310
Santos 161, 188, 189
Sao Paulo 20, 21, 106, 186, 188, 189, 304; future of 226; pollution in 93, 94; population of 310, 311, 339
Sauvy, A. 335
savings 274, 275
Schofield, R. 336
Schumpeter, J.A. 131, 137, 277, 278
Schumpeterian clock 132
Schwind, P.J. 336
sectoral disaggregation 38, 39
Seers, D. 285
self-interest 41, 42, 248, 249
Seoul 20, 21, 106, 185, 186, 188, 266, 307; pollution in 53, 94; population of 310, 339
services sector 64, 107–8, 115, 117, 127
Shanghai 20, 21, 94, 106, 186, 266, 304, 317; growth rate in 318; population of 310, 312, 313, 314, 316, 339
Sheffi, Y. 345
Shell, K. 277, 350
Shenyang 314, 316, 317, 318, 340
shocks: environmental 140; exogenous 132; technological 139
signal processing 132
Simmons, J.W. 336
Simon, H. 212
Simon, J.L. 336
Singapore 90, 99, 262, 350
Sipes, R.G. 337
Sixth of October (Cairo) 357
Skellam, J.G. 335
slaving principle 4, 44, 140
slums 85, 157, 255
Smale, S. 269
Smith, A. 271, 272, 282, 330
social actions 8, 34–5, 194, 196, 202, 245, 247, 301; theories of 39, 302
social agents 10, 14, 36, 39, 201, 205, 223, 236–7; behavioral attributes of 244
social aggregates 38
social choice 258–9
social Darwinism 290
social development 26–8
social disaggregation 38, 39, 40, 41, 42, 233, 261
social equity 263
social events 9, 11, 27, 36, 47, 337, 338; in cities 22; as games 235–8; interpretations of 11–13
social evolution 10, 291, 297
social science 25–6, 38, 332
social segregation 261
social systems 13, 154, 234–5, 251, 332–3; anthropology and 298–302; and games 235–8
social welfare 276
socio-economic development 61, 115
socio-spatial dualism 57–8, 205, 261, 262–8
socio-spatial dynamics 9, 17
socio-spatial evolution 23, 154
socio-spatial-temporal system variables 17, 28, 40–1, 44, 231
sociological forces 31
sociological kinship bonding 31
sociology 25
Solow, R.M. 275, 348
Sonis, M. 31, 207, 278, 328, 329, 335, 344
Sonora 128
Spain 77

Sparta 355
spatial cycles 134, 135, 137, 139, 141, 159
spatial disaggregation 38–9, 40, 42
spatial economics 88
spatial interaction 3–4, 25, 57, 59–62, 78
spatio-temporal diffusion 25, 349
specialization 90, 106, 283
spectators 236, 237
speculation 113, 124, 286, 329, 343; collective 108–9; in land 111, 112
speculative action 250–1, 256
speculative behavior 5, 13, 108, 153, 166, 329; in cycles 132, 138, 139; and ecological determinism 34–7
speculative spectator 237, 239
speculators 36, 233, 251, 252–3, 332; Keynes and 289
Spencer, H. 290, 296, 335
Spengler, J.J. 335, 347
Speigelman, M. 335
spiral dynamic equilibrium 294, 337
Sraffa, P. 271, 277, 350
start-up processes 164, 168
status 104
steady state 224, 326–7
Steedman, I. 271
Stewart, H.B. 278
stochastic models 330
stochasticity 217, 222
stock exchanges 11, 332, 352, 356
strange containers 23, 329
structural anthropology 300
structural Marxism 298–9
subsidies 88; to informal currency nations 100, 101
suburbanization 112, 115
suitability 156
superstructure 299
supply/demand 34, 205, 243–4, 246; of population 184; and public policy 257
Surabaya 340
surplus value 286
survival 88, 249, 250
Suwarno, B. 352
Sweet, J.A. 336

Tabuchi, T. 260, 336, 354
Taeuber, K.E. 336
Taipei 104, 340, 351–2
Taiwan 90, 99, 233, 349
Taiyuan 313, 315, 317
take-off 106
Talien 315, 317
Tampico 127
tariffs 88
Tashkent 220
taxes 88
technological change 64, 245, 274, 277–80
technological innovation 3, 22, 55, 61, 103, 112, 231; cycles and 137, 139, 140–1; and economic growth 279–80; and manufacturing decline 106; primate cities and 91; and urban dynamics 153
Teheran 20, 21, 161, 177, 186, 188, 306, 310, 340
Teitelbaum, M.S. 345
temporal disaggregation 38, 39, 40, 42
Tenth of Ramadan (Cairo) 357
Texas 121–5, 128
Third World 85, 86, 169, 171, 262
Thom, R. 328
Thomas, B. 336
Thompson, W.S. 52, 335, 346
thresholds 19, 79–80, 213; developmental 161, 176, 186, 188, 218
Thunen, J.H. von 282, 325
Tianjin 162, 209, 310, 313, 314, 316, 317, 318, 340
Tiebout, C.M. 282
time horizon 34, 35, 183, 256
time lags 112, 124, 202
Tlalnepantle 358
Todaro, M.P. 49, 172, 280
Tokyo 20, 21, 106, 146, 161, 306, 311, 339
topography 82, 156
Toynbee, A. 355
trade *see* international trade

trading blocs 78, 81
transaction costs 93, 134, 137, 194, 195
transportation 107, 258; costs of 130, 134–5, 137, 163–4
Tucson 108, 128
Tultitlan 358
turbulence 5, 137–8; *see also* chaos
turnpike dynamics 276

umpires 236
underclass 38, 256, 269; currencies and 99, 101, 102; dualisms and 104, 262, 266; growth of 105
uniformity 230, 231, 235
United Kingdom 72, 77, 281
United Nations 19, 26, 175, 209, 253; data from 15, 309, 319, 356
United States of America: economy of 63–73, 356; foreign trade of 65–73, 77, 82, 348; GDP of 281; GNP of 63, 65; population of 63, 210, 211, 232; regions of 108–9, 128; urban agglomerations in 110, 111–14, 339; urban dynamics in 153, 154–5
United States Bureau of the Census 120, 122, 232, 341, 357; data from 309, 319
USSR 72, 77, 210, 211, 291, 325
urban agglomerations 1, 5, 89, 91, 156, 176, 339; changes in size of 174; control of 104; interdependencies among 29–33; macrodynamics of 18–23, 83, 190, 209–12; population dynamics of 4, 18, 162, 209–12; relative parity and 206
urban centers 127, 156, 160, 170–1
urban decline 26, 174, 176
urban development 82, 83, 106–9, 189; and agglomeration forces 165; currency and 102; and national growth 84–90; in USA 111–14
urban dynamics 3, 61, 177, 212–20
urban ecology 137, 300, 325
urban economics 139, 234; industrial structure of 105, 106–7
urban evolution 6, 105, 189, 191–2
urban futures 191
urban growth 26, 83, 176, 267–8
urban hierarchy 8, 57, 81, 83, 91; dynamic code of 160–220; interdependencies and 207–8
urban hyperconcentrations 127, 161
urban landscapes 56–7, 164
urban macrodynamics 141–5, 148, 158–9, 161, 180–1; 225, 226–7; unstable 82–95
urban wealth 146–7
urbanization 107, 112
utility function 41, 42, 193–4, 195, 248, 252
utility theory 167, 198, 236, 267

Vacalopoulos, A.E. 355
vacuums 94–5, 205, 208, 226
value: labor theory of 246; surplus 286
variables 3, 193, 197; in anthropological analysis 301; and carrying capacities 199; in defining urban development 189; socio-economic 25; socio-spatial 40–1, 44; state 43; *see also*l economic variables; exogenous variables; macrovariables; socio-spatial-temporal system variables
variability 27, 45; inter-urban 110
Vasko, T. 278
Veblen, T. 342
Venezuela 114, 129
Vera Cruz 127
Verdoorn, P.J. 260
Verdoorn's law 260
Verhulst, P. 342
Verhulst equation 30, 207, 278, 342
Viedma 108
Vienna 106
viewpoints 236
Vining, D.R. 338
volatility 233
Volterra–Lotka formalism 30, 79, 150–1, 152, 207, 292, 342

wage rates 85–6
Walras, L. 272

Walrasian model 97, 98
war 147, 338
wealth 16, 26, 27, 133, 211, 224, 228; concentration of 169; difference in 232–3, 286; and population dynamics 17, 32, 48–53, 180–9; redistribution of 87, 100; *see also* per capita income
Weber, A. 282
Weeks, J.R. 336
Weidlich, W. 165, 330, 344
Western Europe 70, 114, 233
Wheeler, J. 236
White, H. 350
Williamson, J.G. 84, 85, 86, 337
Winter, J.M. 345
Winter, S.G. 272, 278
World Bank 100, 254, 341, 359
World War II 115
Wuhan 314, 317, 340
Wuxi City 319
Wycherley, R.E. 355

Xian 314, 317

Yamakawa, T. 331
Yangtze River 313
yen (Japanese) 58, 100, 123, 350
Young, L.S. 328
Yucatan Peninsula 27, 144, 354–5

Zaire 2, 221, 339
Zeeman, E.C. 328
zero-sum game 36, 235
Zhang, W.B. 356
Zipf, G.K. 212